BEYOND MIMESIS AND CONVENTION

BOSTON STUDIES IN THE PHILOSOPHY OF SCIENCE

VOLUME 262

For further volumes:
http://www.springer.com/series/5710

BEYOND MIMESIS AND CONVENTION

Representation in Art and Science

Edited by

ROMAN FRIGG
London School of Economics, London, England

and

MATTHEW C. HUNTER
California Institute of Technology, USA

Springer

Editors
Roman Frigg
Department of Philosophy, Logic
and Scientific Method
London School of Economics
and Political Science
Houghton Street
London WC2A 2AE
United Kingdom
r.p.frigg@lse.ac.uk

Matthew C. Hunter
California Institute of Technology
Division of Humanities and Social Sciences
MC 101-40
Pasadena, CA 91125
USA
mchunter@caltech.edu

ISBN 978-90-481-3850-0 e-ISBN 978-90-481-3851-7
DOI 10.1007/978-90-481-3851-7
Springer Dordrecht Heidelberg London New York

Library of Congress Control Number: 2010924549

Printed on acid-free paper

Springer is part of Springer Science+Business Media (www.springer.com)

Preface

This volume has grown out of a conference that the editors organized at the London School of Economics and the Courtauld Institute of Art in June 2006. The aim of the conference was to bring together philosophers of science and historians of art to discuss representation. A topic of central importance to both the arts and the sciences, representation has generated similar conceptual problems in both fields, largely unbeknownst to the other community. Opening dialogue, we thought, would be productive and timely. In response to the call for papers, we received over eighty submissions, thirty of which were chosen for presentation by the program committee. As the present volume took shape, we sought to complement the conference's focus on visual art by soliciting further contributions. Thus, seven of the papers included here were presented in early form at the conference in 2006, while four have been added subsequently.

In organizing the conference and putting this book together, we have incurred many debts. We would like to thank Peter Ainsworth, Elisabeth Schellekens, Christine Stevenson, and Sabine Wieber for serving on the conference's program committee. The conference itself would not have been possible without the support of the Courtauld Institute of Art's Research Forum, and especially its former director, Pat Rubin; the Institute of Philosophy of the University of London; and the London School Economics. While we were still working on the program, Ingrid van Laarhoven of Springer encouraged us to submit a book proposal, and her continued enthusiasm for the project has been crucial. We have been lucky enough to be able to count on Lucy Fleet whose guiding hand and sustained support have helped keep the project on course. We would like to thank all of the speakers who made the 2006 conference such a memorable event and, especially, the contributors to this volume for their stimulating work. Each essay in the collection was read by two anonymous referees, whose input made an invaluable contribution. Finally, we would like to thank Andrew Goldfinch and Daphne Kouretas for their excellent assistance in organizing the event and preparing the manuscript.

Contents

Contributors

Nancy Cartwright London School of Economics, London, UK; University of California, San Diego, CA, USA, N.L.Cartwright@lse.ac.uk

Anjan Chakravartty Institute for the History and Philosophy of Science and Technology, University of Toronto, Toronto, ON M5S IK7, Canada, anjan.chakravartty@utoronto.ca

David Davies McGill University, Montreal, QC, Canada, david.davies@mcgill.ca

Catherine Z. Elgin Harvard University, Cambridge, MA, USA, catherine_elgin@harvard.edu

James Elkins School of the Art Institute of Chicago, Chicago, IL, USA, jameselkins@fastmail.fm

Roman Frigg London School of Economics, London, UK, r.p.frigg@lse.ac.uk

Manuel García-Carpintero LOGOS-Universitat de Barcelona, Barcelona, Spain, m.garciacarpintero@ub.edu

Matthew C. Hunter California Institute of Technology, Pasadena, CA, USA, mchunter@caltech.edu

John Hyman University of Oxford, Oxford, UK, john.hyman@queens.ox.ac.uk

Dawna Schuld Indiana University Bloomington, Bloomington, IN, USA, dlschuld@indiana.edu

Adam Toon University of Bielefeld, Bielefeld, Germany, adam.toon@uni-bielefeld.de

About the Authors

Nancy Cartwright is Professor of Philosophy at the London School of Economics and at the University of California at San Diego. She specializes in the philosophy of natural and social science and has worked extensively on modeling in science, especially in physics and economics. Her most recent work is on the nature and use of evidence for evidence-based policy. She is a Fellow of the British Academy, a member of the American Academy of Arts and Sciences, of the German Society of Science (Leopoldina), the American Philosophical Society and a former MacArthur Fellow. She is currently president of the Philosophy of Science Association.

Anjan Chakravartty is Associate Professor and Director of the Institute for the History and Philosophy of Science and Technology at the University of Toronto. His research focuses on central issues in the epistemology of science and metaphysics, including topics in the philosophy of physics and biology. He is a winner of the biennial Canadian Philosophical Association Book Prize for *A Metaphysics for Scientific Realism: Knowing the Unobservable* (Cambridge University Press 2007), and has published widely on scientific realism, causation, laws of nature, and metaphysics and empiricism, as well as on models, abstraction and idealization, and scientific representation.

David Davies is Associate Professor of Philosophy at McGill University. He is the author of *Art as Performance* (Blackwell, 2004), *Aesthetics and Literature* (Continuum, 2007), and *Philosophical Foundations of the Performing Arts* (Blackwell, forthcoming), and the editor of *The Thin Red Line* (2008) in the Routledge series *Philosophers on Film*. He has published widely in the philosophy of art on topics relating to ontology, artistic value, literature, film, music, theatre, and the visual arts. He has also published articles on topics in metaphysics, philosophy of language, philosophy of mind, and philosophy of science.

Catherine Z. Elgin is professor of philosophy of education at Harvard Graduate School of Education. She is the author of *Considered Judgment, Between the Absolute and the Arbitrary, With Reference to Reference*, and co-author (with Nelson Goodman) of *Reconceptions in Philosophy and Other Arts and Sciences*. She is editor of *The Philosophy of Nelson Goodman*, and co-editor (with Jonathan E. Adler) of *Philosophical Inquiry*. She has received fellowships from the National

Endowment of the Humanities, the American Council of Learned Societies, the John Dewey Foundation, the Spencer Foundation, the Andrew Mellon Foundation and the Bunting Institute.

James Elkins is E.C. Chadbourne Chair in the Department of Art History, Theory, and Criticism at the School of the Art Institute of Chicago. His writing focuses on the history and theory of images in art, science, and nature. Among his books on scientific images are *Six Stories from the End of Representation: Images in Painting, Photography, Microscopy, Astronomy, Particle Physics, and Quantum Mechanics, 1985–2000* (2008) and *The Domain of Images* (1999). In addition to editing the seven-volume *The Art Seminar* series (2005–2008), his current projects include a book called *The Project of Painting: 1900–2000*, a series called *Theories of Modernism and Postmodernism in the Visual Arts*, and a book written against *Camera Lucida*.

Roman Frigg is a Senior Lecturer in Philosophy at the London School of Economics and Deputy Director of the Centre for Natural and Social Science (CPNSS). He holds a PhD in Philosophy from the University of London and an MSc in Theoretical Physics from the University of Basel, Switzerland. His main research interests are in general philosophy of science and philosophy of physics. He has published papers on scientific modeling, quantum mechanics, the foundations of statistical mechanics, randomness, chaos, complexity theory, probability, and computer simulations. Further information can be found on his website at www.romanfrigg.org.

Manuel García-Carpintero is Professor at the Department of Logic, History and Philosophy of Science, University of Barcelona, and Director of the Master and PhD Program *Analytic Philosophy*. He works on the philosophy of language and he is preparing a book on the nature of speech acts, focusing on assertion and ancillary speech acts such as presupposition and reference.

Matthew C. Hunter is Weisman Postdoctoral Instructor in Art History at California Institute of Technology. His research examines interactions of art and science in early modern Europe, and he is currently preparing a book entitled *Wicked Intelligence: Visual Art and the Science of Experiment in Restoration London*. He has received fellowships from institutions including the Samuel H. Kress Foundation, the Social Science Research Council and the Whiting Foundation. He is also co-organizer of "The Clever Object Research Project" at the Courtauld Institute of Art, London.

John Hyman is a Fellow of The Queen's College, Oxford, Professor of Aesthetics in the University of Oxford, and Editor of *The British Journal of Aesthetics*. In 2001–2002 he was a Getty Scholar at the Getty Research Institute, Los Angeles, and in 2002–2003 he was a Fellow of the Wissenschaftskolleg zu Berlin. His most recent book is *The Objective Eye* (University of Chicago Press, 2006). In 2010–2012, he will hold a Leverhulme Major Research Fellowship, which was awarded to enable him to complete a book about action and cognition, entitled *After the Fall*.

Dawna Schuld teaches Modern and Contemporary American Art in the Department of the History of Art at Indiana University. Trained as an artist, her focus remains the intersections between artistic practice and questions of perception and cognition. She received her PhD at the University of Chicago.

Adam Toon studied for his PhD at the Department of History and Philosophy of Science at the University of Cambridge and is now a Postdoctoral Research Fellow in the Department of Philosophy at the University of Bielefeld. His other publications include "The ontology of theoretical modelling: models as make-believe", *Synthese* (2010).

Introduction

Roman Frigg and Matthew C. Hunter

Representation is a concern crucial to the sciences and the arts alike. Scientists devote substantial time to devising and exploring representations of all kinds. From photographs and computer-generated images to diagrams, charts, and graphs; from scale models to abstract theories, representations are ubiquitous in, and central to, science. Likewise, after spending much of the twentieth century in proverbial exile as abstraction and formalist aesthetics reigned supreme, representation has returned with a vengeance to contemporary visual art. Representational photography, video and ever-evolving forms of new media now figure prominently in the globalized art world, while this "return of the real" has re-energized problems of representation in the traditional media of painting and sculpture. If it ever really left, representation in the arts is certainly back.

Central as they are to science and art, these representational concerns have been perceived as different in kind and as objects of separate intellectual traditions. Scientific modeling and theorizing have been topics of heated debate in twentieth century philosophy of science in the analytic tradition, while representation of the real and ideal has never moved far from the core humanist concerns of historians of Western art. Yet, both of these traditions have recently arrived at a similar impasse. Thinking about representation has polarized into oppositions between mimesis and convention. Advocates of mimesis understand some notion of mimicry (or similarity, resemblance or imitation) as the core of representation: something represents something else if, and only if, the former mimics the latter in some relevant way. Such mimetic views stand in stark contrast to conventionalist accounts of representation, which see voluntary and arbitrary stipulation as the core of representation. Occasional exceptions only serve to prove the rule that mimesis and convention govern current thinking about representation in both analytic philosophy of science and studies of visual art.

This conjunction can hardly be dismissed as a matter of mere coincidence. In fact, researchers in philosophy of science and the history of art have increasingly found themselves trespassing into the domain of the other community, pilfering ideas and approaches to representation. Cognizant of the limitations of the accounts of representation available within the field, philosophers of science have begun to look outward toward the rich traditions of thinking about representation in the visual and literary arts. Simultaneously, scholars in art history and affiliated fields like visual studies have come to see images generated in scientific contexts as not merely interesting illustrations derived from "high art", but as sophisticated visualization techniques that dynamically challenge our received conceptions of representation and aesthetics.

Beyond Mimesis and Convention: Representation in Art and Science is motivated by the conviction that we students of the sciences and arts are best served by confronting our mutual impasse and by recognizing the shared concerns that have necessitated our covert acts of kleptomania. Drawing leading contributors from the philosophy of science, the philosophy of literature, art history and visual studies, our volume takes its brief from our title. That is, these essays aim to put the evidence of science and of art to work in thinking about representation by offering third (or fourth, or fifth) ways beyond mimesis and convention. In so doing, our contributors explore a range of topics—fictionalism, exemplification, neuroaesthetics, approximate truth—that build upon and depart from ongoing conversations in philosophy of science and studies of visual art in ways that will be of interest to both interpretive communities. To put these contributions into context, the remainder of this introduction aims to survey how our communities have discretely arrived at a place wherein the perhaps surprising collaboration between philosophy of science and art history has become not only salubrious, but a matter of necessity.

Before doing so, one qualifying remark is in order. In recent decades, interactions between art and science have commanded substantial attention in the humanities and social sciences. This stimulating work has often employed representation to advance broader theses about the nature of art and science.[1] The aim of our introduction is not to provide an exhaustive survey of that ever-expanding literature or the range of social, political and other contacts it has elaborated.[2] Because the concerns of the essays gathered here are largely conceptual in their focus on representation, our aim is to indicate the major trends in understanding representation in both scientific and artistic domains, emphasizing salient cross-disciplinary connections between them.

From Science to Art

Modern philosophy of science has its roots in the empiricist philosophy that emerged at the end of the nineteenth century in the works of Ernst Mach, Henri Poincaré, and Pierre Duhem, and which found its culmination in the logical positivism of the Vienna Circle and the Berlin Group.[3] This tradition understood

---

[1] Influential examples of this approach include Fyfe and Law (1988); and Lynch and Woolgar (1990).

[2] For a capacious survey of recent humanities-based scholarship on art/science interactions in the twentieth century, see Henderson (2004). More broadly, see Galison and Jones (1998), and Latour and Weibel (2002).

[3] The history of this movement is discussed in Kraft (1953) and Stadler (2001).

scientific representation as linguistic: scientific theories are descriptions of their subject matter articulated in a concise formal language. More specifically, logical positivism advocated what is now commonly referred to as the "syntactic view of theories".[4] According to this view, the backbone of a scientific theory is a formal calculus, consisting of axioms and rules of inference. This calculus contains both logical and non-logical terms. The former are connectives such as "and" and "or", and quantifiers like "for all" and "there exists". These are provided by the formal apparatus and are taken for granted in the context of empirical science. The latter are terms that provide the empirical content of a theory. Newtonian mechanics, for instance, contains the terms "*a*" and "*F*", which are interpreted as standing for acceleration and force respectively. Since the logical terms are assumed to be unproblematic, the main issue facing this paradigm is to explain in what way terms like "*a*" and "*F*" come to stand for something. Considerable efforts have been made to answer this question, and various different proposals have been put forward. The detail of these, as well as their relative advantages and weaknesses, need not occupy us here. The important point is that the problem of scientific representation was conceived to be a special case of a more general problem: the relation of language to reality. Accordingly, understanding the semantics of scientific theory was considered by logical positivists to be a problem pertaining to the philosophy of language.

Scientific models, which are now seen as a central concern for questions of representation in science, had a rather fluctuating fate in the philosophical debate about science. In the logical positivist picture of science, models were regarded as otiose in a systematic exposition of a scientific theory. Rudolph Carnap famously remarked that "the discovery of a model has no more than an aesthetic or didactic or at best heuristic value, but it is not at all essential for a successful application of the physical theory" (1938, 210). Similarly, Carl G. Hempel held that "all reference to analogies or analogical models can be dispensed with in the systematic statement of scientific explanations" (1965, 440). Although some writers, in particular Richard Braithwaite (1953, Chapter 4) and Ernest Nagel (1961, Chapter 6) tried to canvass a more favourable picture of the use and function of models in science, in particular by emphasizing their heuristic function, the positivist take on the subject matter remained deflationary.

The tides changed in the 1960s, when the syntactic view of theories came under attack from various sides. The main tenor of these criticisms was that the syntactic view did not only get the details wrong; it in fact started off on the wrong foot. Indeed, the very idea of the syntactic view—that theories are linguistic entities providing a description of the theory's subject matter—was increasingly deemed untenable.[5] By 1970, the syntactic view had largely been surmounted by a new analysis of theories, the so-called "semantic view of theories". On this view, a scientific

---

[4] Canonical statements of the syntactic view are Carnap (1938, 1956), Braithwaite (1953), and Nagel (1961).

[5] For survey of these criticisms see Suppe (1977).

theory is a collection of models rather than sentences, where models are construed as non-linguistic entities. This move is important for two reasons. First, by construing theories as families of models, the semantic view assigned models a central role in the edifice of science, thereby paving the way for a substantive discussion of the roles and functions that models perform in science. Secondly, by emphasizing the non-linguistic character of models, the semantic view had come to pose the problem of understanding scientific representation in a completely different way. The problem was no longer a matter of understanding the language of science, but rather of cashing out how something non-linguistic can represent a part or aspect of the real world. The question had become: how does a model represent its target system?

Over the years, the semantic view has been developed in different ways. Details aside, these approaches can be divided into two classes according to their understanding of the ontology of models and the representational relation between model and target. Originating with Patrick Suppes and now held by most writers in the field, this first category takes models to be mathematical structures, which represent their target systems by being isomorphic to them.[6] According to this view, a mathematical structure S is a collection of objects that enter into certain relations. The structure required by this account is a *mathematical* structure insofar as nothing is assumed about either the nature of the objects it contains or about the nature of the relations between those objects. These objects are taken to be featureless dummies: all that we can say about them is that they are objects. Not assuming anything about the nature of a relation means simply that it is stipulated to hold between a certain number of things but without assuming anything about what the relation itself is. For instance, if we have three objects a, b, and c, a relation R is the set consisting of the ordered pairs $\langle a, b \rangle$ and $\langle b, c \rangle$. Thus, the relation R holds between a and b, and b and c, but not between, say, a and c. Whether this relation in itself is "being in love with" or "standing to the left of" is irrelevant as far as mathematics is concerned.

Structures thus understood are not in themselves "about" anything in the world. According to the semantic view, they acquire representational power if an isomorphism is established between such a structure and the part of the real world in which we are interested.[7] This involves identifying objects in the world and pairing them up with the objects in the structure so that two conditions are satisfied. First, the pairing has to be one-to-one, meaning that to each object in the given structure corresponds exactly one object in the world, and *vice versa.* Second, these pairings have to be such that their relations are preserved. In other words, if a relation R holds between certain objects $a, b, c, \ldots$ in the structure, there must be a relation R' in the world which holds between (and only between) those objects in the world that have been paired up with $a, b, c, \ldots$ The relations in the structure have to mirror

[6] See Suppes (1960). Further proponents of this view include Suppe (1989), van Fraassen (1980), French and Ladyman (1999), Da Costa and French (1990), and with a different emphasis by Balzer et al. (1987).

[7] Some versions of the semantic view postulate other mappings such as embedding (Redhead 2001) or partial isomorphism (French and Ladyman 1999).

relations in the world. Thus, the structural isomorphism demanded by this version of the semantic view of theories is strongly mimetic in nature.

This first, formal iteration of the semantic view stands in contrast to the work of philosophers like Ronald Giere (1988) who take models to be abstract objects in a rather different sense. Instead of viewing them as structures in the abstract mathematical sense, Giere understands models to be idealized objects. For instance, in mechanics when we want to calculate the frequency of a pendulum bob, we do not make calculations on the real bob. Rather, we neglect air resistance, assume the bob is an ideal sphere, assume the spring has no friction, and so on. The object we thus construct—the object consisting of an ideally spherical bob and so on—is the model. According to Giere's view, this model represents its target by being similar to it in certain respects and to certain degrees. Like the isomorphism sought in the mathematical version of the semantic view, then, Giere's analysis of similarity envisions a mimetic conception of representation. Thus, both prominent versions of the semantic view of scientific theories presents us with an approach to representation that is squarely located within a time-honoured tradition of analyzing representation in terms of mimesis.

Yet, this conception of models and representation has not been universally accepted. Particularly, it has come under attack by writers who stand in a tradition of thinking about models and theories that is driven by a focus on scientific practice and whose method is based on case studies rather than rational reconstruction and formal analysis. In general, these writers have shared the semantic view's dismissal of the syntactic view and agreed that models have to occupy center stage in a tenable analysis of scientific theorizing. However, in a tradition that dates back to the 1960s, these philosophers have disagreed with the semantic view's analysis of models and, in particular, its claim to universality. Peter Achinstein (1968), for example, pointed out that there are many different kinds of models; while some models are irreducibly linguistic, no overarching theory can account for all of them. Focusing on examples like wooden models of cars tested in wind tunnels, Max Black (1960) demonstrated the importance of material models—models that are actually built and used in the laboratory. Mary Hesse (1963), meanwhile, emphasized the many different analogical relations models can hold to their target systems, showing that no one single relation accounts for the representational function of all models. The more recent work of Nancy Cartwright (1983), Margaret Morrison (1998), Mary Morgan (1997), and others in the "models-as-mediators" project (Morgan and Morrison 1999) have argued that both the relations between models and theories and between models and their target systems are far more complex than the semantic view has allowed. It is precisely because models are autonomous from theory and the world alike, according to this approach, that they can meaningfully function as mediators between the two. For this reason, this group has rejected isomorphism and similarity views of representation, emphasizing that models relate to the world in much more complex ways.

But, the utility of isomorphism and similarity to the analysis of representation has had other critics. A long line of thought in the Western tradition has sought to explain pictorial representation in terms of mimesis: a picture represents its target because it resembles the target. If an almost equally long tradition has criticized this analysis, few have done so more powerfully than the modern *locus classicus*: Nelson Goodman's *Languages of Art* (1976). Goodman points out that, for an analysis of representation, similarity is a red herring: it is neither necessary nor sufficient for representation. Goodman's arguments have sparked repeated debate in discussions of the nature of pictorial representation—debates that are ongoing in philosophical aesthetics. Indeed, arguments that have emerged in this debate have recently been brought to bear on scientific representation. Roman Frigg (2002, 2006) and Mauricio Suárez (2003, 2004) have aimed to show that mimetic conceptions of scientific representation based on either similarity or isomorphism are blind alleys. In response to these criticisms, revised similarity and isomorphism accounts have been proposed by Giere (2004) and Bas van Fraassen (2004), yet they remain controversial. An elegant way around the problem seems to be to opt for the other extreme end of the spectrum and declare that conventional stipulation is the core of representation. On such a view, nothing but a voluntary act of stipulation is involved in making something represent something else. Although this view is the foil against which many accounts of representation have been formulated, it is rarely carefully articulated.[8] Craig Callender and Jonathan Cohen (2006) give an explicit endorsement of this view—an argument that Adam Toon's contribution to this volume claims to be untenable.

If neither strongly mimetic nor rigorously conventionalist views can satisfactorily account for the complex, variegated field of scientific representations now studied by philosophers of science, the moment has arrived for us to re-examine our conceptions of representation more comprehensively. As the foregoing criticisms of the isomorphism account demonstrate, work towards such an expanded analysis has drawn parallels between representation in art and science as a way to think through the relations of the mimetic and the conventional. However, where salient parallels have traditionally been identified between science and pictorial representation, more recent work has emphasized the crucial comparison with literature. Relations between storytelling and modeling have become particularly important to this conversation. Donald McCloskey (1990) has drawn attention to the parallels of economic modeling and storytelling; Stephan Hartmann (1999) and Morgan (2001) have emphasized that stories are an integral part of models that cannot be omitted from an analysis of modeling; and Till Grüne-Yanoff and Paul Schweinzer (2008) argue that stories are crucial to applying abstract models to real-world scenarios. Nancy Cartwright takes the parallels between models and literature particularly seriously, and has developed an account of representation by likening them to literary fables. First proposed in her (1999), Cartwright's view is further elaborated in her contribution to this book. Similarly, Toon's contribution to this book takes the

[8] Such a view is often attributed to Goodman himself on the basis that he held that denotation was the core of representation. While there is a grain of truth in this, Goodman's view seems to have been more nuanced because he recognized that denotation is not always rooted (solely) in act of conventional stipulation. Goodman's views are discussed in Elgin's and Chakravartty's contributions to this book.

argument into a different direction. By his reading, Kendall Walton's (1990) pretense theory of fiction offers promising resources for elaborating a powerful account of representation in science.

In the wake of the critique of the semantic view of scientific theories, an account of modeling now faces two central questions: what are models and how do they represent? If most of the available literature has focused upon the latter, representational question, Frigg (2003, 2010) and Peter Godfrey-Smith (2006) have argued that literary fiction also provides the clue for an answer to the former, ontological question. Models, in this account, should be seen as the same kind of entities as imaginary places and characters in literary fiction. This basic idea can be cashed out in different ways. In his contribution, Frigg develops an account of models that, like Toon's, draws on Walton's theory of fiction. Manuel García-Carpintero shares the view that the ontology of literary fiction and models are identical, which he defends in his contribution through an account of fiction based on Stephen Yablo's theory of metaphor.

This renewed interest in exploring contacts between artistic and scientific representation does not stop at semantics and ontology. Catherine Elgin (1996) has argued that science and art share important epistemic practices in common. In her contribution to this volume, she builds upon this approach and presents an account of the acquisition of knowledge based on the notion of exemplification. Few scientists would claim that even our best theories are true; but most would submit that they get essential elements right. In other words, our best theories are approximately true. Anjan Chakravartty sets out to analyze the notion of approximate truth in science by drawing attention to representational practices in the arts. Commensurately, while thought experiments have played an important role in science at least since Galileo, David Davies' contribution to this volume demonstrates that there is much to be learned about how such experiments work by examining their similarity to the plots of literary fiction. What fictions are and how we learn from them, so these contributors suggest, are questions that now need to be shared between students of representation in science and art.

From Art to Science

Contemporary to and often conversant with later nineteenth century philosophers of science, the founders of academic art history looked askance upon a venerable tradition of thinking about representation in art.[9] According to that tradition, the visual arts shared a common root with literature, music and a vast array craft practices in their mutual derivation from imitation. Classical Greek philosophers had designated such arts as *mimesis*, a term that would occupy a central but conflicted

[9] On contacts between science and art history's disciplinary formation in nineteenth century Germany, see Mallgrave and Ikonomou (1996). A standard intellectual history of key figures in art history is Podro (1984).

place in the Western tradition (Auerbach 1953, Halliwell 2002). Writing in the wake of Greek art's naturalistic efflorescence of the fifth century BCE, Plato's philosophy keenly registered this vexed position. In the infamous argument set out in *The Republic*, Plato's (1961) Socrates reasons that because works of mimetic art are but second-hand simulacra—imitations made from the material copies of their ideal Forms—painters, sculptors and poets amount to dangerous dissemblers who should be banned from the *kallipolis*, the ideal state. An ostensibly more sympathetic account of art's mimetic nature and its transformative capacities was advanced by Plato's student Aristotle. In the *Poetics*, for example, Aristotle (1982) noted how a visual art like theatre represented men as better than they really are (as in tragedy) or worse than they really are (as in comedy), thereby yielding versions of human action that depart from reality. These creative (and therefore non-*re*presentational) dimensions of art were significantly expanded by some theorists of the European Renaissance who advocated a new conception of art as the product of a divinely-gifted subject: the artist of genius (Panofsky 1968, Koerner 1993, Belting 1994). But, for many Renaissance writers, the imitation of nature by art was a matter of progressive, observable, and almost miraculous fact. Heir to the reclamation of one-point pictorial perspective, the deployment of oil as a painting medium and a host of other ingenious innovations, Renaissance art would be narrated by theorists like Giorgio Vasari (1998) in the mid-sixteenth century as moving progressively toward the perfection of imitative skill.

For nineteenth century Germanic academics keen to establish the credentials of art history as a science, neither this privilege of naturalistic European art nor the narration of mimetic ascent (or decline) could satisfactorily constrain analysis. Alois Riegl, Heinrich Wölfflin and their art-historical contemporaries understood their project to demand the interpretation of the diverse, but equally-valid, styles of representation through which the art of geographically and historically varying cultures developed from its own, autonomous causes. Instead of assuming some universal standard against which a work's imitative accomplishment could be measured, the intellectual credentials of art history would be established through its ability to historicize the mode of representation in which an artwork was made and to elucidate the desires and cognitive demands expressed by it. So Wölfflin would famously put it: "Every artist finds certain visual possibilities before him, to which he is bound. Not everything is possible at all times. Vision itself has a history, and the revelation of these visual strata must be regarded as the primary task of art history" (1950, 11). Even if mimesis could still then be assumed as a guiding intention for much of the high art produced in the Western tradition, imitative "content" counted less than the stylistic form in which it was materialized.

By the first decades of the twentieth century, however, the demolition of even this diminished role for mimesis was well under way. Systematically, modernist artists had dispensed with the clever modulations of painterly tone, the perspectival constructions of space developed by Renaissance painters and the even the fundamental assumption that a work of art would serve some representational capacity. These were challenges that historians of art could hardly ignore. Indeed, when publishing his seminal *Art and Illusion: A Study in the Psychology of Pictorial Representation* in the heyday of the non-figurative art of Abstract Expressionism, Ernst Gombrich acknowledged the need to justify studying the traditions of pictorial representation that had been so ruthlessly negated by modernism. Citing then-recent psychological research and its revelation of what he called "a radical reorientation of all traditional ideas about the human mind, which cannot leave the historian of art unaffected", Gombrich catalogued the force of formulas and schemata in the production of convincingly representational images (1961, 27). In the sympathetic reading that he sought to give it, such illusionistic representation would be understood as the product of conventions projected onto the visible world, not copying data received from it. Beginning "not with his visual impression but with his idea or concept", Gombrich argued, the artist selectively introduces information from to the observed target "as it were, upon a pre-existing blank or formulary. And, as often happens with blanks, if they have no provisions for certain kinds of information we consider essential, it is just too bad for the information" (1961, 73). Writing at the apex of High Modernism, Gombrich could recuperate the artistic and intellectual credibility of representational art not by appeal to mimesis, but by elaborating the evolving conventions underpinning it.

Reviewing *Art and Illusion* in 1960, philosopher Nelson Goodman found much to admire in Gombrich's work. Goodman emphasized the book's insight into what he called "the nature of vision and of representation, and the problem of reconciling the objectivity of the latter with its conventionality and the relativity of vision" (1972, 142). Although they parted company over the extent to which Renaissance perspective constituted a convention, Goodman integrated Gombrich's work into the devastating critique of mimetic or resemblance theories of representation that he outlined in *Languages of Art* (1968), a work which stands as one of the most powerful examples of a conventionalist reading of representation.[10] "The plain fact", Goodman claimed therein:

> is that a picture, to represent an object must be a symbol for it, stand for it, refer to it; and that no degree of resemblance is sufficient to establish the requisite relationship of reference. Nor is resemblance *necessary* for reference; almost anything may stand for almost anything else. A picture that represents—like a passage that describes—an object refers to and, more particularly, *denotes* it. Denotation is the core of representation and is independent of resemblance (1976, 5).

Far from following from some heightened degree of resemblance, "realism" in Goodman's iconoclastic analysis turned out to be a residue of habit, a symptom of a representation's adherence to acculturated stereotype. Moreover, by analyzing representation in the arts as systems of symbols, Goodman's work suggested significant possibilities for studying varieties of images that deployed conventions utterly foreign to those of the canon of western art.

Although often in ways contrary to the rigorous analytic tenor of his work, Goodman's "conventionalism" and his attention to non-canonical imagery are broadly instructive of the direction of much recent work on representation in art

[10] For Gombrich's response to Goodman's reading of perspective, see Gombrich (1972).

history. Since the 1970s, contact with structuralist linguistics, semiotics and related interpretive frameworks has transformed art-historical thinking about artistic representations, calling attention to the codes and conventions of socio-economic, political, racial or other interests embodied in them.[11] Simultaneously, the discipline has expanded dynamically outward; art historians have come to recognize the necessity of placing the canonical core of European aesthetic objects in dialogue with both the art of "non-Western" cultures and non-elite, non-art images native to the Western tradition itself.[12] If interdisciplinary fields like visual culture and media studies that privilege these questions have had their detractors, one of the most productive topics in this ambit has been the humanities-based study of scientific imagery. Because our own volume touches upon some of the questions this literature has asked, it is instructive to briefly consider how problems of representation have been approached therein.

To several leading scholars, mimetic ambition has stood as a crucial point of conjunction between science and visual art. As the artistic ability to draw empowered Leonardo da Vinci or Galileo to perceive scientific features of natural entities which remained completely unintelligible to their contemporaries[13], so scholars like Svetlana Alpers (1983), Martin Kemp (1990) and Pamela Smith (2004) have argued for strong continuities between the "mirroring of nature" in art and science ca. 1400–1850. An instructively different approach has been taken in historian of science Peter Galison's *Image and Logic*, which analyzes twentieth century particle physics as a struggle between "two competing traditions" (1997, 19). On one side, Galison plots the "image tradition," or those theories and experimental instruments designed to produce representations that are "presented, and defended, as *mimetic*—they purport to preserve the form of things as they occur in the world" (1997, 19). The opposing "logic" tradition, meanwhile, is organized around theories and instruments engineered to yield statistical data, constituting what Galison calls "'homologous' representation" (1997, 19). This strategy of narrating scientific visualization through the opposition of mimetic and conventional representations has recently been developed further by art historian David Freedberg in his *The Eye of the Lynx*. Seventeenth century Italian natural history, Freedberg argues, can be interpreted as a decline of pictures and the rise of conventional diagrams as mimesis was effectively outpaced by the needs of science: "The graphic description of the surfaces of things could not yield the principles of order; these could only be achieved by penetrating beneath the surface, by counting, and by reducing the fullness of pictorial description to their essential geometrical abstractions" (2002, 4). For Freedberg, pictures and diagrams not only map respectively onto the

---

[11] For work that has specifically appealed to Goodman for these ends, see Mitchell (1986). Although this broader art-historical literature is massive, an indicative range of approaches to representation and leading scholars thereof is Bryson et al. (1991).

[12] See, for example, Levenson (1991); and Farago (1995).

[13] See Panofsky (1954, 1962); Edgerton Jr. (1984); and Bredekamp (2000).

resemblance-based epistemological order of the Renaissance and the representational signs of Enlightenment knowledge as theorized by Michel Foucault, but they constitute a "clear, serious, and instructive" polarity (2002, 476 footnote 1). Thus, mimesis and convention have come to be seen not only as different ways of representing natural targets, but as opposing strategies that signal broader intellectual (or other) commitments.

Importantly, the need to think beyond such an opposition of mimesis and convention is one that has already registered within this literature. Especially in studies of the photographic technologies used increasingly in the sciences by the end of the nineteenth century, researchers have aimed to theorize the resulting images in terms of their "indexicality". As influentially articulated by art historian Rosalind Krauss based upon the writings of C.S. Peirce, photography could be understood to produce indexical signs that exceed the mimetic relations of "icons" and conventional relations of "symbols" by means of their causal relation to target objects (1977a, b). "Every photograph", Krauss claimed, "is the result of a physical imprint transferred by light reflections onto a sensitive surface. The photograph is thus a type of icon, or visual likeness, which bears an indexical relationship to its object" (1977a, 75). If the limits of indexical relations upon scientific photography have now been vigorously argued (Snyder 2007, Ellenbogen 2008), attention to the index has developed less in relation to scientific images than in conversations about the implications of photographic aesthetics (Saltzman 2006). More expansive approaches beyond the mimesis/convention opposition—and indeed beyond the art/science binary—have been suggested in the pioneering work of James Elkins (see for example 1999, 2007, 2008). Central to Elkins' work in this direction and as argued in his essay included here, is a contention that the artistic images privileged by humanities-based scholarship possess nothing like interpretive purchase or theoretical hegemony imagined by art historians and visual theorists. So Elkins argues—and as the exhibition and book reported on in his essay sought to enact—humanities-based researchers can only begin to truly theorize our "increasingly visual society" by listening to and engaging in technical detail with the profuse, complicated ways in which visual materials are produced and accorded representational values in the sciences.

The contributions of Matthew C. Hunter, Dawna Schuld and John Hyman all engage with available studies of relations between art and science. Hunter's essay focuses upon the material models and broader visual activities of Robert Hooke in later seventeenth century London. Trained as a painter but best known for his numerous accomplishments as an experimental scientist, Hooke has stood for humanities-based interpreters as an arch example of the mutual hold of mimesis upon early modern art and science. Drawing upon recent work from the philosophy of science, Hunter demonstrates how Hooke's material models frustrate mimetic readings in departing not only from the natural targets they were intended to represent, but from the theories they ostensibly aimed to elucidate. Theorizing this complexity of Hooke's models, Hunter calls attention to the devilish sophistication of thinking and working with representations in art and science at the cusp of the Enlightenment. Although examining a case from some three hundred years later, Dawna Schuld's essay also considers visual practices generated through the direct interaction between artists and scientists. Schuld shows how the artistic activities developed through a collaboration between experimental psychologists and artists Robert Irwin and James Turrell in late 1960s Los Angeles need to be seen as offering a powerful critique of the Formalist models of modernist aesthetic experience which continue to inform the interpretation of their work. For, drawing upon their experiences in sensory-deprivation chambers, Irwin and Turrell made "conditional art" by eliminating the aesthetic object and manipulating the conventional gallery space in which it would appear. As Schuld argues, this artistic project not only discloses compelling alignments between Formalism and behaviorist psychology, but shows how the work of Irwin and Turrell speaks instructively to recent research in cognitive neuroscience. A suggestive juxtaposition to this approach is offered by John Hyman who critically examines recent studies of visual art by neuroscientists. Considering the work of leading figures in "neuro-aesthetics" like V.S. Ramachandran and Semir Zeki, Hyman analyzes what scientific concepts like "peak shift" can or cannot tell us about artistic representation and assesses the broader prospects of a "neurobiological definition of art".

For readers from the history and theory of art, the essays by contributors like Anjan Chakravartty and Nancy Cartwright may well come as a pleasant surprise. Chakravartty outlines the need to develop a theory of "approximate truth" capable of answering to the significant departures scientific representations make from their target systems. Distinguishing between scientific representations that abstract and those that idealize their targets, Chakravartty argues for different "conditions of approximation" by which each type of representation can be evaluated—and does so by appealing to works of twentieth century art as cognitive resources. Commensurately, Cartwright's essay explores what she calls "highly idealized" models used in the sciences—models that are markedly unlike the real world entities and systems they ostensibly represent. Cartwright turns to the theory of the fable proposed by G.E. Lessing, comparing and contrasting the interpretations required of models to those of fables and parables. Echoing the strong interest in the philosophy of literature that marks our collection as a whole, these essays exemplify the broader desire of the project to put works of art and theories of science "to work" in the shared enterprise of thinking representation beyond mimesis and convention.

Problems and Prospects

It goes without saying that substantial work remains to be done in rethinking our familiar stories about representation. To scholars coming from the humanities, the conceptions of representation to be found in the pages that follow may seem extremely foreign. The visual features that we like to attribute to scientific photographs or illustrations (precision, meticulous attention to detail, and "realism" in numerous variants) are thrown into abeyance, while even the fundamental privilege of visualization that we have come to envision as central to science—an iconophilia seen to be meaningfully coextensive with visual art—is brought into question. Likewise, philosophers of science may find the conceptions of models, representation, truth, and learning suggested in this volume eccentric, if not outlandish. Formidable though such challenges are, it is our conviction that the vitality of our conversations demands that we look beyond binary categories, our discrete intellectual traditions, and our comfortable pathways. The aim of this volume is to make the concerns we share salient, and to suggest how they might best be addressed through collaborative enterprise. If studies of art and science are now moving from contraband traffic to officially-sanctioned trade in our parallel but discrete disciplinary zones, our call is for a more global expansion of trading alliances. Granting amnesty to pirates and honor to brave privateers, the aim of *Beyond Mimesis and Convention: Representation in Art and Science* is to demonstrate the necessity and advantage of rethinking representation together.

Acknowledgments Thanks to Josh Ellenbogen and Allison Morehead for comments on earlier drafts of this text.

References

Achinstein, P. (1968), *Concepts of Science: A Philosophical Analysis*. Baltimore: Johns Hopkins Press.

Alpers, S. (1983), *The Art of Describing: Dutch Art in the Seventeenth Century*. Chicago: University of Chicago Press.

Aristotle (1982), *Aristotle's Poetics*, trans. J. Hutton. New York: W.W. Norton.

Auerbach, E. (1953), *Mimesis: The Representation of Reality in Western Literature*, trans. W. R. Trask. Princeton: Princeton University Press.

Balzer, W., Moulines, C. U. and Sneed, J. D. (1987), *An Architectonic for Science: The Structuralist Program*. Dordrecht: D. Reidel.

Belting, H. (1994), *Likeness and Presence: A History of the Image before the Era of Art*, trans. E. Jephcott. Chicago: University of Chicago Press.

Black, M. (1960), "Models and Archetypes", in M. Black (ed.), *Models and Metaphors: Studies in Language and Philosophy*, Ithaca and New York: Cornell University Press, 219–243.

Braithwaite, R. (1953), *Scientific Explanation*. Cambridge: Cambridge University Press.

Bredekamp, H. (2000), "Gazing Hands and Blind Spots: Galileo as Draftsman", *Science in Context* 13, 3–4: 423–462.

Bryson, N., Holly, M. A. and Moxey, K. (eds.) (1991), *Visual Theory: Painting and Interpretation*. Cambridge: Polity Press.

Callender, C. and Cohen, J. (2006), "There Is No Special Problem About Scientific Representation", *Theoria* 55: 7–25.

Carnap, R. (1956), "The Methodological Character of Theoretical Concepts", in H. Feigl and M. Scriven (eds.), *The Foundations of Science and the Concepts of Phsychologz and Psychoanalysis*, Minneapolis: University of Minnesota Press, 38–76.

Carnap, R. (1938), "Foundations of Logic and Mathematics", in O. Neurath, C. Morris and R. Carnap (eds.), *International Encyclopaedia of Unified Science. Vol. 1.*, Chicago: University of Chicago Press, 139–213.

Cartwright, N. (1983), *How the Laws of Physics Lie*. Oxford: Oxford University Press.

Cartwright, N. (1999), *The Dappled World: A Study of the Boundaries of Science*. Cambridge: Cambridge University Press.

Da Costa, N. and French, S. (1990), "The Model-Theoretic Approach to the Philosophy of Science", *Philosophy of Science* 57: 248–265.

Edgerton Jr., S. Y. (1984), "Galileo, Florentine 'Disegno,' and the 'Strange Spottednesse' of the Moon'", *Art Journal* 44, 3: 225–232.

Elgin, C. Z. (1996), *Considered Judgment*. Princeton: Princeton University Press.
Ellenbogen, J. (2008), "Camera and Mind", *Representations* 101: 86–115.
Elkins, J. (1999), *The Domain of Images*. Ithaca: Cornell University Press.
Elkins, J. (2007), *Visual Practices Across the University*. Munich: Wilhelm Fink Verlag.
Elkins, J. (2008), *Six Stories from the End of Representation: Images in Painting, Photography, Astronomy, Microscopy, Particle Physics, and Quantum Mechanics, 1980–2000*. Stanford: Stanford University Press.
Farago, C. (ed.) (1995), *Reframing the Renaissance: Visual Culture in Europe and Latin America 1450–1650*. New Haven: Yale University Press.
Freedberg, D. (2002), *The Eye of the Lynx: Galileo, His Friends and the Beginnings of Modern Natural History*. Chicago: University of Chicago Press.
French, S. and Ladyman, J. (1999), "Reinflating the Semantic Approach", *International Studies in the Philosophy of Science* 13: 103–121.
Frigg, R. (2002), "Models and Representation: Why Structures Are Not Enough", *Measurement in Physics and Economics Project Discussion Paper Series, DP MEAS 25/02*.
Frigg, R. (2003), *Re-presenting Scientific Representation*, PhD Thesis. London: University of London.
Frigg, R. (2006), "Scientific Representation and the Semantic View of Theories", *Theoria* 55: 49–65.
Frigg, R. (2010), "Models and Fiction", *Synthese* 172(2): 251–268.
Fyfe, G. and Law, J. (eds.) (1988), *Picturing Power: Visual Depiction and Social Relations*. London: Routledge.
Galison, P. (1997), *Image and Logic: A Material Culture of Microphysics*. Chicago: University of Chicago Press.
Galison, P. and Jones, C. (eds.) (1998), *Picturing Science, Producing Art*. London: Routledge.
Giere, R. N. (1988), *Explaining Science: A Cognitive Approach*. Chicago: Chicago University Press.
Giere, R. N. (2004), "How Models Are Used to Represent Reality", *Philosophy of Science* 71, 4: 742–752.
Godfrey-Smith, P. (2006), "The Strategy of Model-Based Science", *Biology and Philosophy* 21: 725–740.
Gombrich, E. (1961), *Art and Illusion: A Study in the Psychology of Pictorial Representation*. London: Phaidon.
Gombrich, E. (1972), "The 'What' and the 'How': Perspective Representation and the Phenomenal World", in R. Rudner and I. Scheffler (eds.), *Logic and Art: Essays in Honor of Nelson Goodman*, New York: Bobbs-Merrill, 129–149.
Goodman, N. (1972), "Review of Goodman's *Art and Illusion*," in *Problems and Projects*, New York: Bobbs-Merrill: 141–146.
Goodman, N. (1976), *Languages of Art*. 2nd ed., Indianapolis and Cambridge: Hackett.
Grüne-Yanoff, T. and Schweinzer, P. (2008), "The Roles of Stories in Applying Game Theory", *Journal of Economic Methodology* 15, 2: 131–146.
Halliwell, S. (2002), *The Aesthetics of Mimesis: Ancient Texts and Modern Problems*. Oxford: Princeton University Press.
Hartmann, S. (1999), "Models and Stories in Hadron Physics", in M. Morgan and M. Morrison (eds.), *Models as Mediators. Perspectives on natural and social science*, Cambridge: Cambridge University Press, 326–346.
Hempel, C. G. (1965), *Aspects of Scientific Explanation and Other Essays in the Philosophy of Science*. New York: Free Press.
Henderson, L. D. (2004). "Editor's Introduction: I. Writing Modern Art and Science – An Overview", *Science in Context* 7, 4: 423–466.
Hesse, M. (1963), *Models and Analogies in Science*. London: Sheed and Ward.
Kemp, M. (1990), *The Science of Art: Optical Themes in Western Art from Brunelleschi to Seurat*. New Haven: Yale University Press.

Koerner, J. L. (1993), *The Moment of Self-Portraiture in German Renaissance Art*. Chicago: University of Chicago Press.
Kraft, V. (1953), *The Vienna Circle: The Origins of Neo-Positivism*. New York: Philosophical Library.
Krauss, R. (1977a), "Notes on the Index: Seventies Art in America", *October* 3: 68–81.
Krauss, R. (1977b), "Notes on the Index: Seventies Art in America. Part 2", *October* 4: 58–67.
Latour, B. and Weibel, P. (2002), *Iconoclash: Beyond the Image Wars in Science, Religion and Art*. Cambridge: MIT Press.
Levenson, J. A. (ed.) (1991), *Circa 1492: Art in the Age of Exploration*. New Haven: Yale University Press.
Lynch, M. and Woolgar, S. (eds.) (1990), *Representation in Scientific Practice*. London: MIT Press.
Mallgrave, H. F. and Ikonomou, E. (1996), "Introduction", in *Empathy, Form, and Space: Problems in German Aesthetics, 1873–1893*, Santa Monica: Getty Research Center, 1–66.
McCloskey, D. N. (1990), "Storytelling in Economics", in C. Nash (ed.), *Narrative in Culture: The Uses of Storytelling in the Sciences, Philosophy, and Literature*, London: Routledge, 5–22.
Mitchell, W. J. T. (1986), *Iconology: Image, Text, Ideology*. Chicago: University of Chicago Press.
Morgan, M. (2001), "Models, Stories and the Economic World", *Journal of Economic Methodology* 8, 3: 361–384.
Morgan, M. and Morrison, M. (eds.) (1999), *Models as Mediators: Perspectives on Natural and Social Science*. Cambridge: Cambridge University Press.
Morgan, M. S. (1997), "The Technology of Analogical Models: Irving Fisher's Monetary Worlds", *Philosophy of Science* 64, Supplement: S304–S314.
Morrison, M. (1998), "Modelling Nature: Between Physics and the Physical World", *Philosophia Naturalis* 35:65–85.
Nagel, E. (1961), *The Structure of Science*. London: Routledge and Keagan Paul.
Panofsky, E. (1954), *Galileo as a Critic of the Arts*. The Hague: M. Nijhoff.
Panofsky, E. (1962), "Artist, Scientist, Genius: Notes on the 'Renaissance-Dämmerung'", in W. K. Ferguson et al. (eds.), *The Renaissance: Six Essays*, New York: Harper & Row, 121–182.
Panofsky, E. (1968), *Idea: A Concept in Art Theory*, trans. J.S. Peake. New York: Harper & Row.
Plato (1961), *The Collected Dialogues of Plato*, in E. Hamilton and H. Cairns (ed.), Princeton: Princeton University Press.
Podro, M. (1984), *The Critical Historians of Art*. New Haven: Yale University Press.
Saltzman, L. (2006), *Making Memory Matter: Strategies of Remembrance in Contemporary Art* Chicago: University of Chicago Press.
Smith, P. H. (2004), *The Body of the Artisan: Art and Experience in the Scientific Revolution*. Chicago: University of Chicago Press.
Snyder, J. (2007), "Pointless", in J. Elkins (ed.), *Photography Theory*, New York: Routledge, 369–385.
Stadler, F. (2001), *The Vienna Circle: Studies in the Origins, Development and Influence of Logical Empiricism*. Berlin and New York: Springer.
Suárez, M. (2003), "Scientific Representation: Against Similarity and Isomorphism", *International Studies in the Philosophy of Science* 17, 3: 225–244.
Suárez, M. (2004), "An Inferential Conception of Scientific Representation", *Philosophy of Science (Supplement)* 71: 767–779.
Suppe, F. (1989), *The Semantic View of Theories and Scientific Realism*. Urbana and Chicago: University of Illinois Press.
Suppe, F. (ed.) (1977), *The Structure of Scientific Theories*. Urbana and Chicago.
Suppes, P. (1960), "A Comparison of the Meaning and Uses of Models in Mathematics and the Empirical Sciences", in P. Suppes (ed.), *Studies in the Methodology and Foundations of Science: Selected Papers from 1951 to 1969*, Dordrecht: Reidel 1969, 10–23.
van Fraassen, B. C. (1980), *The Scientific Image*. Oxford: Oxford University Press.
van Fraassen, B. C. (2004), "Science as Representation: Flouting the Criteria", *Philosophy of Science* 71, Supplement: S794–S804.

Vasari, G. (1998), *The Lives of the Artists*, trans. J.C. Bondanella and P. Bondanella. New York: Oxford World Classics.

Walton, K. L. (1990), *Mimesis as Make-Believe: On the Foundations of the Representational Arts*. Cambridge, MA: Harvard University Press.

Wölfflin, H. (1950), *Principles of Art History: The Problem of the Development of Style in Later Art*. trans. M.D. Hottinger. New York: Dover.

Telling Instances

Catherine Z. Elgin

Science, we are told, is (or at least aspires to be) a mirror of nature, while art imitates life. If so, both disciplines produce, or hope to produce, representations that reflect the way the mind-independent world is. Scientific representations are supposed to be complete, accurate, precise and distortion-free. Although artistic representations are granted more leeway, they too are supposed to resemble their subjects. Underlying these clichés is the widespread conviction that representations are intentional surrogates for, or replicas of, their objects. If so, a representation should resemble its referent.

This stereotype is false and misleading. It engenders unnecessary problems in the philosophy of science and the philosophy of art. It makes a mystery of the effectiveness of sketches, caricatures, scientific models, and representations with fictional subjects. Indeed, the stereotype strongly suggests that there is something intellectually suspect about such representations. Caricatures exaggerate and distort. Sketches simplify. Models may do all three. Many pictures and models flagrantly fail to match their referents. Representations with fictional subjects have no hope of matching, since they have no referents to match. The same subject, real or fictive, can be represented by multiple, seemingly incongruous representations. These would be embarrassing admissions if representations were supposed to accurately reflect the facts.

Mimetic accounts of representation fail to do justice to our representational practices. Many seemingly powerful and effective representations turn out on a mimetic account to be at best flawed, at worst unintelligible. Nor is it clear why we should want to replicate reality. As Virginia Woolf allegedly said, "Art is not a copy of the real world. One of the damn things is enough!"[1] To replicate reality would simply be to reproduce the blooming buzzing confusion that confronts us. What is the

C.Z. Elgin (✉)
Harvard University, Cambridge, MA, USA
e-mail: catherine_elgin@harvard.edu

[1] Goodman (1968, 3). Goodman was not able to find the original source for this quotation. Although a number of sources credit Woolf with it, I have found none that knows where in her work it is to be found.

R. Frigg, M.C. Hunter (eds.), *Beyond Mimesis and Convention*, Boston Studies in the Philosophy of Science 262, DOI 10.1007/978-90-481-3851-7_1,

value in that? Our goal should be to make sense of things—to structure, synthesize, organize, and orient ourselves toward things in ways that serve our ends.

Nominalism is of no help with this task, for it is undiscriminating. According to nominalism, there are no natural kinds. Since, except for paradoxically self-referential cases, every collection of entities constitutes an extension, every two or more objects resemble each other in virtue of their joint membership in some extension. Thus mere resemblance cannot serve as a ground for representation, else everything would represent everything else. This is true but unhelpful. That there are no natural kinds tells us virtually nothing about how representations function.

The problem lies in the metaphor of the mirror and the ideal of replication. Neither art nor science is, can be, or ought to be, a mirror of nature. Rather, I will argue, effective representations in both disciplines embody and convey an *understanding* of their subjects. Since understanding is not mirroring, failures of mirroring need not be failures of understanding. Once we recognize the way science affords understanding, we see that the features that look like flaws under the mirroring account are actually virtues. A first step is to devise an account of scientific representations that shows how they figure in or contribute to understanding. It will turn out that an adequate account of scientific representation also affords insight into representation in the arts.

Representation

The term "representation" is irritatingly imprecise. Pictures represent their subjects; graphs represent the data; politicians represent their constituents; representative samples represent whatever they are samples of. We can begin to regiment by restricting attention to cases where representation is a matter of denotation. Pictures, equations, graphs, charts, and maps represent their subjects by denoting them. They are representations *of* the things that they denote.[2] It is in this sense that scientific models represent their target systems: they denote them. But, as Bertrand Russell notes, not all denoting symbols have denotata (Russell 1968, 41). A picture that portrays a griffin, a map that maps the route to Mordor, a chart that records the heights of Hobbits, and a graph that plots the proportion of caloric in different substances are all representations, although they do not represent anything. To be a representation, a symbol need not itself denote, but it needs to be the sort of symbol that denotes. Griffin pictures are representations then because they are animal pictures, and some animal pictures denote animals. Middle Earth maps are representations because they are maps and some maps denote real locations. Hobbit height charts are

[2]This use of "denote" is slightly tendentious, both because denotation is usually restricted to language and because even within language it is usually distinguished from predication. As I use the term, predicates and generic non-verbal representations denote the members of their extensions; see Elgin (1983, 19–35).

representations because they are charts and some charts denote magnitudes of actual entities. Caloric proportion graphs are representations because they are graphs and some graphs denote relations among real substances. So whether a symbol is a representation is a question of what kind of symbol it is. Following Goodman, let us distinguish between representations *of p* and *p*-representations. If *s* is a representation *of p,* then *p* exists and *s* represents *p*. But *s* may be a *p*-representation even if there is no such thing as *p* (Goodman 1968, 21–26). Thus, there are griffin-pictures even though there are no griffins to depict. There is an ideal-gas-description even though there is no ideal gas to describe. There are also mixed cases. The class of dog-representations includes both factual and fictional representations. Factual dog-representations are representations of dogs; fictional dog-representations lack denotata.

Denoting symbols with null denotation may seem problematic. Occasionally philosophers object that in the absence of griffins, there is no basis for classifying some pictures as griffin pictures and refusing to so classify others. Such an objection supposes that the only basis for classifying representations is by appeal to an antecedent classification of their referents. This is just false. We readily classify pictures as landscapes without any acquaintance with the real estate—if any—that they represent. I suggest that each class of *p*-representations constitutes a small genre, a genre composed of all and only representations with a common ostensible subject matter. There is then a genre of griffin-representations and a genre of ideal-gas-representations. And we learn to classify representations as belonging to such genres as we study those representations and the fields of inquiry that devise and deploy them. This is no more mysterious than learning to recognize landscapes without comparing them to the terrain they ostensibly depict.

Some representations denote their ostensible objects. Others do not. Among those that do not, some—such as caloric-representations—simply fail to denote. They purport to denote something, but there is no such thing. They are therefore defective. Others, such as ideal-gas-representations are fictive. They do not purport to denote any real object. So their failure to denote is no defect. We know perfectly well that there is no such animal as a griffin, no such person as Othello, no such gas as the ideal gas. Nonetheless, we can provide detailed representations *as if* of each of them, argue about their characteristics, be right or wrong about what we say respecting them and, I contend, advance understanding by means of them.

Representation As

x is, or is not, a representation *of y* depending on what *x* denotes. And *x* is, or is not, a *z*-representation depending on its genre. This enables us to form a more complex mode of representation in which *x* represents *y as z*. In such a representation, symbol *x* is a *z*-representation that as such denotes *y*. Caricature is a familiar case of representation-as. Winston Churchill is represented as a bulldog; George W. Bush is represented as a deer in the headlights. According to R. I. G. Hughes, representation-as is central to the way that models function in science (Hughes 1997). This excellent idea needs elaboration.

Representation-of can be achieved by fiat. We simply stipulate: let *x* represent *y* and *x* thereby becomes a representation of *y*. This is what we do in baptizing an individual or a kind. It is also what we do in ad hoc illustrations as, for example, when I say (with appropriate accompanying gestures), "If that chair is Widener Library, and that desk is University Hall, then that window is Emerson Hall" in helping someone to visualize the layout of Harvard Yard. We could take any *p*-representation and stipulate that it represents any object. We might, for example, point to a tree-picture and stipulate that it denotes the philosophy department. But our arbitrary stipulation does not bring it about that the tree-representation represents the philosophy department as a tree.

Should we say then that representation-as requires similarity? In that case, what blocks seemingly groundless and arbitrary cases of representation-as is the need for resemblance between the representation and the referent. But as Goodman, Suárez, and others argue, similarity does not establish a referential relationship (Goodman 1968, 4, Suárez 2003). Representation is an asymmetrical relation; similarity is symmetrical. Representation is irreflexive; similarity is reflexive. One might reply that this only shows that similarity is not sufficient for representation-as. Something else determines direction. Then it is the similarity between symbol and referent that brings it about that the referent is represented as whatever it is represented as. The problem is this: Via stipulation, we have seen, pretty much anything can represent pretty much anything else. So nothing beyond stipulation is required to bring it about that one thing represents another. But similarity is ubiquitous. This is the insight of nominalism. For any *x* and any *y*, *x* is somehow similar to *y*. Thus if all that is required for representation-as is denotation plus similarity, then for any *x* that represents *y*, *x* represents *y* as *x*. Every case of representation turns out to be a case of representation-as. In one way or another, the philosophy department is similar to a tree-picture, but it is still hard to see how that fact, combined with the stipulation that a tree-picture represents the department, could make it the case that the department is represented as a tree-picture, much less as a tree. Suppose we add that the similarity must obtain between the content of the *p*-representation and the denotation. Then for any *x-representation* and any *y*, if the *x-representation* denotes *y*, it represents *y* as *x*. In that case, a tree that represented the philosophy department would not represent it as a tree. But a tree-picture that represented the philosophy department would represent it as a tree.

The trouble is that contentful representations, as well as chairs and desks, can be used in ad hoc representations such as the one I gave earlier. If the portrait of the dean on the wall represents Widener Library, and the graph on the blackboard represents University Hall, then the map represents Emerson Hall. This does not make the dean's portrait represent Widener Library as the dean. Evidently, it takes more than being represented by a tree-picture to be represented as a tree. Some philosophy departments can be represented as trees. But to bring about such representation-as is not to arbitrarily stipulate that a tree picture shall denote the department, even if we add a vague intimation that somehow or other the department is similar to a tree. The question is, what is effected by such a representation?

To explicate representation-as, Hughes discusses Sir Joshua Reynolds' painting, *Mrs. Siddons as The Tragic Muse*. The painting denotes its subject and represents her as the tragic muse. How does it do so? It establishes Mrs. Siddons as its denotation. It might represent Mrs. Siddons, a person familiar to its original audience, in a style that audience knows how to interpret. Then, without further cues, they could recognize that the picture is a picture of her. But the painted figure need not bear any particular resemblance to Mrs. Siddons. *We* readily take her as the subject even though we have no basis for comparison. (Indeed, we even take Picasso's word about the identities of the referents of his cubist portraits, even though the figures in them do not look like anyone on earth.) Captioning the picture as a portrait of Mrs. Siddons suffices to fix the reference. So a painting can be connected to its denotation by stipulation. The painting is a tragic-muse-picture. It is not a picture of the tragic muse, there being no such thing as the tragic muse. But it belongs to the same restricted genre as other tragic-muse-representations. To recognize it as a tragic-muse-picture is to recognize it as an instance of that genre. Similarly in scientific cases. A spring is represented as a harmonic oscillator just in case a harmonic-oscillator-representation as such denotes the spring. The harmonic-oscillator-representation involves idealization. So it is not strictly a representation *of* a harmonic oscillator, any more than the Reynolds is a picture *of* the tragic muse.

In both cases a representation that does not denote its ostensible subject is used to denote another subject. Since denotation can be effected by stipulation, there is no difficulty in seeing how this can be done. The difficulty comes in seeing why it is worth doing. What is gained by representing Mrs. Siddons as the tragic muse, or a spring as a harmonic oscillator, or in general by representing an existing object as something that does not in fact exist? The quick answer is that the representation affords epistemic access to features of the object that are otherwise difficult or impossible to discern. To make this out requires resort to another Goodmanian device—exemplification.

Exemplification

Consider a mundane case. Commercial paint companies provide sample cards that instantiate the colors of the paints they sell. The cards also instantiate innumerable other properties. They are a certain size, shape, age, and weight. They are at a certain distance from the Eiffel Tower. They are excellent bookmarks but poor insulators. And so on. Obviously, there is a difference between the colors and these other properties. Some of the properties the cards instantiate, such as their distance from the Eiffel Tower, are matters of complete indifference. Others, such as their size and shape, facilitate but do not figure in the cards' standard function. Under their standard interpretations, the cards serve exclusively as paint samples. They are mere instances of their other properties, but telling instances of their colors. A symbol that is a telling instance of a property exemplifies that property. It points up, highlights, displays or conveys the property. Since it both refers to and instantiates the property, it affords epistemic access to the property (Goodman 1968, 45–68, Elgin 1996, 171–183).

Because exemplification requires instantiation as well as reference, it cannot be achieved by stipulation. Only something that is colored dusky rose can exemplify that shade. Moreover, exemplification is selective. An exemplar can exemplify only some of its properties. It highlights those properties by marginalizing, downplaying, or overshadowing other properties it instantiates. It may exemplify a cluster of properties, as a fabric swatch exemplifies its colors, texture, pattern and weave. But it cannot exemplify all its properties. Moreover, an exemplar is selective in the degree of precision with which it exemplifies. A single splotch color that instantiates dusky rose, rose, and pink may exemplify any of these properties without exemplifying the others. Although the color properties it instantiates are nested, it does not exemplify every property in the nest. Exemplars are symbols that require interpretation.

Paint samples and fabric swatches belong to standardized, regimented exemplificational systems. But exemplification is not restricted to such systems. Any item can serve as an exemplar simply by being used as an example. So items that ordinarily are not symbols can come to function symbolically simply by serving as examples. A teacher might use one student's work as an example of what she wants (or does not want) her other students to do. Moreover, in principle, any exemplar can exemplify any property it instantiates, and any property that is instantiated can be exemplified.

But what is feasible in principle is not always straightforward in practice. Exemplification of a particular property is not always easy to achieve, for not every instance of a property affords an effective example of it. The roof of a crocodile's mouth is a distinctive shade of yellowish pink. Nevertheless, a paint company would be ill advised to recommend that potential customers peer into a crocodile's mouth order to see that color. Crocodiles are so rare and so dangerous that any glimpse we get of the roof of one's mouth is unlikely to make the color manifest. We could not see it long enough or well enough and would be unlikely to attend to it carefully enough or survive long enough after our investigation to decide whether it was the color we want to paint the hall. It is far better to create a lasting, readily available, easily interpretable sample of the color—one whose function is precisely to make the color manifest. Such a sample should be stable, accessible, and have no properties that distract attention from the color. Effective samples and examples are carefully contrived to exhibit particular features. Factors that might otherwise predominate are omitted, bracketed, or muted. This is so, not only in commercial samples, but in examples of all kinds. Sometimes elaborate stage setting is required to bring about the exemplification of properties that are subtle, scarce, or tightly intertwined.

Scientific experiments are vehicles of exemplification. They do not purport to replicate what happens in the wild. Instead, they select, highlight, control and manipulate things so that features of interest are brought to the fore and their relevant characteristics and interactions made manifest. To ascertain whether water conducts electricity, one would not attempt to create an electrical current in a local lake, stream or bathtub. The liquid found in such places contains impurities. So a current detected in such a venue might be due to the electrical properties of the impurities, not those of water. By experimenting on distilled water, scientists bring it about that the conductivity of water is exemplified. But distilled water is nowhere to be found in nature.

Experiments are highly artificial.[3] They are not slices of nature, but contrivances, often involving unnaturally pure samples tested under unnaturally extreme conditions. The rationale for resorting to such artifices is plain. A natural case is not always an exemplary case. A pure sample that is not to be found in nature, tested under extreme conditions that do not obtain in nature, may exemplify features that obtain but are not evident in nature. So by sidelining, marginalizing, or blocking the effects of confounding factors, experiments afford epistemic access to properties of interest.

Not all confounding factors are easily set aside. Some clusters of properties so tightly fuse that they cannot be prized apart. In such cases, we cannot devise a laboratory experiment to test one in the absence of the others. This is where idealizations enter. Factors that are inseparable in fact can be separated in fiction. Even though, for example, every actual swinging bob is subject to friction, we can represent an idealized pendulum that is not. We can then use that idealization in our thinking about pendulums, and (we hope) understand the movement of swinging bobs in terms of it. The question though is how something that does not occur in nature can afford any insight into what does. Here again, it pays to look to art.

Fiction

Like an experiment, a work of fiction selects and isolates, manipulating circumstances so that particular properties, patterns, and connections, as well as disparities and irregularities are brought to the fore. It may localize and isolate factors that underlie or are interwoven into everyday life or natural events, but that are apt to pass unnoticed because other, more prominent factors typically overshadow them. This is why Jane Austen maintained that "three or four families in a country village is the very thing to work on" (Austen 2005). The relations among the three or four families are sufficiently complicated and the demands of village life sufficiently mundane that the story can exemplify something worth noting about ordinary life and the development of moral personality. By restricting her attention to three or four families, Austen in effect devises a tightly controlled thought experiment. Drastically limiting the factors that affect her protagonists enables her to elaborate the consequences of the relatively few that remain.

If our interests are cognitive though, it might seem that this detour through fiction is both unnecessary and unwise. Instead of resorting to fiction, wouldn't it be

---

[3]See Cartwright (1999, 77–104).

cognitively preferable to study three or four real families in a real country village? Probably not, if we want to glean the insights that Austen's novels afford. Even three or four families in a relatively isolated country village are affected by far too many factors for the social and moral trajectories that Austen's novels exemplify to be salient in their interactions. Too many forces impinge on them and too many descriptions are available for characterizing their interactions. Any such sociological study would be vulnerable to the charge that other, unexamined factors played a non-negligible role in the interactions studied, that other forces were significant. Austen evades that worry. She omits such factors from her account and in effect asks: Suppose we leave them out, then what would we see? Similarly, the model pendulum omits friction and air resistance, allowing the scientist in effect to ask: Suppose we leave them out, then what would we see?

Models, like other fictions, can simplify, omitting confounding factors that would impede epistemic access to the properties of interest. They can abstract, paring away unnecessary and potentially confusing details. They can distort or exaggerate, highlighting significant aspects of the features they focus on. They can augment, introducing additional elements that focus attention on properties of interest. They can insulate, screening off effects that would otherwise dominate.

The question is how this is supposed to inform our understanding of reality. That Elizabeth Bennet and Mr. Darcy, who do not exist, are said to behave thus and so does not demonstrate anything about how real people really behave. That an idealized pendulum, which also does not exist, is said to behave thus and so does not demonstrate anything about how actual pendulums behave.

Let us return to the paint company's sample cards. Most people speak of them, and probably think of them as samples *of* paint—the sort of stuff you use to paint the porch. They are not. The cards are infused with inks or dyes of the same color as the paints whose colors they exemplify. It is a fiction that they are samples of paint. But since the sole function of such a card is to convey the paint color, the fiction is no lie. All that is needed is something that is the same color as the paint. A fiction thus conveys the property we are interested in because in the respect that matters, it is no different from an actual instance. The exemplars need not themselves be paint. Similarly in literary or scientific cases. If the sole objective is to exemplify particular properties, then in a suitable context, any symbol that exemplifies those properties will do. If a fiction exemplifies the properties more clearly, simply, or effectively than a strictly factual representation, it is to be preferred to the factual representation.

Still there is a worry.[4] Many scientific models are not capable of instantiating the properties they apparently impute to their targets. If they cannot instantiate a range of properties, they cannot exemplify them. Suppose we model a pendulum as a simple harmonic oscillator. Since exemplification requires instantiation, if the model is to represent the pendulum as having a certain mass, the model must have that mass. But, not being a material object, the model does not have mass. So it

[4]I am grateful to an anonymous referee for pressing this point.

cannot exemplify the mass of the pendulum. This is true. Strictly, the model does not exemplify mass. Rather it exemplifies an abstract mathematical property, the magnitude of the pendulum's mass. Where models are abstract, they exemplify abstract patterns, properties, and/or relations that may be instantiated by physical target systems. It does no harm to say that they exemplify physical magnitudes. But this is to speak loosely. Strictly speaking, they exemplify mathematical (or other abstract) properties that can be instantiated physically.

Both literary fictions and scientific models exemplify properties and afford epistemic access to them. By omitting or downplaying the significance of confounding factors (the Napoleonic wars in the case of *Pride and Prejudice*, intermolecular attraction in the ideal gas, friction in the model pendulum), they constitute a cognitive environment where certain aspects of their subjects stand out. They thereby facilitate recognition of those aspects and appreciation of their significance. They thus give us reason to take those aspects seriously elsewhere.

Of course this does not justify a straightforward extrapolation to reality. From the fact that Elizabeth Bennet was wrong to distrust Mr. Darcy, we cannot reasonably infer that young women in general are wrong to distrust their suitors, much less that any particular young woman is wrong to distrust any particular suitor. But the fiction exemplifies the grounds for distrust and the reasons those grounds may be misleading. Once we have seen them clearly there, we may be in a better position to recognize them in everyday situations. Nor can we reasonably infer from the fact that ideal gas molecules exhibit no mutual attraction, that neither do helium molecules. But the behavior ideal gas molecules exemplify in the model may enable us to recognize such behavior amidst the confounding factors that ordinarily obscure what is going on in actual gases.

Epistemic Access

Let us return to Reynolds' representation of Mrs. Siddons as the tragic muse. The tragic muse is a figure from Greek mythology who is supposed to inspire works of tragedy—works that present a sequence of events leading inexorably from a position of eminence to irrecoverable, unmitigated loss, thereby inspiring pity and terror (Aristotle 1973, 677). A tragic muse representation portrays a figure capable of inspiring such works, one who exemplifies such features as nobility, seriousness, inevitability, and perhaps a somber dramaticality, along with a capacity to evoke pity and terror. To represent a person as the tragic muse is to represent her in such a way as to reveal or disclose such characteristics in her or to impute such characteristics to her.

The ideal gas law is an equation ostensibly relating temperature, pressure, and volume in a gas. To satisfy that equation, a gas would have to consist of perfectly elastic spherical particles of negligible volume and exhibiting no mutual attraction. The law thus defines a model that mandates specific values for size, shape, elasticity, and attraction. With these parameters fixed, the interdependence of the values of temperature, pressure, and volume is exemplified. The law and the model it defines are fictions. There is no such gas. Nevertheless, the model advances our understanding of gas dynamics. It exemplifies a relation that is important, but hard to discern in the behavior of actual gases. Hughes maintains that the relation between a model and its target is representation-as. The model is a representation—a denoting symbol that has an ostensible subject and portrays its ostensible subject in such a way that certain features are exemplified. It represents its target (its denotatum) as exhibiting those features. So to represent helium as an ideal gas is to impute to it features that the ideal gas model exemplifies. By setting the parameters to zero, it in effect construes the actual size, shape, inelasticity, and mutual attraction of the molecules as negligible. Strictly, of course, in helium the values of those parameters are not zero. But the imputation allows for a representation that discloses regularities in the behavior of helium that a more faithful representation would obscure. The model then foregrounds the interdependence of temperature, pressure, and volume, making it and its consequences manifest.

Representing a philosophy department as a tree might exemplify the ways the commitments of the various members branch out of a common, solid, rooted tradition, and the way that the work of the graduate students further branches out from the work of their professors. It might intimate that some branches are flourishing while others are stunted growths. It might even suggest the presence of a certain amount of dead wood. Representing the department as a tree then affords resources for thinking about it, its members and students, and their relation to the discipline in ways that we otherwise would not.

I said earlier that when x represents y as z, x is a z-representation that *as such* denotes y. We are now in a position to cash out the "as such". It is because x is a z-representation that x denotes y as it does. x does not merely denote y and happen to be a z-representation. Rather in being a z-representation, x exemplifies certain properties and imputes those properties or related ones to y. "Or related ones" is crucial. A caricature that exaggerates the size of its subject's nose, need not impute an enormous nose to its subject. By exemplifying the size of the nose, it focuses attention, thereby orienting its audience to the way the subject's nose dominates his face or the way his nosiness dominates his character. The properties exemplified in the z-representation thus serve as a bridge that connects x to y. This enables x to provide an orientation to its target that affords epistemic access to the properties in question.

Of course there is no guarantee that the target has the features the model exemplifies, any more than there is any guarantee that a subject represented as the tragic muse has the features that a painting representing her as the tragic muse exemplifies. This is a question of fit.

A model may fit its target perfectly or loosely or not at all. Like any other case of representation-as, the target may have the features the model exemplifies. Then the function of the model is to make those features manifest and display their significance. We may see the target system in a new and fruitful way by focusing on the features that the model draws attention to.

In other cases, the fit is looser. The model does not exactly fit the target. A target that does not instantiate the precise properties its model exemplifies may instantiate more generic properties that subsume the exemplified properties. If gas molecules are roughly spherical, reasonably elastic and far enough apart, then we may gain insight into their behavior by representing them as perfectly elastic spheres with no mutual attraction. Perhaps we will subsequently have to introduce correction factors to accommodate the divergence from the model. Perhaps not. It depends on what degree of precision we want or need. Sometimes, although the target does not quite instantiate the features exemplified in the model, it is not off by much. Where their divergence is negligible, the models, although not strictly true of the phenomena they denote, are true enough of them (Elgin 2004, 113–131). This may be because the models are approximately true, or because they diverge from truth in irrelevant respects, or because the range of cases for which they are not true is a range of cases we do not care about, as for example when the model is inaccurate at the limit. Where a model is true enough, we do not go wrong if we think of the phenomena as displaying the features that the model exemplifies. Obviously whether such a representation is true enough is a contextual question. A representation that is true enough for some purposes or in some respects is not true enough for or in others. This is no surprise. No one doubts that the accuracy of models is limited.

In other cases, of course, the model simply does not fit. In that case, the model affords little or no understanding of its target. Not everyone can be informatively represented as the tragic muse. Nor can every object be informatively represented as a perfectly elastic sphere.

Earlier I dismissed resemblance as the vehicle of representation. I argued that exemplification is required instead. But for *x* to exemplify a property of *y*, *x* must share that property with *y*. So *x* and *y* must be alike in respect of that property. It might seem then that resemblance in particular respects is what is required to connect a representation with its referent.[5] There is a grain of truth here. If exemplification is the vehicle for representation-as, the representation and its object resemble one another in respect of the exemplified properties. But resemblance, even resemblance in a particular, relevant respect, is not enough, as the following tragic example shows.

On January 28, 1986, the space shuttle Challenger exploded because its O-rings failed due to cold weather. The previous day, engineers involved in designing the shuttle had warned NASA about that very danger. They faxed data to NASA to support their concern. The printouts contained complex descriptions conveying vast amounts of information about previous shuttle flights. They included measurements of launch temperatures for previous flights and measurements of six types of O-ring degradation after each flight. Had loss of elasticity been plotted against temperature, the danger would have been clear. The evidence that the O-rings were vulnerable in cold weather was contained in the data. But it was obscured by a melange of other information that was also included (Tufte 1997, 17–31). So although the requisite

[5]This is the position Giere (1999) takes about the relation between a model and its target system.

resemblance between model and target obtained, it was overshadowed in the way that a subtle irregularity in an elaborate tapestry might be. As they were presented, the data instantiated but did not exemplify the correlation between degradation of elasticity and temperature. They did not represent the O-rings as increasingly inelastic as the temperature dropped. Because the correlation between O-ring degradation and temperature was not perspicuous, the NASA decision makers did not see it. The launch took place, the shuttle exploded, and the astronauts died. When the goal of a representation is to afford understanding, its merely resembling the target in relevant respects is not sufficient. The representation must make the resemblance manifest.

Problems Evaded

The account I have sketched evades a number of controversies that have arisen in recent discussions of scientific models. Whether models are concrete or abstract makes no difference. A tinker toy model of a protein exemplifies a structure and represents its target as having that structure. An equation exemplifies a mathematical relation between temperature and pressure and represents its target as consisting of molecules whose temperatures and pressures are so related. Nor does it matter whether models are verbal or non-verbal. One could represent Mrs. Siddons as the tragic muse in a picture, as Reynolds did, or in a poem as Russell did (Russell 2006).

In all cases, models are contrived to exemplify particular features. Theoretical models are designed to realize the laws of a theory (Giere 1999, 92). But we should not be too quick to think that they are therefore vacuously true. For by exemplifying features that follow from the realization of the laws, the models may enhance understanding of what the realization of the laws commits the theory to. They may, for example, show that any system that realizes the laws has certain other unsuspected properties as well. A model then can provide reasons to accept or reject the theory. Such a model is a mediator between the laws and the target system (Morrison and Morgan 1999). It in effect puts meat on the bare bones of the theory, makes manifest what its realization requires, and exemplifies properties that are capable of being instantiated in and may be found in the target system. In discussing theoretical models, we should be sensitive to the ambiguity of the word "of". Such a model is a model *of* a theory because it exemplifies the laws of the theory. It is a model *of* the target because it denotes the target. It thus stands in different referential relations to the two systems it mediates between.

Not all models are models of laws or theories. There are phenomenological models as well. These too exemplify features they ascribe to their target systems. They are streamlined, simplified representations that highlight those properties and exhibit their effects. The difference is that the features phenomenological models exemplify are not captured in laws.

Data models regiment and streamline the data. They impose order on it, by smoothing curves, omitting outliers, grouping together readings that are to count as the same, and discriminating between readings that are to count as different. They thereby bring about the exemplification of patterns and discrepancies that are apt to be obscured in the raw data.

There is evidently no limit on what can be a target. It is commonplace that scientists rarely if ever test theoretical models or phenomenological models against raw data. At best, they test such models against data models. Only data models are apt to be tested against raw data. A theoretical model might take as its target a phenomenological model or a less abstract theoretical model (Suárez 2003, 237). Then its accuracy would be tested by whether the features it exemplifies are to be found in the representations that other model provides, and its adequacy would be tested by whether the features found are scientifically significant. We can and should insist that eventually models in empirical sciences answer to empirical facts. But there may be a multiplicity of intervening levels of representation between the model and the facts it answers to.

Because models depend on exemplification, they are selective. A model makes some features of its target manifest by overshadowing or ignoring others. So different models of the same target may make different features manifest. Where models are thought of as mirrors, this seems problematic. It is hard to see how the nucleus of an atom could be mirrored without distortion as a liquid drop and as a shell structure.[6] Since a single material object cannot be both fluid and rigid, there might seem to be something wrong with our understanding of the domain if both models are admissible. But if what one model contends is that in some significant respects the nucleus behaves like a liquid drop, and another model contends that in some other significant respects it behaves as though it has a shell structure, there is in principle no problem. There is no reason why the same thing should not share some significant properties with liquid drops and other significant properties with rigid shells. It may be surprising that the same thing could have both sets of features, but there is no logical or conceptual difficulty. The models afford different perspectives on the same reality. And it is no surprise that different perspectives reveal different aspects of that reality. There is no perfect model for the same reason that there is no perfect perspective (Teller 2001). Every perspective, in revealing some things, inevitably obscures others.

Nothing in this account favors either scientific realism or anti-realism. One can be a realist about theoretical commitments, and take the success of the models to be evidence that there really are such things as, for example, charmed quarks. Or one can be an anti-realist and take the success of the models to be evidence only of the empirical adequacy of representations that involve charmed-quark-talk. Where models do not exactly fit the data, one can take an instrumentalist stance to their function. Or one can take a realist stance and say that the phenomena are a product of signal and noise, and that the models just eliminate the noise. I am not claiming that there are no real problems here, only that the cognitive functions of models that I have focused on do not favor either side of the debates.

---

[6] I owe this example to Roman Frigg.

Objectivity

The close affinity I find between scientific and artistic representations may heighten anxieties about the objectivity of science. I do not think this is a real problem, but I need to say a bit about objectivity to explain why.

We need to distinguish between objectivity and accuracy. A representation is accurate if things are the way it represents them to be. A hunch may be accurate. My wild guess that it is raining in Rome may be correct. But there is no reason to believe it, since it is entirely subjective and utterly ungrounded. A portrait portraying Aristotle as blue-eyed may be accurate. But there is no reason to think so, since we have no evidence of the color of Aristotle's eyes. An objective representation may be accurate or inaccurate. Its claim to objectivity turns not on its accuracy, but on its relation to reasons. A representation is objective to the extent that it admits of interpretations that are assessable by reference to intersubjectively available and evaluable reasons, where a reason is a consideration favoring a contention that the other members of the community cannot intellectually responsibly reject.[7]

In the first instance then objectivity attaches to interpretations. For it is interpretations that are (or are not) directly backed by reasons. To say that a representation is objective then is to say that it admits of objective interpretations. Whether this is so depends on the norms governing the institutional framework within which it functions.

Where we are concerned with science, the relevant community is a scientific community. So scientific objectivity involves answerability to the standards of a scientific community. According to these standards, among the factors that make a scientific result objective are: belonging to a practice which regards each of its commitments as subject to revision or refinement on the basis of future findings; being grounded in evidence; being subject to confirmation by further testing; being corroborated or capable of being corroborated by other scientists; being consistent with other findings; and being delivered by methods that have been validated. And generating objective results is what makes a model or method objective.

My characterization of scientific objectivity is plainly schematic. What counts as evidence, and what counts as being duly answerable to evidence, and who counts as a member of the relevant community are not fixed in the firmament. Answers to such questions are worked out with the growth of a science and the refinement of its methodology. This is not the place to go into the details of such an account of objectivity.[8] What is important here is that to be duly answerable to evidence is not necessarily to be directly answerable to evidence. A representation may be abstract. Then it needs multiple levels of mediating symbols to bring it into contact with the facts. A representation may be indirect. It may involve idealizations, omissions,

---

[7]See Scanlon (1998, 72–75). I say "assessible by reference to reasons" rather than "supportable by reasons" because an objective judgment may not stand up. If I put forth my judgment as an objective judgment, submit it to a (real or hypothetical) jury of my peers, it is objective, even if my peers repudiate it.

[8]For the start of such an account, see Scheffler (1982).

and/or distortions that have to be acknowledged and accommodated, if we are to understand how it bears on the facts. But if it is objective, then evidence must bear on its acceptability and the appropriate scientific community must be in at least rough accord about what the evidence is (or would be) and how it bears or would bear on the representation's acceptability.

Since the same representation might be deployed by communities governed by different norms, a single representation may be objective when functioning in one context and subjective when functioning in another. This result is welcome. Leonardo's scientific drawings are frequently exhibited in both science museums and art museums. When an illustration of a machine functions as a scientific representation, as it does in a science museum, features such as gear ratios are exemplified. When it functions as a work of art, as it does in an art museum, features like shading and delicacy of line are exemplified. The representation has all of the features each interpretation focuses on. But when interpreted against a background where different interests and values predominate, different features stand out.

A difference between art and science emerges from this characterization of objectivity. Aesthetic interpretations, unlike scientific ones, are endlessly contestable. There are relatively few reasons that no member of a community of connoisseurs could reasonably reject. A single work can bear multiple correct interpretations. Velázquez's *Infanta Margarita Teresa in Blue* is a portrait of the 8-year-old infanta. One interpretation might construe the work politically. The pose is regal; the accoutrements, opulent. But the infanta is trapped in an immobilizing gown, unable to move. She is completely unfree, nothing but a pawn in a political game. Foreign policy considerations dictate whom she will marry and when. On this interpretation, the portrait is not unsympathetic to her plight, but the sympathy is entirely general. It applies to anyone fated to play such a role. Another interpretation is more personal. It focuses on the fact that, despite the accoutrements, it is a portrait of a little girl—a specific little girl. According to this interpretation, the painting exemplifies her fragility and the poignant tragedy inherent in her position. It notes the tenderness with which she is portrayed. The picture is not just a portrait of *an* infanta in the Spanish Court, but of a particular person—Infanta Margarita Teresa. The one interpretation looks outward, interpreting the portrait and its subject in light of dynastic politics. The other looks inward, inviting us to consider what the experience of this particular child might be. Arguably, both interpretations are correct. Each affords an understanding of the picture, its subject, and the forces of circumstance that constrict lives and fetter freedom. But because one is public and political while the other is private and personal, the understandings that emerge are different. A viewer in the grip of one might reasonably reject the perspective the other affords.

To say that there is no consensus about how to interpret works of art is not to say that reasons are inert. One can give reasons for one's interpretation, and both the reasons and the interpretation are open to public scrutiny (Kant 1968, 183–184). Objectivity and subjectivity belong to a continuum. There is no sharp dividing line. So rather than asking whether an interpretation or a representation is objective, it is preferable to ask how objective (or subjective) it is. Typically some aspects of an interpretation of a work of art are backed by reasons that no members of a community of connoisseurs can reasonably reject. All, for example, are apt to agree that *Infanta Margarita Teresa in Blue* is a portrait of a little girl. But to the extent that interpretations outrun the prospect of community consensus, to the extent that the reasons adduced to support them are contestable, they lack objectivity. The finest differences can make a difference to the interpretation of a work of art. Competent viewers discern, focus on, and weigh the significance of aspects of a work differently. So the reasons supporting an interpretation of a work of art are apt to be inconclusive.

The differences in objectivity suggest that the understandings we glean from the arts may be more tentative and tenuous than those we glean from science. There is far less agreement about the adequacy of the interpretations they generate. But all understanding is provisional and fallible. Even the most well established claim may be revised or rejected on the basis of further findings. So we should not repudiate the cognitive deliverances of art merely because they are tentative, controversial, and subject to revision.

I said that the outset that science and art embody understandings. An understanding is a grasp of a general body of information that is and manifests that it is responsive to reasons. It is a grasp that is grounded in fact, is duly answerable to evidence, and enables inference, argument, and perhaps action regarding the subject the understanding pertains to. This entails nothing about the way the body of information is encoded or conveyed. Whether symbols are qualitative or quantitative, factual or fictional, direct or oblique, they have the capacity to embody an understanding. To glean an understanding requires knowing how to interpret the symbols that embody it. So although scientific models and fictional portrayals do not accurately mirror anything in the world, they are capable of figuring in an understanding of the world.

Acknowledgments I would like to thank Israel Scheffler, Nancy Nersessian, John Hughes, the participants in the 2006 Workshop on Scientific Representation at the Universidad Complutense de Madrid and the Conference "Beyond Mimesis and Nominalism: Representation in Art and Science" in London, and two anonymous referees for helpful comments on earlier drafts of this paper.

References

Aristotle (1973), "Poetics", in R. McKeon (ed.), *Introduction to Aristotle*. Chicago: University of Chicago Press.

Austen, J. (2005), "Letter to Her Niece, Anna Austen Lefroy", September 9, 1814, in *Letters of Jane Austen*, Bradbourn Edition. URL = www.pemberley.com/janeinfo/brablets.html. Consulted May 4, 2005.

Cartwright, N. (1999), "Aristotelian Natures and the Modern Experimental Method", in *The Dappled World*. Cambridge: Cambridge University Press, 77–104.

Elgin, C. Z. (1983), *With Reference to Reference*. Indianapolis: Hackett.

Elgin, C. Z. (1996), *Considered Judgment*. Princeton: Princeton University Press.

Elgin, C. Z. (2004), "True Enough", *Philosophical Issues* 14: 113–131.

Giere, R. (1999), *Science Without Laws*. Chicago: University of Chicago Press.
Goodman, N. (1968), *Languages of Art*. Indianapolis: Hackett.
Hughes, R. I. G. (1997), "Models and Representation", in *PSA 1996*, vol. 2. East Lansing, MI: Philosophy of Science Association, S325–S336.
Kant, I. (1968), *Critique of Judgment*. New York: Hafner.
Morrison, M. and Morgan, M. S. (1999), "Models as Mediating Instruments", in M. Morgan and M. Morrison (eds.), *Models as Mediators: Perspectives on Natural and Social Science*, Cambridge: Cambridge University Press, 10–38.
Russell, B. (1968), "On Denoting", in *Logic and Knowledge*. New York: Capricorn.
Russell, W. (2006), "The Tragic Muse: A Poem Addressed to Mrs. Siddons", 1783. URL = www.dulwichpicturegallery.org.uk/collection/search/display.aspx?im=252. Consulted January 12, 2006.
Scanlon, T. M. (1998), *What We Owe to Each Other*. Cambridge, MA: Harvard University Press.
Scheffler, I. (1982), "Epistemology of Objectivity", in *Science and Subjectivity*. Indianapolis: Hackett, 93–124.
Suárez, M. (2003), "Scientific Representation: Against Similarity and Isomorphism", *International Studies in the Philosophy of Science* 17: 225–243.
Teller, P. (2001), "Twilight of the Perfect Model Model", *Erkenntnis* 55: 393–415.
Tufte, E. R. (1997), *Visual and Statistical Thinking: Displays of Evidence for Making Decisions*. Cheshire, Connecticut: Graphics Press.

Models: Parables v Fables

Nancy Cartwright

How Fables and Parables Help Us Understand the Use of Models: A Short Survey of This Paper

Models of different kinds appear throughout the natural and social sciences serving a variety of different ends. This paper will discuss one particular kind of model whose purpose is opaque: the "highly idealized" model, prevalent in physics and economics but widely used elsewhere as well. Models of this kind study the behavior of stripped-down systems in unrealistic circumstances. The models may study balls rolling down totally frictionless totally stable planes (Galileo 1914, 61–69), or laborers of only two kinds—old and young—concerned only with leisure and income (Pissarides 1992), or, as in Thomas Schelling's famous model, black and white checkers moving according to artificial rules on a checkerboard, ending up in clumps of similar color (Cartwright 2009a, Schelling 2000). The objects and situations pictured in these models are very unlike real objects in the real world of interest to the sciences. Yet they are supposed to teach something, indeed something important, about that real world. How?

I am going to defend the use of descriptions of highly unrealistic situations to learn about real-life situations. That, I maintain, is just what Galileo did in his famous rolling-ball experiments. He honed his planes to make them as smooth as possible, and bolted them down, to learn about the effects of gravity acting on its own. Models I urge are often experiments *in thought* about what would happen in a real experiment like Galileo's if only it could be conducted: What would happen were we able to create just the right artificial situation to see the feature under study acting all on its own, without any other causes interfering to mask its effect?

That however is not enough. Doing what Galileo did sounds a good thing. But Galileo's results are still results about the behavior of balls rolling down totally frictionless planes. We don't have any such planes and anyway what we really want to know is about canon balls and rocket ships. How do we get from a Galilean

N. Cartwright (✉)
London School of Economics, London, UK; University of California, San Diego, CA, USA
e-mail: N.L.Cartwright@lse.ac.uk

R. Frigg, M.C. Hunter (eds.), *Beyond Mimesis and Convention*, Boston Studies in the Philosophy of Science 262, DOI 10.1007/978-90-481-3851-7_2,

conclusion: "The pull of the earth induces an acceleration of 32 ft/sec/sec in balls rolling down totally frictionless totally stable planes"—to a result about cannonballs, teetering coffee cups or rocket ships? I shall here repeat an earlier answer of mine, that these models are like fables, for instance like this fable that I shall discuss below:

A marten eats the grouse;
A fox throttles the marten; the tooth of the wolf, the fox.
Moral: the weaker are always prey to the stronger.

Like the characters in the fable, the objects in the model are highly special and do not in general resemble the ones we want to learn about. Just as I have never seen a frictionless plane or a worker interested only in leisure and income, I don't think I have ever seen a marten, and seldom a wolf. But the conclusion of the model, like the moral of the fable, can be drawn in a vocabulary abstract enough to describe the things we do want to learn about. For instance, we conclude from Galileo's experiment, "The pull of the earth induces a downwards acceleration in massive objects of 32 ft/sec/sec". This is a correct way of describing what we see in the rolling-ball experiments, just as my earlier description is. But this more abstract description also applies to cannonballs, coffee cups and rocket ships since they too are massive objects. Similarly, the moral of Schelling's model might be, "A group of individuals moving not under the rule, 'Create segregated neighbourhoods' but only under the rule 'Move from a neighbourhood where almost everyone is different from you to one where that is not so' will almost always end up in clumps in each of which everyone is alike". In both cases a description of what happens in the model that does not fit the target gets recast as one that can, just as the moral of the fable can apply to a broader range than the kinds of individuals pictured in the fable. To underline this, I shall employ an idea and a slogan from Menno Rol: Climbing up the ladder of abstraction can take one from falsehood to truth.

There is a problem, however. Fables, like those of Aesop or Lafontaine, typically have the moral built right in. Many of our most familiar parables do not. What is the moral of the parable of the prodigal son[1] or of the laborers in the vineyard?[2] It

[1]Luke 15: 11–32.

[2]Matthew 20: 1–16 (The Holy Bible, King James Version): 20:1 For the kingdom of heaven is like unto a man that is an householder, which went out early in the morning to hire laborers into his vineyard. 20:2 And when he had agreed with the laborers for a penny a day, he sent them into his vineyard. 20:3 And he went out about the third hour, and saw others standing idle in the marketplace, 20:4 And said unto them; Go ye also into the vineyard, and whatsoever is right I will give you. And they went their way. 20:5 Again he went out about the sixth and ninth hour, and did likewise. 20:6 And about the eleventh hour he went out, and found others standing idle, and saith unto them, Why stand ye here all the day idle? 20:7 They say unto him, Because no man hath hired us. He saith unto them, Go ye also into the vineyard; and whatsoever is right, [that] shall ye receive. 20:8 So when even was come, the lord of the vineyard saith unto his steward, Call the laborers, and give them their hire, beginning from the last unto the first. 20:9 And when they came that were hired about the eleventh hour, they received every man a penny. 20:10 But when the first came, they supposed that they should have received more; and they likewise received every man a penny. 20:11 And when they had received it, they murmured against the goodman of the house,

is not written into these parables as morals are written into Aesop and Lafontaine's fables. Consider for instance the laborers in the vineyard. I always thought that the parable teaches about God's intentions to accept late repentants into the Kingdom of Heaven and that those who live virtuously their entire lives just have to put up with that, with perhaps some knock-on moral that we too should indulge in generosity at the cost of seeming fairness. But can its moral cover this case, described to me by a lawyer friend?

> Lloyd's of London is an insurance market where individuals ("Names") underwrite insurance risks through syndicates. A syndicate normally consists of several hundred Names. In the late 80s and early 90s Names lost a great deal of money—so much so that the stability of the market was threatened. The Society of Lloyd's, the body which runs the market, became involved in trying to settle the Names' claims which arose, said the Names, not from bad luck but from the negligence of their agents—the market practitioners who actually underwrote the insurance risks. Some Names sued their agents for damages—something which had rarely, if ever, been done before.
>
> When it came to offering a settlement the question arose whether the offer should be to those who had sued or to all the Names on a syndicate. Those who had sued argued that they had borne the heat and burden of the day and the offer would not be made if they had not banded together and taken legal action. Those who had not sued said they had the same claims and had refrained from suing for reasons of their own—perhaps because they did not believe it to be the right thing to do to de-stabilize Lloyd's. There were only limited funds available to make the offer so if it was made to all Names then each would get less.

My lawyer friend says "yes", the parable does apply to the Lloyd's case. The relevant issues are not so much the actions or intentions of the parties or whether they are deserving or not but their legal rights and the probability of their enforcement. Here is his read on what the parable teaches:

> The laborers in the vineyard is a lawyers' parable. Never mind the slightly confusing stuff about the kingdom of God. ... The point of the parable for me is that you get what you contract for, and shouldn't complain if others get more. It is a market-based capitalist parable, opposed to socialist ideas of "fairness". The laborers union would not have liked it but the Chamber of Viticulturalists would get the point immediately...

I shall argue here that my earlier defense of "idealized" models was overly optimistic. Many of the highly artificial, "idealized" models of economics are like this parable and are unlike the fable of the grouse and the marten where the lesson is written right in. A variety of morals can be attributed to the models, expressed in a variety of different vocabularies involving abstractions of different kinds and at different levels. Importantly, these morals can point in different directions, implying opposite predictions for the real-life situations to which we want to apply them. Climbing up the ladder of abstraction, whether in a fable, a parable or a model, can

---

20:12 Saying, These last have wrought but 1 hour, and thou hast made them equal unto us, which have borne the burden and heat of the day. 20:13 But he answered one of them, and said, Friend, I do thee no wrong: didst not thou agree with me for a penny? 20:14 Take that thine is, and go thy way: I will give unto this last, even as unto thee. 20:15 Is it not lawful for me to do what I will with mine own? Is thine eye evil, because I am good? 20:16 So the last shall be first, and the first last: for many be called, but few chosen.

take us from falsehood to truth, but only if we know which ladder to climb up. That knowledge, I will propose in closing, comes, if it does at all, not from the model itself but from the rich context of the science in which it is imbedded.

The Problem of Unrealistic Assumptions, Round 1: Valid Arguments but False Premises

The models I discuss here, found typically in physics and economics, offer descriptions of imaginary situations or systems using a combination of mathematics and natural language. The descriptions are thin: Not much about the situation is filled in. They are often unrealistic as well in that what is filled in is not true of many real situations. Yet in many cases we want to use the results of these models to inform our conclusions about a range of actually occurring (so-called ***target***) situations.

I am also going to restrict my attention to models in which results are derived by deduction. The whole point of these formal models is rigor, which is why they are preferred by physicists and economists alike over more informal reasonings that merely make results plausible. Deduction is a key ingredient in this rigor. We are assured that the consequences drawn from the models are genuine because they follow deductively from the starting descriptions; these consequences must occur whenever these descriptions are satisfied.

The "unrealistic" assumptions that are offered in a model's descriptions are no problem so long as they play no role in deducing the intended results of the model. But this is seldom the case. In fact quite the contrary. They are often necessary to the deductions offered in the model.[3] This gives rise to the canonical **problem of unrealistic assumptions:** How can a result that must occur given characteristics different from those in the target inform conclusions about what will happen in the target? The conclusion is supposed to be guaranteed because it follows deductively from the premises. How does that provide information about what conclusions to expect when the premises are different?

The Plan

In tackling the problem of unrealistic assumptions this paper will rely on three different strands of enquiry:

- previous work of mine on Galilean thought experiments
- previous work of mine on models as fables
- Menno Rol's insight that abstraction can turn falsehoods into truths.

[3] The requisite deduction will sometimes not be literally on offer in the model but rather presumed.

It will wind through an increasingly narrow spiral:

- from **problem, Round 1:** the first broad *problem of unrealistic assumptions* just described
- to **solution, Round 1,** a solution that should work for certain specific kinds of models—those that can count as *Galilean thought experiments*
- to a twist at which the problem of unrealistic assumptions re-emerges in narrower form, as **problem, Round 2**: the problem of *overconstraint*
- to another twist provided by Menno Rol that offers **solution, Round 2** to this new version of the problem, a solution for an even more restricted set of models—by *abstraction* of the kind seen in the moral of a fable
- finally ending in yet another **problem, Round 3**, which arises because models are far more like *parables* than fables.

In the end, this final problem for models with unrealistic assumptions, I hazard, cannot be solved within the model itself nor by philosophy. The problem in the end demands that the model be located in a strong, rich scientific network that can pick out the right abstract concepts with which to formulate the model's results.

Solution, Round 1: Galilean Thought Experiments

Unrealistic assumptions do not always stand in the way of drawing lessons about real situations from models. Some models function as Galilean thought experiments and for these unrealistic assumptions are not a hindrance but a necessity. A real Galilean experiment (a Galilean experiment really conducted), as I use the term, isolates a single factor as best possible to observe its natural effect when it operates "on its own" with no other causes at work. In a thought experiment we just imagine the situation and what would happen in it if it were conducted. In a real Galilean experiment the effect is produced in accord with the laws of nature. In a model that pictures a Galilean thought experiment it is the principles built into the model that determine what the effect must be. So real experiments and thought experiments have complementary virtues. In the real experiment we can never be sure that we have eliminated all confounding factors but we can be sure the effect is produced in accord with nature's laws. By contrast the situation described in the thought experiment has only the factors in it that we put there. So we can be sure that confounders are absent but we cannot be sure the effect is right because that depends on the principles we provide in the model.

Typical economics models, and many in physics as well—especially those set as problems to work out in physics texts—can certainly be taken to be Galilean thought experiments, isolating a single factor to study its effect when no other causes of that effect are there to interfere. This is clear not only from the practice in both cases; it is explicit in many economics discussions and in some from physics as well. With respect to those models that serve as Galilean thought experiments, unrealistic assumptions that suppose the factor is at work all on its own, with no alternative causes at work, are no more of a problem than they are for real Galilean experiments. If we can learn about target situations with more "realistic" arrangements from actual Galilean experiments despite the "unrealistic" assumptions necessary to the experiment, the same is true for Galilean thought experiments (so long as the basic principles used in the model to drive the consequences are accurate enough).[4] So at least for some models and some kinds of unrealistic assumptions, unrealistic assumptions pose no problem.

The Problem of Unrealistic Assumptions, Round 2: Overconstraint

This is a rather too happy conclusion however. That's because a good many of the models that can be cast as Galilean thought experiments have a number of "unrealistic" assumptions beyond those necessary for them to count as Galilean experiments—that is beyond those that eliminate all the other causes of the same effect. This is generally for two interacting reasons.

First: Many kinds of causes, unlike gravity, cannot just act without involvement of the specific setting in which they are placed. They need a concrete situation in which to play out. Consider for example a model to study the effect of skill loss during unemployment on future employment levels. If it is to be a Galilean model there must be no further causes of employment or unemployment at work in it: no downturns in consumer spending, no shift to a war economy, and so forth; no motives that differ between when employers invest to open future jobs from when they don't other than the difference in profit they expect due to a difference in the efficiency of the workers. Still, the model needs employers in it in order to study what happens given their different expectations; and it needs to have workers who have lost skills and workers who have not to create these different expectations. How many workers, how many employers? What ratio of employed to unemployed? Etc. These factors are not properly thought of as alternative causes of employment or unemployment, as alternative mechanisms to those of skill loss that can affect future employment levels independently. Hence the answer to what form they should take is not dictated by the rules for a Galilean experiment, i.e., the demand to eliminate all alternative causes of the effect studied. Still they must take some form or other, otherwise the skill-loss mechanism cannot be set operating.

Second is the well-rehearsed reason that matters must often be set in very particular ways if calculations and deductions are to be at all possible. So often mathematically more tractable descriptions are substituted for descriptions that are more true to the target situations that we want the model results to bear on.

---

[4]In either case exporting from the Galilean experiment requires both more and stronger assumptions than those supplied in the experiment. My own view (1989, 2009b) is that exporting often employs the logic of capacities, where the assumption that a factor has a capacity to study in the first place takes a great deal of highly varied independent evidence.

Indeed it is often the need for mathematical tractability that solves the first problem by settling how to fix the concrete setting in which the isolated cause will play out.

So most Galilean thought experiments have many more "unrealistic" assumptions than those they should. Again, this would not be a problem if these assumptions did not play a role in deducing the final results. But of course generally they do—that is the point of including them in the first place. Just by inspection we can see that they are a necessary part of the deduction offered by the model.[5]

In these cases I say that the results of the model are *overconstrained.* All the conditions sufficient to insure that the model describes a Galilean experiment are met. The results are constrained to be ones brought about by the cause operating without any other independent causes of the same effect present. So (pace mistakes in the driving principles) the results must be ones we would see in a real Galilean experiment. The problem is that the Galilean experiment takes place in a very special and unusual setting. What we see is indeed the result of the cause acting on its own without other causes interfering but it is a very special result that we cannot expect in all other Galilean experiments. The setting *overconstrains* the results—it constrains them to a narrower set than those permitted by just the assumptions necessary to ensure that no independent causes are at work. We know we cannot expect the overconstrained result to occur in other different settings for the Galilean experiment because we can see by inspection that the description of the special setting plays a necessary role in the derivation offered. So unrealistic assumptions that overconstrain the results are a problem for learning lessons that apply elsewhere even if the model does function as a Galilean thought experiment.[6]

In order to explain the proposed solution to this new problem of overconstraint, I first turn to another topic altogether, that of fables and their morals. I shall spend quite a bit of time on this topic because doing so will make it easy to see Rol's proposal, which I summarize in the slogan: "Abstraction can turn falsehoods into truths".

---

[5]Explicit attempts to deal with this problem often involve so-called "robustness" investigations: Vary these extra assumptions in different ways to see if the results are still more or less the same. Then, I suppose, we are supposed to do a quick induction to the conclusion that the results will be the same under the conditions that hold in the target situations. Not only is this inductive inference dicey but usually the variation is not very great. Also often the interest is not so much in varying the "extra-Galilean" assumptions but rather in adding in some further causes to see how the results are affected when a more realistic arrangement of causes occurs. This latter offers some help with the problem of whether the results are exportable from the experiment to other situations—the question "Can an induction be done at all?"—but not with the problem of *which* results to export.

[6]For a more detailed description of Galilean thought experiments and the problem of over constraint see my (2007, Chapter 15).

Fables and Models, Their Morals and Lessons

Many models, I argue, are like *fables,* and the lesson derived from the model is its *moral.* I say this in order to stress that the relationship between the description of the result that can be exported to new situations and the description of the result using the language of the model is often that of the *abstract* to the *concrete*. It is a truism that scientific terms are often abstract. My claims here involve one very specific way in which one description used in science—like "...is a source of utility" or "...is a laborer"—is more abstract than another—like "income" and "leisure" or "...is an 'older' laborer in a setting containing only two generations of laborers". This is a sense of "abstract" that I take from the theory of the fable defended by Gotthold Ephraim Lessing, the great critic and dramatist of the German Enlightenment.[7]

Lessing argues (1759, Section 1, 100), "In order to give a general symbolic conclusion all the clarity of which it is capable, that is in order to elucidate it as much as possible, we must reduce it to the particular in order to know it intuitively". For him this is in part a matter of *Anschaulichkeit*—intuitive understanding. "Income" for instance is probably more intuitively understandable than "utility". It is also a matter of ontology: "The general exists only in the particular..." (1759, Section 1, 73). This is the aspect I want to stress about the conclusions derived in models that function like Galilean thought experiments.

I illustrate the relation of the abstract to the concrete, following Lessing, with a fable of his own, which I introduced at the start of this paper:

> A marten eats the grouse;
> A fox throttles the marten; the tooth of the wolf, the fox.
> Moral: the weaker are always prey to the stronger.

As I described in *The Dappled World,* Lessing makes up this story as part of his argument to show that a fable is no allegory. Allegories say not what their words seem to say, but rather something similar. But where is the allegory in the fable of the grouse, the marten, the fox and the wolf: "What similarity here does the grouse have with the weakest, the marten with the weak and so forth? Similarity! Does the fox merely *resemble* the strong and the wolf the strongest or *is* the former the strong, the latter the strongest. He *is* it" (1759, Section 1, 73). For Lessing similarity is the wrong idea to focus on. The relationship between the moral and the fable is that of the general to the more specific and it is "a kind of misusage of the word to say that the special has a similarity with the general, the individual with its type, the type with its kind" (1759, Section 1, 73). Each particular *is* a case of the general under which it falls.

The point comes up again when Lessing protests against those who maintain that the moral is hiding in the fable or disguised there. Lessing argues: "How can one disguise (*verkleiden*) the general in the particular... If one insists on a similar word here it must at least be *einkleiden* rather than *verkleiden*". *Einkleiden* is to fit out, as when you take the children to the shops in the autumn to buy them new school

---

[7]My discussion of Lessing here is taken from my (1999, 35–48).

clothes. The children are not disguised by their clothes, hidden in them. Dressed up one way or another, they are still the same children, visible as such. But when dressed up, they are filled out and have more to them. And the very same children could be, and out of school generally are, dressed up, filled out, in a different ways. So the moral is to be "fitted out" by the fable. The moral describes just what happens in the fable; but the fable fits it out in a special way—a way true to the moral but not necessarily shared by all cases of which the moral is true.

The account of abstraction that I borrow from Lessing provides two necessary conditions.

- A concept or claim that is abstract relative to a set of more concrete descriptions or more concrete claims never applies unless one of the more concrete descriptions or claims also applies. These are the descriptions/claims that can be used to fit out the abstract description or claim on any given occasion.
- Satisfying the associated concrete description/claim that applies on a particular occasion is what satisfying the abstract description/claim consists in on that occasion.

What I want to take away from Lessing's account of the fable and its moral is the idea of how the model relates to the abstract lesson that might be drawn from it. Like fables and their morals, the lesson we might hope to export from the model may be *abstract* relative to the more concrete conclusion derived in the model using the more concrete descriptions provided by the model. Like the fable, the model "fits out" the more abstract lesson; and when a situation satisfies the more concrete result expressed in the language of the model, that is what it is for that situation to satisfy the more abstractly expressed result.

Solution, Round 3: From Falsehood to Truth via Abstraction

The problem of models with unrealistic assumptions is one of the standard worries both in the philosophy of economics and in economics itself. Philosopher of economics Menno Rol (2008) has a nice account of why it need not always be a problem. One can, he argues, go from falsehood to truth by climbing up the ladder of abstraction. Rather than delving into economics, let me illustrate his point with a physics example that I think will be familiar to everyone.

Suppose we perform a careful real Galilean experiment to see how bodies move *inertially*, that is, subject to no forces. We do it perfectly; we succeed in stripping (or calculating) away all forces. So we have eliminated all the other independent causes of motion besides inertia, as we are supposed to do in a Galilean experiment. But we do our experiments on a Euclidean plane. From this we conclude that bodies moving inertially follow a Euclidean straight line. This conclusion is entirely correct in the setting of the experiment. But it need not be true elsewhere. In particular this will not describe correctly inertial motion in a spherical geometry, where a body subject to no forces will move on a great circle. To use my earlier language, we succeed in carrying out a Galilean experiment but the results are overconstrained. The solution, following Rol, is to move away from the overconstrained result and describe the results of the experiment *equally correctly* in more abstract vocabulary: The bodies in the experiment travel on *geodesics*—that is, they take the shortest distance between any two points in the relevant geometry. This conclusion is true both in the experiment we conduct and (putatively) everywhere else as well.

So, suppose then that we want to learn from our experiment how a body subject to no forces will travel in our target case, which is a body in a spherical geometry. If the result seen in the experiment is expressed too concretely—"Bodies subject to no forces follow Euclidean straight lines"—then the conclusion of the experiment is false of the target we had hoped to learn about from the experiment. But if the conclusion is expressed more abstractly, we get a prediction from the experiment that is true of the target. That is the sense in which climbing up the ladder of abstraction in describing the results of the experiment can take us from falsehood to truth: Stating the lesson of a model using more abstract concepts than those directly involved in presenting the model can generate true predictions about behaviors in a target.

This account dovetails with the image of models as fables. The lesson of the model is, properly, more abstract than what is seen to happen in the model and that can be described in the concepts introduced there. In the model the marten eats the grouse; the body moves along a Euclidean line. The lesson is that the weaker are prey to the stronger; that inertial bodies move on geodesics. The abstract lesson can be true of a variety of new, different situations where the more concrete behavior will fail.

The advantage of thinking of what happens here in terms of Lessing's account of morals and fables is that it makes clear that there is *nothing wrong with the initial experiment*. What is wrong vis-à-vis applicability elsewhere is the level at which the conclusion is described. Moreover, no experiment could have done better. Experiments must be performed in some geometry or other. That is the point of invoking Lessing's theory of the relation of the abstract to the concrete. The abstract can exist *only* in the concrete. You can't get it unless it is fitted out in one way or another. What the abstract consists in given one filling out will be very different from what it consists in given another. For the marten and the grouse, the grouse's being weaker consists in being slow and not having sharp teeth, claws or a hard shell; being prey is being eaten. For a worker vis-à-vis employer, being weaker can consist in having no union, no transferrable in-demand skills and no wealth; and being prey to equals working long hours in bad conditions for little pay. Still, both are cases where the weaker are prey to the stronger. And in any case, it cannot just be true that someone is weaker and prey to another. In every case there must be something more concrete—and thus less generalizable—that this consists in.

My topic here is thought experiments not real experiments. But the same lessons apply. A thought experiment can succeed perfectly in isolating the factor under study and observing—correctly—what it does on its own, without impediment. But if the results are overconstrained they will not readily generalize. Yet, just as with the "real" Galilean experiment I described for inertia, there may be no alternative. The experiment must be performed in some geometry or other. Similarly, the model to study the effects of skill loss during unemployment on future unemployment rates may have only two generations of workers and one employer, where these affect the outcome though they could not properly be counted as alternative causes of unemployment, as confounding factors that must be eliminated in a Galilean experiment. Yet all situations have some generational structure among the workers and some number of employers. "Real" economic experiments cannot eliminate them either and so will also be overconstrained.

Thinking of thought experiments as fables, then, points out two important methodological lessons

- Though the results of an experiment or of a thought experiment may be overconstrained, this may be inevitable since the abstract exists only in the concrete.
- To get a conclusion that is true both in the model and in a variety of other cases, it may well be necessary to follow Rol's advice and climb up the ladder of abstraction.

The Problem of Unrealistic Assumptions, Round 3: Not Fables but Parables

Consider the parable of the prodigal son, of the good Samaritan or of the workers in the vineyard that I discussed in the introduction. As I pointed out there these parables differ from Aesop's and Lafontaine's fables and from Lessing's fable of the marten and the grouse in that no moral appears as part of the parable itself. The moral is not written in but must be supplied from elsewhere. Defending a moral as the correct one requires a great deal of outside work, including much interpretation of other parts of the available text and of the world itself and how it operates.

So too with "unrealistic" models. Many of these may be Galilean thought experiments and so rightly have "unrealistic" assumptions. And in many cases the correct lessons to be drawn may be more abstract than those described immediately in the concrete situation of the model. But seldom can we really cast the models as fables because the moral is not written in. They are rather like parables, where the prescription for drawing the right lesson must come from elsewhere. Theory can help here, as can a wealth of other cases to look to, and having a good set of well-understood more abstract concepts to hand will play a big role. So the good news that one can move from falsehood in a model to truth by climbing up the ladder of abstraction is considerably dampened by the fact that the model generally does not tell us which ladder of abstraction to use and how far to climb.

I should stress that this problem is not peculiar to thought experiments. As I have mentioned, real experiments can be overconstrained too. As with thought experiments this need not be a problem since, as with fables and their morals, what results in the (correctly conducted) overconstrained experiment will be what the generalizable result consists in for that situation; it will be an instance of the generalizable conclusion. But the experiment does not show what the generalizable conclusion is, how far up which ladder of abstraction one must climb to reach a result that will be true of new target situations as well or whether we can do so at all. This is, I think, clearly recognized in physics and in much of economics as well, even though not articulated in this way. I stress it because I think that it has not been taken on board in the new drive for experiments in evidence-based policy, where practitioners are trying to draw general conclusions without the aid of theory or appeal to a set of well-understood abstract concepts whose reliability has been established elsewhere. So it is important to stress that real experiments, just like thought experiments, are far more often parables than fables.

Still, it may be harder to notice this problem in the case of models and thought experiments because these come—indeed must come—in some specific vocabulary or other, using some particular concepts or other. If we are to use Rol's ladder to derive from the model conclusions true of the various targets we are concerned with, the trick is not to get stuck in that vocabulary but to climb (if possible!) to one sufficiently abstract to be true both of the model situation and of our target situations. The bigger trick of course is to figure out which ladder, if any, to climb.

Conclusion

If we are to use Galilean thought experiments to inform ourselves about target situations we had better recognize that these models are more like parables than they are like fables. So constructing the model and deriving its consequences are just a small step towards drawing a lesson from it. In order to know what the parable means we need to study a great deal of text, reading both the theory that imbeds the model and reading the world itself.

Acknowledgments This work was carried out during a modeling project at the Institute for Advanced Study at the University of Durham and I am very grateful for their support. I would also like to thank two anonymous referees for helpful comments as well as the participants in the conference "The Experimental Side of Modeling" at San Francisco State University, March 2009.

References

Cartwright, N. (1989), *Nature's Capacities and Their Measurement*. Oxford: Oxford University Press.

Cartwright, N. (1999), *The Dappled World. A Study of the Boundaries of Science*. Cambridge: Cambridge University Press.

Cartwright, N. (2007), *Hunting Causes and Using Them. Approaches in Philosophy and Economics*. Cambridge: Cambridge University Press.

Cartwright, N. (2009a), "If No Capacities Then No Credible Worlds. But Can Models Reveal Capacities?", in T. Gruene-Yannof (ed.), Economic Models as Isolating Tools or as Credible Worlds? *Erkenntnis* 70, 1: 45–58.

Cartwright, N. (2009b), "What is This Thing Called 'Efficacy'?", to appear in C. Mantzavinos (ed.), *Philosophy of the Social Sciences. Philosophical Theory and Scientific Practice*. Cambridge: Cambridge University Press.

Galilei, G. ([1564–1642] 1914), *Dialogues Concerning Two New Sciences*, trans. H. Crew and A. de Salvio. New York: Dover.

Lessing, G. E. ([1759] 1967), *Abhandlungen Uber die Fable*. Stuttgart: Philipp Reclam.

Pissarides, C. A. (1992), "Loss of Skill During Unemployment and the Persistence of Unemployment Shocks", *Quarterly Journal of Economics*, 107: 1371–1391.

Rol, M. (2008), "Idealization, Abstraction, and the Policy Relevance of Economic Theories", *Journal of Economic Methodology*, 15, 1: 69–97.

Schelling, T. (1978), *Micromotives and Macrobehavior*. New York: Norton.

The Holy Bible, King James Version: http://www.biblegateway.com/passage/?search=matthew% 2020:1-16&version=9;

[illegible]

Truth and Representation in Science: Two Inspirations from Art

Anjan Chakravartty

Varieties of Truth in Art and Science

Not so long ago, pursuing the notion that the philosophies of art and science can inform one another in mutually productive ways might have been considered a cultured but fringe activity. Recently, however, philosophers more generally have awoken to the import of provocative and substantive analogies between practices of representation in art and science, and it is in the spirit of exploring such analogies that this essay is intended. My primary concern is to understand the nature of truth in the scientific context, and it will be my contention that this understanding requires an appreciation of a distinction between two different conventions of representation—one associated with practices of what I will call "abstraction", and the other with practices of what is often called "idealization". I believe that analogies to practices of representation in art can serve as valuable heuristics towards understanding how and in what manner scientific representations can be true.

The term "scientific representation" is commonly applied to many things, and would benefit from a more precise consideration than I can give it here. For present purposes, let me simply take such representations to include the usual items traditionally held to have representational status in the sciences, *viz.* theories and models, however these things are best defined, inhabiting the usual variety of ontological categories commonly associated with them: linguistic and mathematical entities; computer simulations; concrete objects; and so on. Other key concepts here will of course include those of abstraction and idealization, and I will have something to say about both and my reasons for focusing on these concepts in particular in due course. Let me begin, however, with the central concept whose explication this essay is intended to serve. Clearly, not all philosophers of science believe that the sciences are in the truth business, but an impressive diversity do, including different kinds of realists and empiricists. The former take the truths of science to include facts about

A. Chakravartty (✉)
Institute for the History and Philosophy of Science and Technology, University of Toronto, Toronto, ON M5S IK7, Canada
e-mail: anjan.chakravartty@utoronto.ca

R. Frigg, M.C. Hunter (eds.), *Beyond Mimesis and Convention*, Boston Studies in the Philosophy of Science 262, DOI 10.1007/978-90-481-3851-7_3,

unobservable entities and processes, and some of the latter acknowledge only truths about the observable, but all believe that scientific knowledge involves or at the very least aspires to substantive truths about the world, in some form or other. This is the first of two assumptions I will make here, at the outset.

The second assumption is that descriptions of entities and processes afforded by scientific representations are generally false, strictly speaking. I will not argue for this presently, but neither do I take it to be controversial. Even realists and empiricists who think that the sciences are in the truth business readily admit the hyperbole involved in suggesting that current representations (however circumscribed) are generally perfectly and comprehensively true. The history of the sciences has made a mockery of that suggestion in the past, and no doubt there is further mockery to come. It seems that anyone who endorses the idea of scientific truth as a reasonable aspiration requires some means of making sense of the idea that inaccurate representations can be close to the truth, and perhaps even get better with respect to truth over time. In the literature this requirement has motivated several accounts of "approximate truth", in terms of which, it is argued, one may understand such improvements. There would seem to be a widely held intuitive platitude concerning the notion of approximate truth, and Stathis Psillos (1999, 277) summarizes it nicely: "A description *D* ... is approximately true of [a state] *S* if there is another state S' such that *S* and S' are linked by specific conditions of approximation, and *D* ... is true of S'." By itself, however, the helpfulness of this statement is impaired by the vagueness of the phrase "conditions of approximation". The remainder of this essay is essentially an attempt to clarify this phrase.

It will be my contention that the clarification required is wonderfully illuminated by drawing analogies to certain practices of representation in art, and as a final foreshadowing remark, let me mention briefly the path-breaking work in this area that informs several of the thoughts to follow. Nelson Goodman (1976) is celebrated for presenting a detailed analysis of the "symbol systems" in terms of which different forms of art express their content. At the end of his book on the subject, Goodman (1976, 262) says something particularly striking about the comparison between representations in art and in science:

> ... have I overlooked the sharpest contrast: that in science, unlike art, the ultimate test is truth? Do not the two domains differ most drastically in that truth means all for the one, nothing for the other? ... Despite rife doctrine, truth by itself matters very little in science.

It should be noted immediately that Goodman does not of course think that truth is unimportant in the sciences. Important truths about the natural world are indeed of great interest to scientists, and while one may admit that scientific laws are seldom true as they stand, we have an interest in "arriving at the nearest approximation to truth that is compatible with our other interests" (1976, 263). Ultimately, says Goodman, truth can be understood in terms of "a matter of fit" between theories and facts, and as it turns out, just this sort of "fitting" is characteristic of the relationship between art and the world (1976, 264). Truth in both domains should be understood in terms of approximating reality by means of representations. What more precisely "fitting" and "approximating" mean in these contexts, however, is something I hope to explore here.

One reason Goodman suggests that truth *by itself* matters little, is that truth amounts to nothing unless one, in addition to having truthful representations, is properly acculturated with the conventions of representation in terms of which they express their content. It is precisely these conventions that I take to constitute the "conditions of approximation" whose explication is required in order to make sense of the notion of approximate truth, and so by exploring the former, I aim to shed light on the latter. In the following I will suggest that understanding two central features of scientific knowledge are crucial to illuminating these conditions of approximation, and it is here that analogies to representation in art may prove useful. The first of these features is a distinction between abstraction and idealization in connection with scientific representation, and the second concerns the pragmatics of scientific practice.

In the following section, "Preliminaries on Approximate Truth", I will briefly review extant approaches to the issue of approximate truth in the context of scientific knowledge. My goal here is modest: to convey in summary fashion the gist of these approaches and some reasons philosophers have worried about them. Leaving aside the details of proofs and potential resolutions of these worries, the main function of this discussion is to illuminate, by way of contrast, the approach taken here, which pays greater attention to the conventions of representation whereby scientific knowledge departs from truth at the outset, in its construction. The next section, "Truth in the Context of Abstraction and Idealization", considers the most central of these conventions recognized in contemporary philosophy of science—abstraction and idealization—and the ways in which one might articulate conceptions of approximate truth in either case. Idealization is the greater challenge here, and in the next section, "Denotation in Art, Reference in Science", Goodman's reflections on the nature of realistic and non-realistic artistic representation are exploited in furnishing a proposal for understanding the truth content of idealizations. In the fifth and final section, "Representations and Practice as Products and Production", a second analogy to representation in art, this time to the work and reception of some twentieth century avant-gardes, furnishes a poetic insight into the nature of scientific work, and its reception by philosophers in the latter twentieth century. In the process, I will make reference to pieces by Pablo Picasso, Jackson Pollock, and Yoko Ono.

Preliminaries on Approximate Truth

A moment ago I suggested that generally speaking, the knowledge contained in scientific representations is usually understood as approximately true at best. Three main families of accounts of approximate truth have emerged in the literature since the 1960s, and before considering the nature and relevance of abstraction and idealization in this context, it will serve us to have a synoptic overview of these approaches. I will refer to them respectively as the verisimilitude approach, due to Karl Popper, the possible worlds approach, formulated in different ways by philosophers including Pavel Tichý, Ilkka Niiniluoto, and Graham Oddie, and the type hierarchy approach, offered by Jerrold Aronson, Rom Harré, and Eileen Cornell Way.

Popper was the first to give a definition of what he called "verisimilitude" or "truth-likeness". On his (1972, 231–236) view, scientific theories within a domain may exhibit increasing levels of verisimilitude over time, and this relative ordering can be expressed as follows. Consider a temporally-ordered sequence of theories concerning the same subject matter: $T_1, T_2, T_3, \ldots$. Now for each of these theories, consider the set of all of its true consequences (for example, $T_1{}^{\mathrm{T}}$) and the set of all of its false consequences ($T_1{}^{\mathrm{F}}$). A comparative ranking of the verisimilitude of any two theories can be given, suggests Popper, by comparing their true and false consequences. For any theory T_{n}, and any previous theory $T_{<\mathrm{n}}$, T_{n} has a higher degree of verisimilitude than $T_{<\mathrm{n}}$ if and only if either of the following statements is true ("$\subseteq$" here stands for set-theoretic inclusion, and "$\subset$" for proper inclusion):

1. $T^{\mathrm{T}}_{<\mathrm{n}} \subset T^{\mathrm{T}}_{\mathrm{n}}$ and $T^{\mathrm{F}}_{\mathrm{n}} \subseteq T^{\mathrm{F}}_{<\mathrm{n}}$
2. $T^{\mathrm{T}}_{<\mathrm{n}} \subseteq T^{\mathrm{T}}_{\mathrm{n}}$ and $T^{\mathrm{F}}_{\mathrm{n}} \subset T^{\mathrm{F}}_{<\mathrm{n}}$

Though intuitive, this account is unfortunately afflicted by fatal difficulties, first described by David Miller (1974) and Tichý (1974). As these and other authors have proven formally, one can show that in neither the case of 1 or 2 above can both conjuncts be satisfied together. It turns out that, on Popper's definition, in order that T_{n} have greater approximate truth than $T_{<\mathrm{n}}$, T_{n} would have to be true *simpliciter*. Thus, on this view, one false theory cannot have more approximate truth than another, and this rather defeats the aim of providing an understanding of what it could mean to rank false theories with respect to truth.

The precise formulation of the possible worlds approach (also called the "similarity" approach) varies between different authors, but what follows is a general characterization of it. The basic idea is first to identify the truth conditions of a theory with the set of possible worlds in which it is true, and then to calculate what one may call "truth-likeness" by means of a function that measures the average "distance" between the actual world and the worlds in that set. In this way, one may generate an ordering of theories with respect to truth-likeness. For example, consider the class of atomic propositions entailed by a theory, each attributing a specific state to a particular; possible worlds here are described by distributions of truth values across these atomic propositions. The greater the agreement between a given theory and a theory correctly describing the actual world, the greater the former's truth-likeness.[1]

Though less clearly undermined by objections than Popper's account, the possible worlds approach is itself subject to two important controversies. First, Miller (1976) argues that on this view, measures and relative orderings of truth-likeness

[1]See Tichý (1974, 1976, 1978), Niiniluoto (1984, 1987, 1999), and Oddie (1986a, b, 1990). Niiniluoto (1998) summarizes the different formulations associated with this approach.

are language-dependent: logically equivalent theories may have different degrees of truth-likeness depending on the language in which they are expressed, and the relative truth-likeness of two theories may be reversed when translated into another language with logically equivalent predicates. Second, Aronson (1990) shows that on this view, the truth-likeness of a proposition (whether true or false) depends on the number of atomic states under consideration, and it is at least questionable whether the truth-likeness of propositions concerning states of affairs *other* than that described by the proposition at issue should be relevant in this way. Disputes concerning these charges continue to surround this approach today.

Finally, a third approach to approximate truth analyzes truth-likeness in terms of similarity relationships between nodes in type hierarchies: tree-structured graphs of types and subtypes.[2] The nodes represent concepts or things in the world, and links between them represent relations between concepts or things. As an illustration, consider the standard biological taxonomy of organisms divided into kingdoms, phyla, and so on down to species. Similarity here is defined with respect to locations within type hierarchies. In order to show that dolphins are more similar to whales than tuna, for example, one calculates their degrees of similarity to one another by means of a weighted difference measure, comparing the properties these types share and those in which they differ. Now consider an analogous comparison between a node in a theoretical type hierarchy and a corresponding node in the actual type hierarchy of the world. Truth-likeness is measured by the "distance" between a theoretical claim about a type and the correct description of that type, reflecting the degree of similarity of the nodes with which these descriptions are associated. The most striking difficulty with this approach is that it appears to require the existence of a unique type hierarchy of the world. Lacking this, it seems there is no determinate answer to the question of what a node in a theoretical type hierarchy should be compared *to*. As Psillos (1999, 277) observes, on this view, significantly different type hierarchies would lead to different measures of approximate truth, and the assumption that nature admits of only one correct taxonomy is controversial at best.

Each of the three approaches to approximate truth just outlined face interesting challenges, and it is not my intention here to see whether these challenges can be met. My goal in reviewing them has been rather to set the stage for what I take to be a more general complaint, to be leveled against all three. None of these established approaches to approximate truth pays much if any explicit attention to the qualitative dimensions of the concept, which concern the ways in which theories and models typically diverge from truth in the first place. It is precisely a better understanding of these details, I contend, that is crucial to understanding the nature and truth content of scientific representations, and it is in this context that I will take inspiration from certain analogies to practices of representation in art. Understanding *how* scientific representations give inaccurate accounts of their subject matter is an important precursor to thinking about approximate truth, for as we shall see, there are different *kinds* of representational inaccuracy, and as a consequence, I will suggest, the

[2]See Aronson (1990), and Aronson et al. (1994, 15–49).

concept of approximate truth is best explicated differently in different circumstances, depending on the relevant modes of inaccuracy. Extant approaches to the concept place no emphasis on these differences, which I take to be central. In the following section, I will attempt to explain what these differences are, and why they are so important.

Truth in the Context of Abstraction and Idealization

Let us turn now to consider the ways in which scientific theories and models are constructed so as to deviate from the truth. There are, I believe, two such ways, and I will call them *abstraction* and *idealization* respectively. In focusing on these two practices specifically, I am drawing on a recently established tradition in the philosophy of science that regards them as the primary means by which scientific representations are constructed and related to the worldly phenomena they target. I will not explore the nuances of this rapidly growing literature here; let it suffice to say that while there are, of course, idiosyncratic differences in various presentations of the relevant concepts, the fundamental ideas are widely shared.[3] In the following, I will sketch my own view of them, which is faithful to the central tenets of this recent tradition in thinking about scientific representation.

Roughly put, an abstract representation is the result of a process of abstraction; that is, one in which only some of the potentially many factors that are relevant to the behavior of a target system are built into the representation. In such a process other parameters are ignored, either intentionally or unwittingly, so as to permit the construction of a tractable representation. A commonly discussed example of this is the model of the simple pendulum. Here, among other simplifying assumptions made in the construction of the model, one simply omits the factor of frictional resistance due to air. The reason such omissions are thought to compromise the truth of resultant representations is not merely the fact that they leave out theoretically important aspects of the systems they represent, but that even more importantly, in doing so, they also generate predictions regarding these systems that deviate from reality. By omitting factors that play a causal role in determining the values of certain parameters, for example, abstractions often fail to be accurate in their estimations of them. The greater the discrepancy between the output of an abstract representation and the behavior of its target system, the less approximately true it may seem.

On the other hand, an idealized representation is the result of a process of idealization; that is, one in which at least one of the parameters of the target system is represented in a way that constitutes a distortion or a simplification of its true nature. In such a process, one is not excluding parameters, as in abstraction, but incorporating them, again either intentionally or unwittingly, in such a manner as to represent them in ways they are not—indeed, as I shall use the term, in ways they could not

[3]For some of the most influential and comprehensive discussions, see McMullin (1985), Cartwright (1989, Chapter 5), Suppe (1989, 82–83, 94–99), and Jones (2005).

possibly be. Idealized representations thus furnish strictly false descriptions of their counterparts in the world. For example, in the *Principia*, Newton assumes that the sun is at rest in his derivation of Kepler's laws of planetary motion. According to his own theory, however, this would require that the sun have infinite mass, and Newton clearly did not believe this to be the case. On his understanding, the sun experiences small amounts of motion as a consequence of the attractions of other bodies, and so, the "attribution" of infinite mass constitutes an idealization. Abstraction and idealization are not mutually exclusive processes, and consequently, representations are often both abstract and idealized. A model of a frictionless plane, for instance, is an abstraction because it leaves out frictional forces associated with the plane, and if no surfaces are totally frictionless, the model also incorporates an idealization. Elements of idealization enter the picture when representations describe systems in ways they could not be, given the laws of nature that obtain in our world.

Now, given these two practices of deviation from the truth, how should one think about the approximate truth of scientific representations? It seems to me that there is a straightforward answer to this question in cases of pure abstraction—that is, in cases of abstraction which incorporate no idealization—and a less obvious answer in cases of idealization. Regarding the former, there would seem to be no impediment to thinking of a pure abstraction as true *simpliciter*, if only in connection with a certain class of target systems. Pure abstractions correctly describe certain features of things in the world, even if they do not describe all of the properties and relations potentially relevant to the phenomena at issue. It is natural, of course, to view abstractions as yielding false descriptions because of their omissions and resultant inaccuracies of prediction. But in the case of pure abstractions, where no idealization is involved, target systems in which only those parameters represented contribute to the behavior of the system are presumably possible; if they were not, this would indicate the presence of an idealization. Therefore, pure abstractions are perfectly accurate representations of *some* nomically possible target systems (that is, ones that could exist, given the laws of nature that obtain in our world), even if they are impoverished representations of other, more complex ones. Clearly, one may apply a pure abstraction to a more complex system it does not correctly describe, but this should not be taken to discredit the truth of the representation in connection with systems it *does* correctly describe. Neglecting air resistance usually counts as an error of omission, but it would not be in connection with a system in a vacuum.

This suggests a straightforward articulation of the notion of approximate truth *qua* abstraction. Consider all of the parameters potentially relevant to the behavior of a particular target system. Degrees of approximate truth are correlated here with the extent to which representations incorporate these parameters. The greater the number of factors built into the representation, the greater its approximate truth. This suggestion for assessing relative degrees of approximate truth does justice, I think, to the intuition that higher degrees of abstraction may correspond to lesser degrees of truth, but without failing to appreciate that abstractions may yet characterize some things perfectly accurately. Pure abstractions yield correct descriptions of certain classes of target systems while being more or less approximately true in application to others, and here we have our first insight into what "conditions of approximation" means in the analysis of approximate truth. In these cases, conditions of approximation can be understood simply in terms of how much information a representation yields, or its comprehensiveness, relative to a specific kind of target system, or class of systems.

In cases of idealization, however, one requires a rather different understanding of the relevant conditions of approximation. For here, unlike in cases of pure abstraction, one does not have the luxury of representations that accurately characterize at least some nomically possible phenomena. Idealizations are more egregiously fictional than abstractions; they constitute not mere omissions, but distortions of things in the world. Models in classical mechanics, for example, generally treat the masses of bodies as though they are concentrated at extensionless points, but given the nature of mass as we understand it, in accordance with the laws of nature, it cannot be concentrated this way in any world such as ours, where particulars with masses exist. What information about the world is contained in fictions such as these?

A failure to grapple seriously with the qualitative nature of idealization is, I believe, a defect of extant accounts of approximate truth. Consider my illustration earlier of the possible worlds approach, in which one considers the class of atomic propositions entailed by a theory, each attributing a specific state to a particular, as a means towards evaluating its approximate truth. In cases of pure abstraction, one may justifiably claim here that the greater the extent to which a representation yields true descriptions of systems in the world, the greater its truth-likeness. Idealizations, however, do not generally give true descriptions of atomic states of affairs, for they are constructed in such a way as to characterize their objects in a distorted manner. Likewise, consider again the type hierarchy approach, where one calculates degrees of similarity between theoretical propositions and true ones by performing weighted difference measures involving the properties these propositions describe in common and those in which they differ. Arguably, however, idealized characterizations may describe few *if any* properties in common with true theories, because they correctly describe fictional properties, not actual ones. The conditions of approximation relevant to assessing approximate truth *qua* idealization must be understood differently, I think, than the relevant conditions of approximation *qua* abstraction.

Denotation in Art, Reference in Science

In order to appreciate how idealizations bear on the notion of approximate truth, let me now return to Goodman, and draw a first analogy to representation in art. In the first section, "Varieties of Truth in Art and Science", I quoted Goodman as suggesting that in both art and the sciences, successful representation is a matter of fitting or approximating things in the world. Let us now consider this suggestion in more detail, beginning with an examination of Goodman's reflections on the nature of realistic and non-realistic representation. It is precisely this distinction, I will argue, that is important to understanding the "truth content" of idealized theories and models. The contrast between the nature of this content in cases of abstraction and idealization will provide crucial insight into how different contexts of representation call for a flexible approach on the part of those seeking to explicate the concept of approximate truth.

In the opening sections of his major work on artistic representation, Goodman (1976, 34) considers the question of how best to understand the nature of realism in this context. His answer appears at first both provocative and negative: "Surely not [in terms of] any sort of resemblance to reality". Goodman does not provide much help with the ambiguous term "resemblance", here, but on any obvious reading, his answer presents a *prima facie* puzzle of interpretation. If one interprets "resemblance" narrowly to mean "similarity in appearance", this might seem a strange claim regarding much art, though not perhaps regarding science; in the latter case one hardly expects mathematical equations (for example) to share similarities in appearance with acids and bases or populations of organisms. Reading "resemblance" more broadly, as "having some feature or features in common", the puzzle of interpretation extends even to the scientific case, since most philosophers of science hold that at least some parts of our best theories and models do, in fact, have features in common with their target systems, such as commonalities in structure (whether concerning observable or unobservable parts of the world). These interpretive puzzles, however, are resolved with the clarification that for Goodman, realism is by no means inconsistent with resemblance in either of the senses just mentioned. His point is rather to emphasize that, as suggested earlier, realism of representation is only achieved in special circumstances, *viz.* those in which agents considering the representation have been acculturated with the system or systems of representation that have been employed in constructing it.

> Consider a realistic picture, painted in ordinary perspective and normal colour, and a second picture just like the first except that the perspective is reversed and each colour is replaced by its complementary. The second picture, appropriately interpreted, yields exactly the same information as the first. And any number of other drastic but information preserving transformations are possible. Obviously, realistic and unrealistic pictures may be equally informative; informational yield is no test of realism. ... The two pictures just described are equally correct, equally faithful to what they represent, provide the same and hence equally true information; yet they are not equally realistic or literal. ... Just here, I think, lies the touchstone of realism: not in quantity of information but in how easily it issues. And this depends upon how stereotyped the mode of representation is, upon how commonplace the labels and their uses have become (Goodman 1976, 35–36).

Goodman is a conventionalist about systems of representation: anything can represent anything else, subject to appropriately internalized conventions. As a consequence, one and the same representation can be realistic or not, depending on whether the relevant conventions have been internalized by the viewer or user.

Several fascinating issues concerning conventionalism and representation are raised by this and surrounding passages, but for present purposes, let me simply extract one key point: an understanding of the relevant and potentially different conventions of reading information from representations is crucial to how one understands that information. This point bears directly on my contention that if one is to have a genuinely informative account of what it means to say that scientific representations are approximately true, one must understand the different conditions of approximation exemplified by abstraction and idealization. These different conditions, I suggest, should be understood in terms of different conventions of representation.

Having already considered the way in which pure abstractions approximate reality, let us do the same now for idealizations. One last point arising from Goodman's discussion of artistic representation will prove useful in this regard. Goodman (1976, 5) suggests that "the core of representation" is denotation. That is, in order for *x* to represent *y*, *x* must be a symbol for, or stand for, or refer to *y*. Symbols here include "letters, words, texts, pictures, diagrams, maps, models, and more" (1976, xi). In Goodman's terms, denotation is simply a particular species of reference, pointing from representations to things represented. And with this in mind, here finally is the first feature of artistic representation that I believe furnishes a provocative analogy for those hoping to understand the nature of approximate truth in science: just as in the case of art, where successful representation can be a function of denotation, in the sciences, successful representation can be a function of reference, even when theories and models offer idealized descriptions of the properties and relations of their target systems. In the first instance, idealizations "fit" or "approximate" reality by latching on and successfully referring to aspects of it. Let us consider this suggestion in some detail.

To be sure, emphasizing reference is hardly novel in the philosophy of science, especially in discussions of scientific realism. Entity realists are especially well known for this, holding that under conditions in which one has significant evidence of an ability to manipulate or otherwise systematically exploit the causal properties of entities, one has good reason to believe that such entities exist, even while withholding belief from the theories that describe them. It is for this reason that entity realism can be cast as a response to challenges to realism posed by the history of science, which provides ample evidence of theoretical descriptions changing over time. Hacking (1983, Chapter 6), for example, contends that one may continue to refer to the same causal entity despite changes in the theories that describe it, and this furnishes a stable point around which realists can organize their knowledge claims regarding unobservables. Despite the fact that theories are false and likely to change, says the entity realist, there are conditions under which one has good reason to think that unobservable terms refer, and will continue to refer. Interestingly, the importance of relations of reference has never really shaped thinking about approximate truth, but it is my contention here that they are clearly relevant to understanding the differential truth content of pure abstractions and idealizations. Insofar as *true* (that is, non-idealized) claims about entities and processes can be extracted from idealizations, these are for the most part claims of successful reference, not the more detailed descriptions of target systems that one may associate with cases of pure abstraction.

This should not be taken to suggest, of course, that merely appealing to reference can save realist blushes. This appeal on behalf of entity realism is controversial, even amongst scientific realists. Many question whether it is coherent to be a realist merely about certain entities described by theories, since causal manipulations and exploitations seem to be based on further and substantial parts of those theories. And as I have argued elsewhere (2007, Chapter 2), there does seem to be something anachronistic in the suggestion that scientists from different periods in the history of scientific investigation into an entity all believe in the same thing. In order to be more compelling, the realist's story must be told at a deeper level, with respect to specific properties and relations on which existential claims are based, and that are likely to survive (if only as limiting cases) in theories over time. Despite what I take to be serious difficulties, however, there is an important insight at the heart of entity realism: degrees of belief in unobservable entities are generally and rightly correlated with the extent of one's causal contact with those entities. There are no stronger grounds for belief in an entity than an impressive ability to systematically exploit its causal properties, and less impressive abilities rightly ground more attenuated beliefs. On the impressive side of this continuum, claims of reference are concomitantly strong. For those of a strict empiricist bent, the same point can be made, *mutatis mutandis*, regarding observable entities.

Having emphasized the notion of reference, and having gestured towards some of the nuances that must be taken into consideration concerning reference in this context, I am now in a position to describe what it means for one theory to be more or less approximately true than another *qua* idealization, and to contrast this with cases of pure abstraction. When it comes to truth, even the best idealizations contribute primarily existential claims. This does not mean, however, that all idealizations are on a par when it comes to the approximate truth of the more substantive descriptions they provide. Some idealizations approximate true descriptions of properties and relations better than others, and this is an important consideration in assessing their relative approximate truth. The relevant notion of approximation here is usually specified mathematically. One can define mathematically how Newtonian descriptions of certain properties approximate those of special relativity, for example, by showing how the equations of Newtonian mechanics are limiting cases of relativistic equations. The ideal gas law assumes that molecules of gas are point particles and that there are no forces of attraction between them, but it is possible to take into account both the space occupied by molecules of gas and small forces of mutual attraction. Thus, while the van der Waals equation generates values for various properties that approach those given by the ideal gas law at lower pressures (larger volumes), it yields different, more accurate values at higher pressures (smaller volumes).[4] The Van der Waals equation, over certain ranges of pressure, volume, and temperature, describes the natures of these properties and their relations more accurately than the ideal gas law.

Earlier I credited Psillos with a precise statement of what I take to be a widely-held intuitive platitude regarding approximate truth, to the effect that a description

[4]McMullin (1985, 259) contains a nice discussion of this and similar cases.

is approximately true of a state if it can be "linked by specific conditions of approximation" to a true description. It was precisely because of a lack of clarity regarding the question of what these "conditions of approximation" might be that I undertook to focus attention on them, with the goal of generating a more satisfying explication of the concept of approximate truth. I believe the various pieces of the puzzle are now in hand. When representations deviate from true characterizations of their target systems, they do so via abstraction, or idealization, or in many cases both. I have argued that insofar as representations are abstract, approximate truth may be gauged in terms of the numbers of potentially relevant features of their target systems they incorporate, so that theories incorporating greater numbers of these features may be thought of as more approximately true than those incorporating fewer. Pure abstractions yield descriptions of properties and relations that are true *simpliciter* of certain classes of target systems, and that may be more or less approximately true in application to others. The notion of approximate truth *qua* abstraction is thus simply the notion of comprehensiveness, generally assessed in connection with a specific target system, and the relevant condition of approximation here is the extent to which the numbers of factors incorporated into a representation match up with those in the target systems to which it is applied.

The notion of approximate truth *qua* idealization is importantly different, for here the issue is not the comprehensiveness of representations, but in the first instance, their successful reference, and thereafter, the accuracy with which they characterize the natures of the specific parameters they represent. Unlike pure abstractions, idealizations do not generally offer true characterizations of the properties and relations they concern, even if they do permit ontological claims, in virtue of successful reference. By reducing the number of idealized assumptions or the extent to which they idealize—by de-idealizing—one describes target systems in ways that admit of greater degrees of approximate truth. Unlike the case of abstraction, however, where improving a representation is simply a matter of increasing the number of potentially relevant factors it incorporates, there is no reason to expect that processes of de-idealization should follow any common pattern from one domain of theorizing to the next. There are many ways of incorporating idealized assumptions into representations, and the ways in which one describes possible de-idealizations may vary in just the way that idealizations do. In any case, whatever these variations, idealized representations may be improved in ways determinable in specific instances. Approximate truth *qua* idealization concerns the degree to which a representation that has successfully latched on to an aspect or aspects of some target system resembles a non-idealized representation of that system, where degrees of resemblance are defined in specific cases. The relevant condition of approximation here is not comprehensiveness, but successful reference, and degrees of distortion or simplification of the specific properties and relations targeted.[5]

---

[5]Interesting questions naturally arise here concerning whether, in the context of scientific (if not artistic) representations, distortions can be so severe as to sever relations of reference, whether in such cases it is reasonable to speak of idealizations *of* target systems at all, and so on. For some thoughts on these issues, see my (2009).

Let me sum up the import of the first analogy furnished by representation in art before turning to a second. When viewing a painting or a sculpture, one may extract more or less information regarding the things it represents, depending on the amount of information it contains, and the extent to which one has mastered the conventions of representation it employs. At one end of this spectrum is what Goodman calls realistic representation in art. Here, the viewer is sufficiently acculturated with some relevant system or systems of representation to derive significant information about the subject matter represented. At the other end of the spectrum representations may convey very little information, but information nonetheless. Consider the representational content of paintings, for example. Just one of the reasons Pablo Picasso's *Guernica* (1937) is one of the most celebrated artworks of the twentieth century is its captivating representational power. Its subject is the bombing of the Basque town of Guernica by Hitler's and Mussolini's air forces, with the complicity of Franco, during the Spanish Civil War. Aspects of the work, such as the figures of a bull, a dead baby in the arms of a screaming woman, a speared horse, the broken body of a soldier, and so on, represent various things with greater and lesser degrees of realism. The painting taken as a whole also has representational content. Among other things, for instance, it represents the rising threat of European fascism.[6] Insofar as the painting represents this, however, it is not depictive, but merely denotative. It does not provide much in the way of "description" beyond the existential "claim" it makes concerning the presence of a terrifying danger.

Scientific representations also yield information about their subject matter, but whether they do so by providing true characterizations of specifically chosen parameters, or by distorting the parameters to which they successfully refer, will depend on how abstract and idealized they are. The contrast between depiction and mere denotation as a central feature of representation in art is an analogy for the contrast between truth and mere reference as a central feature of representation in the sciences. Higher degrees of approximate truth can be understood in terms of improved representations of the natures of target systems in the world, and this improvement can be mapped along two dimensions: how many of the relevant properties and relations one describes (abstraction), and how accurately one describes them (idealization). This simple formula, combined with an understanding of the conditions of approximation involved in the practices it describes, comprises an explication of the principal notions at stake in making sense of the idea of approximate truth.

Representations and Practice as Products and Production

Let me now turn, finally, to the second analogy between representation in art and science I promised at the start. This one also concerns approaches to truth, but in a rather different way than the first. For quite apart from the question of whether one can make sense of the notion of approximate truth, it should be noted that in the

[6]See Suarez (2003, 236), for a discussion of this painting and associated literature.

philosophy of science, there is no consensus regarding what *sorts* of scientific claims one ought reasonably to regard as true or approximately true in the first place. While realists defend the reasonableness of believing scientific claims concerning both observable and unobservable aspects of target systems, various critics, including varieties of empiricists and instrumentalists, accept only the former. My goal in this final section is to employ a second analogy of representation between art and science so as to extend a bridge between these opposing camps. In part because of their obsession with unobservable things, realists are often guilty, I believe, of failing to note the significance of the observable. Certainly, realists like everyone else regard observable consequences as furnishing tests of the accuracy of representations, but I have something else in mind here. In scientific practice, one is often primarily concerned with whether and to what extent theories, models, procedures, tests, etc. *work*. Can we use them to make faster computer chips, manage eco-systems, and successfully complete the astounding variety of tasks associated with laboratories and fieldwork across the globe every day? Success in practice is measured in terms of observable consequences, and there is a strong current of pragmatism built into everyday scientific pursuits. The pragmatist's test of epistemic significance is utility, and utility is assessed by means of observables.

Antirealists often intimate that realist interpretations of scientific knowledge are out of touch with the everyday worlds of real science, as opposed to the rarefied philosophical worlds of imagined science. The prevalence of empirically adequate idealizations and pure abstractions applied to systems they do not correctly describe serves to fuel this skepticism. It is for this reason that the idea of approximate truth, and more specifically, the idea that different sorts of truths may be contained within different sorts of scientific representations, is so important to realism. Armed with an understanding of the truth content of both idealizations and pure abstractions applied to systems they do not correctly describe, one may connect desiderata that skeptics generally believe to be independent of one another: the generation of observable predictions within acceptable margins of error (the goal of much scientific endeavor); and the uncovering of facts regarding unobservables that underlie these predictions. In the first section of this essay, I suggested that a consideration of two important features of scientific knowledge would facilitate an account of the concept of approximate truth. The first of these was the distinction between abstraction and idealization. The second concerns the pragmatic dimensions of scientific practice, and this topic, I believe, is intimately connected to the first. Let me move on to the second, now, by means of a second analogy to representation in art.

The history of twentieth-century art is, to a great extent, the history of the avant-gardes and their forms of "abstraction". Realistic conventions of representation, in Goodman's sense, were supplemented by varieties of experiments seeking to realize different sorts of conventions, both in the service of representation and even, in some cases, in the service of *non*-representational expression. These experiments initiated traditions that we now recognize as familiar artistic movements such as Cubism, Surrealism, Constructivism, and Abstract Expressionism. The disparate approaches of the avant-gardes exemplified a shared rebellion against traditional approaches to representation, and a striking feature of much of this work is an increasing focus on processes of art production, as opposed to the precise visual properties of the products of these processes. This is not to say, of course, that in many or most of these cases, products such as paintings and sculptures ceased to be important to artists and their critics. Rather it is to note that many of these artistic pioneers were self-consciously and sometimes primarily interested in reflecting on the nature of artistic representation itself, paying great attention, for example, to the nature of the canvas as a two-dimensional surface, as opposed to the task of realistically representing three-dimensional subjects. This is one, partial interpretation of the motivations of analytic Cubism, but it is also a recurring theme in other movements.

Consider the emergence of Abstract Expressionism in the 1940s and 1950s. One of the defining features of this work is a commitment to *expressing* emotional and other cognitive states of the artist, as opposed to depicting them as such. The methodology of Jackson Pollock is legendary in this regard: Pollock would drip, fling, and otherwise propel paint onto canvases placed on the ground, by means of controlled and sometimes highly athletic movements. The process of creation here is a central part of the content of the work. The artists of this and other movements increasingly emphasized the materiality of the process of painting, as opposed anything like realistic representation. The surface of the canvas, its shape, the thickness of the paint, and so on, took on a new significance. Co-opting the slogan of the American art critic Clement Greenberg (2003/1939, 539), this is "art for art's sake".

The rise of performance art may exemplify this tendency towards attaching greater significance to processes involved in the creation of art as opposed to its products per se better than anything else. Works associated with the Fluxus movement, for example, such as Yoko Ono's *Cut Piece*, may serve to illustrate the point. *Cut Piece* was performed by Ono four times, in Kyoto (1964), Tokyo (1964), New York (1965), and London (1966). During these events, the artist sat on a stage while members of the audience approached, individually and in succession, to cut pieces of clothing from her body with a pair of scissors. The piece is variously interpreted as engaging with issues of female vulnerability, sexual violence, and gender politics; and as a response to the horrors of war and the threat of nuclear annihilation.[7] Here as in all work in the performance art genre, the idea of a process takes on so much significance that *it* now is the central focus of the artwork. What matters is an event or a series of events. The idea that the value of the performance resides in any further output is completely lost. Of course, photographs of works of performance art are very important for purposes of discussion and art criticism, but such outputs are considered mere documents of the art form, not things that are important in their own right, and certainly not things that are the proper focus of attention when considering the nature or significance of the work.

Keeping in mind this analogy of a transition in focus from products to production, let us now return to the case of the sciences, and give due consideration to the pragmatic dimension of scientific practice. Focusing on processes of production

[7]For an insightful discussion of the variety of interpretation and the literature surrounding it, see Bryan-Wilson (2003).

led artists to a dizzying array of less-than-realistic representations. Analogously, focusing on processes of detection, experiment, and the innumerable tasks constituting everyday scientific work leads scientists to create ingenious abstractions and idealizations. In the successful pursuit of much of this work, one does not require anything like truth *simpliciter*. One of the main reasons abstractions and idealizations are so ubiquitous in the sciences is that they facilitate these tasks so well, within the degrees of accuracy and precision required in the contexts of particular scientific endeavors. Indeed, it is often the case that less approximately true representations are preferred to more approximately true ones. For while both may generate predictions that are adequate to specific endeavors, simpler though less approximately true theories and models are, generally, more easily taught, learned, and used.

Furthermore, the epistemic virtues of inaccurate representations often extend beyond their mere contextual adequacy. Scientists routinely apply pure abstractions and idealizations to phenomena whose properties and relations they do not correctly describe, but that is not to say that in such cases, representations yield no truths.[8] The classical theory of gases idealizes several properties of gas molecules and their relations to one another, but nevertheless has the (putatively) true consequences that there are molecules composing gases, and that they have properties such as mass. These are truths about particulars and properties that follow immediately from successful reference, but other truths stemming from idealization arguably go further. Frictionless surfaces are ideal, but models of spherical objects sliding down frictionless inclined planes correctly describe the motions of these objects as linear. Newtonian models of the earth-moon system are idealized, but correctly represent the mass of the earth as being greater than that of the moon. Idealizations generate substantially less truth *simpliciter* than pure abstractions, but what truths they do yield may nevertheless add to their pragmatic utility.

The emphasis on production as opposed to products in art and science has an echo in the intellectual traditions that study these practices. A delightful symmetry can be found, for example, in the juxtaposition of twentieth-century art criticism and post-positivist philosophy of science. One of the recurring themes of critiques of logical positivism was that it was too absorbed with normative projects based on rational reconstructions of the products of the sciences, such as theories and models, and as a consequence, it is argued, the positivists found themselves out of touch with actual scientific practice. Thus it comes as no surprise that the demise of positivism in the twentieth century was accompanied by the rise of the history of science as an essential tool for philosophers. Much post-positivist philosophy of science takes the details of scientific practice as its focus, thereby de-emphasizing considerations of the epistemic status of its products. And thus, the word "truth" does not appear in Kuhn's iconic essay in the history and philosophy of science, *The*

---

[8] I owe thanks to Martin Thomson-Jones and Juha Saatsi for illustrating this point with some of the following examples.

Structure of Scientific Revolutions, and Hacking (1983) is ultimately more interested in intervening in the natural world than representing it.

This sort of pragmatism is something that realists must take to heart in grappling with the concept of approximate truth. It is a concept that is differently instantiated by means of different representational relationships, involving true descriptions of properties and relations in some cases, and little more than successful reference in others. Some representations are purely abstract, in which case they yield a multitude of true descriptions of certain classes of phenomena. Other representations are heavily idealized, and consequently their truth rests primarily in existential claims, and in the extent to which their descriptions of properties and relations measure up to true descriptions, in ways specifiable in connection with specific target systems. Most cases of scientific representation are neither pure abstractions nor pure idealizations, of course, but rather mixtures of both, in different proportions and to varying degrees. The concept of approximate truth is thus heterogeneous, to be explicated as may be appropriate in particular cases, within the myriad contexts of representation to which it may be applied.

It is thus the conclusion of this paper that in the sciences, approximate truth is best understood as a virtue that is multiply realized by means of different kinds of representational relationships between scientific products such as theories and models on the one hand, and target systems in the world on the other.[9] These different conventions of representation reflect the degrees to which theories and models abstract and idealize, and as a consequence, anyone hoping to understand the ways in which they approximate truth must first subject these conventions to serious consideration. In this and perhaps other ways, those interested in the nature of scientific knowledge may find illumination in positive analogies to the nature of representation in art.

Acknowledgments My thinking about these topics has benefited immensely from discussions with audiences at the Tilburg Centre for Logic and Philosophy of Science, the "Beyond Mimesis and Nominalism: Representation in Art and Science" conference in London, the National University of Singapore, York University, and the Swiss Federal Institute of Technology and University of Zurich. I am grateful to all the participants, and especially thankful for the detailed input of Roman Frigg, Matthew C. Hunter, and two anonymous referees for this volume.

References

Aronson, J. L. (1990), "Verisimilitude and Type Hierarchies", *Philosophical Topics* 18: 5–28.
Aronson, J. L., Harré, R. and Way, E. C. (1994), *Realism Rescued: How Scientific Progress Is Possible*. London: Duckworth.
Bryan-Wilson, J. (2003), "Remembering Yoko Ono's *Cut Piece*", *Oxford Art Journal* 26: 99–123.
Cartwright, N. (1989), *Nature's Capacities and Their Measurement*. Oxford: Clarendon.
Chakravartty, A. (2007), *A Metaphysics for Scientific Realism: Knowing the Unobservable*. Cambridge: Cambridge University Press.

[9]For a more leisurely route to this conclusion, see Chakravartty (2007, Part III).

Chakravartty, A. (2009), "Informational Versus Functional Theories of Scientific Representation", *Synthese* 172: 197–213.
Goodman, N. (1976), *Languages of Art: An Approach to a Theory of Symbols*, 2nd ed. Indianapolis: Hackett.
Greenberg, C. (2003/1939), "Avant-Garde and Kitsch", in C. Harrison and P. Wood (eds.), *Art in Theory, 1900–2000: An Anthology of Changing Ideas*, Oxford: Blackwell.
Hacking, I. (1983), *Representing and Intervening*. Cambridge: Cambridge University Press.
Jones, M. R. (2005), "Idealization and Abstraction: A Framework", in M. R. Jones and N. Cartwright (eds.), *Idealization XII: Correcting the Model, Poznañ Studies in the Philosophy of the Sciences and the Humanities*, vol. 86. 173–217.
McMullin, E. (1985), "Galilean Idealization", *Studies in History and Philosophy of Science* 16: 247–273.
Miller, D. (1974), "Popper's Qualitative Theory of Verisimilitude", *British Journal for the Philosophy of Science* 25: 166–177.
Miller, D. (1976), "Verisimilitude Redeflated", *British Journal for the Philosophy of Science* 27: 363–380.
Niiniluoto, I. (1984), *Is Science Progressive?* Dordrecht: D. Reidel.
Niiniluoto, I. (1987), *Truthlikeness*. Dordrecht: D. Reidel.
Niiniluoto, I. (1998), "Verisimilitude: The Third Period", *British Journal for the Philosophy of Science* 49: 1–29.
Niiniluoto, I. (1999), *Critical Scientific Realism*. Oxford: Clarendon.
Oddie, G. (1986a), "The Poverty of the Popperian Program for Truthlikeness", *Philosophy of Science* 53: 163–178.
Oddie, G. (1986b), *Likeness to Truth*. Dordrecht: D. Reidel.
Oddie, G. (1990), "Verisimilitude by Power Relations", *British Journal for the Philosophy of Science* 41: 129–135.
Popper, K. R. (1972), *Conjectures and Refutations: The Growth of Knowledge*, 4th ed. London: Routledge & Kegan Paul.
Psillos, S. (1999), *Scientific Realism: How Science Tracks Truth*. London: Routledge.
Suárez, M. (2003), "Scientific Representation: Against Similarity and Isomorphism", *International Studies in the Philosophy of Science* 17: 225–244.
Suppe, F. (1989), *The Semantic Conception of Theories and Scientific Realism*. Chicago: University of Illinois Press.
Tichý, P. (1974), "On Popper's Definitions of Verisimilitude", *British Journal for the Philosophy of Science* 25: 155–160.
Tichý, P. (1976), "Verisimilitude Redefined", *British Journal for the Philosophy of Science* 27: 25–42.
Tichý, P. (1978), "Verisimilitude Revisited", *Synthese* 38: 175–196.

Learning Through Fictional Narratives in Art and Science

David Davies

Thought experiments (henceforth, "TE's") in science take the form of short narratives in which various experimental procedures are described. The competent reader understands that these procedures have not been, and usually could not (for some appropriate modality) be, enacted. She is invited, however, to imagine or make believe that these procedures *are* enacted and to conclude that certain consequences would ensue, where this is taken to bear upon a more general question which is the topic of the TE. Perhaps the most famous example of such a device is Galileo's "Tower" TE which aimed both to refute the standing Aristotelean account of the behavior of falling bodies, and to establish the alternative account favored by Galileo himself. The Aristotelean account held that the speed at which a body falls is directly proportional to its weight. Galileo asks us to imagine that we take two bodies, one heavy [H] and one light [L], to the top of a high tower. We strap the bodies together and drop the resulting object [H+L] from the tower. The Aristotelean is then committed to two contradictory claims. First, since [H+L] is heavier than [H], it should fall faster than H. On the other hand, since [L] falls more slowly than [H] it should retard the fall of [H], and since [H] falls more quickly than [L] it should accelerate the fall of [L]. So [H+L] should fall at a speed somewhere between the rate of fall of [H] and the rate of fall of [L]. Since the Aristotelean view leads to an absurdity—that [H+L] will fall both more quickly and more slowly than [H]-rate of fall must be independent of weight. Given this "intermediate" conclusion, Galileo further concludes that (if we remove the resistance of a medium) all bodies fall at the same rate.[1]

It is not only in science that we find the TE playing a substantial cognitive role. Analytic philosophers also accord both critical and constructive roles to this device. Among the more celebrated examples are John Searle's "Chinese Room" TE (1980), directed at claims about the cognitive capacities of certain forms of artificial

D. Davies (✉)
McGill University, Montreal, QC, Canada
e-mail: david.davies@mcgill.ca

[1]For a fuller description of Galileo's "Tower" TE, and for discussion of its philosophical significance, see Brown (1991, 1992), McAllister (1996), Norton (1996), and Gendler (1998).

R. Frigg, M.C. Hunter (eds.), *Beyond Mimesis and Convention*, Boston Studies in the Philosophy of Science 262, DOI 10.1007/978-90-481-3851-7_4,

intelligence, Judith Thompson's "violinist" TE (1971), which challenged some standard approaches to the morality of abortion, and Hilary Putnam's "Twin Earth" TE's (1975), which challenged hitherto entrenched views about mental and linguistic representation.[2]

A number of writers on TE's such as James McAllister (1996), Roy Sorensen (1992), and Nancy Nersessian (1993) have further observed that scientific and philosophical TE's take the form of short *fictional* narratives. This seems to be correct, given the literature on the nature of fiction. Rather than survey this (extensive) literature here, I shall merely offer what I have elsewhere argued to be two plausible necessary and sufficient conditions for the fictionality of a narrative.[3] First, fictional narratives must be products of acts of "fiction-making", where the maker's intention is that we make-believe, rather than believe, the content of the story narrated.[4] Second, the primary constraint on the construction of a fictional narrative must not be what I have termed the "fidelity constraint" ("include only events you believe to have occurred, narrated as occurring in the order in which you believe them to have occurred"), but, rather, some more general purpose in telling a given story, such as entertaining or perhaps instructing readers in certain specific ways.[5] The second condition elaborates upon the first by placing a constraint on what can count as a legitimate act of fiction-making.

TE's, whether scientific or philosophical, clearly seem to involve the construction of fictional narratives so construed. They present the reader with a hypothetical situation in which an event or process is taken to occur. The reader is intended to make-believe, rather than believe, that the hypothetical situation and described sequence of events occur. Thus the first condition for fictionality seems to be met. The second condition also seems to be met, given that the author of a TE doesn't think that the envisaged situation and sequence of events actually occurred, and is therefore not guided by the fidelity constraint.[6]

Not all fictional narratives that play some cognitive part in scientific and philosophical reasoning are usefully classified as TE's. For not all such narratives function as "experiments" in any clear sense. Some only provide indirect support for a scientific or philosophical position by illustrating how certain things are

---

[2]For a detailed discussion of philosophical thought experiments, see Häggqvist (1996).

[3]For a critical overview of the literature on the nature of fiction and a defense of these two conditions, see my (2005, 2007a, Chapter 3).

[4]See, for example, Currie (1990, Chapter 1).

[5]This allows not merely for fictions that are accidentally true, but also for fictions that the author knows or believes to be true, as long as their being true is not what guides the author's constructive activity. See my (2005).

[6]Nersessian (1993, 297) argues that TE's in science differ from the narratives in literary fictions in that "unlike the fictional narrative, ... the context of the scientific thought experiment makes the intention clear to the reader that the situation is one that is to represent a potential real-world situation", one in which "objects behave as they would in the real world" (1993, 295) if the hypothesized circumstances were to obtain. In my (2007b), I argue that literary fictions also standardly satisfy these conditions, and thus that no such principled distinction can be drawn.

possible given that position. Arguably, this applies to Maxwells' "Demon" TE, and to what are sometimes unflatteringly termed "just so stories" that feature in evolutionary accounts of how certain biological features might have emerged through a process of natural selection (see Brown 1992). Other such narratives, especially in philosophy, function as helpful visualizable illustrations of more abstract positions, rather than as arguments for them in any strict sense. For example, in *The Republic* Plato presents the "parable" of the prisoners in the cave to illustrate the "Theory of Forms".

Interestingly, recent treatments of the fictional narratives that feature in many works of literary art[7] also appeal to a connection between fictions and thought experiments. But the import of the connection between fictions and TE's for those whose primary interest is in TE's differs in one striking respect from its import for those whose primary interest is in literary fictions. The claim that TE's are fictional narratives is generally taken to *problematize* the cognitive credentials of TE's. James McAllister, for example, sets up what is usually termed the "epistemological problem" of scientific TE's as follows: how can we make genuine cognitive progress through engaging with TE's if they are *merely* fictional narratives? On the other hand, the claim that at least some literary fictions are TE's has been taken to *un*problematize the cognitive credentials of such fictions. Literary fictions, it is claimed, can have genuine cognitive value and educate us about the extra-fictional world *because* they are elaborate thought experiments and (it is assumed) the latter possess such cognitive virtue. Thus, rather paradoxically, that TE's are fictions has been taken (by some) to call into question the very thing that is supposed to be established (for others) by the fact that fictions are TE's! It should be noted that literary cognitivists—to borrow James Young's term (2001) for those who argue for the cognitive value of works of literary fiction—do not claim that *all* literary fictions are TE's. They claim only that it is because some paradigm works of literary fiction function as TE's that they can share in the cognitive value that the latter possess.

In section "I", I outline the "epistemological problem" presented by TE's in science and provide an overview of responses to that problem. In section "II", I present closely parallel epistemological problems associated with the claims of literary cognitivism. In section "III", I critically examine recent attempts to defend literary cognitivism by drawing analogies between works of literary fiction and TE's in philosophy and science. Finally, in section "IV", I explore a strategy for defending literary cognitivism that takes fuller account of arguments, outlined in section "I", for the cognitive value of TE's themselves.

---

[7]See, for example, Noel Carroll (2002) and Catherine Elgin (2007), discussed in section "III" below.

I

As may be apparent from my earlier remarks, what unites those who reflect on the fictional nature of TE's and those who claim the status of TE's for some literary fictions is an epistemological concern with the claim to cognitive status of certain narratives. In the case of scientific TE's, the focus of this concern is an epistemological puzzle most clearly posed by Thomas Kuhn (1964). The puzzle resides in the apparent tension between the following three claims:

> C1/ Scientific TE's do not rely on or provide any new empirical data concerning the state of the world. Any empirical data upon which we draw must have been known and generally accepted before the TE was conceived.
> C2/ TE's provide us with new information about the physical world.
> C3/ TE's, while they involve reasoning, cannot be reduced without epistemic loss to inferences of any standard kind (deductive, inductive, or abductive).

We need C3 to get a genuine puzzle because we routinely learn new things about the world by *constructing inferences* based on existing knowledge.

This epistemological "puzzle" admits of broadly "deflationary" and "inflationary responses" [8]:

> 1/ A *deflationary* response denies that there is a genuine puzzle, either by denying C2, or by denying C3. (We may take C1 to be true by definition, if "new empirical data" means new evidence about the world derived directly or indirectly from sense experience.)
> 2/ An *inflationary* response accepts C2 and C3, thereby takes TE's to have a distinctive cognitive value, and offers an explanation of how TE's are able to possess that value.

There are extreme and moderate versions of each kind of response. An extreme deflationist simply denies C2. A moderate deflationist retains C2 but casts doubt on the distinctive epistemic virtues of TE's by denying C3. A moderate inflationist supplements C1 by arguing that prior empirical knowledge can be mobilized in a new way by TE's. And an extreme inflationist argues that TE's involve non-empirical modes of intuition.

For extreme deflationists such as Duhem (1914) and Hempel (1965), scientific TE's are of at best heuristic value. They may serve as instruments of discovery, but they cannot provide justified beliefs about the world unless independently tested. A TE may *suggest* that physical reality has a certain feature, and may even provide the idea for a concrete experiment which may itself justify the belief that such a feature exists. But TE's cannot themselves teach us anything about the world. A more modest version of extreme deflationism—what Kuhn terms the "standard view"—holds that the new understandings provided by TE's are not of nature but of the scientist's conceptual apparatus. The TE elicits recognition of contradictions inherent in a scientist's way of thinking. So assessed, scientific TE's help to clarify the nature of our concepts by exploring their implications in counter-factual situations.

The moderate deflationist response is exemplified in the writings of John Norton (1991, 1996) and Andrew Irvine (1991). Both maintain that, insofar as TE's can tell

---

[8] For a fuller discussion of these matters, see my (2007b).

us about the world, they are epistemically unremarkable. They are merely colorful uses of our standard epistemic resources—ordinary experiences and the inferences that are to be drawn from them. TE's are simply picturesque arguments that reorganize and make explicit what we already know about the physical world. Norton expresses this view in his "reconstruction thesis", according to which all TE's can be reconstructed as arguments once we fill in the tacit or explicit assumptions. Belief in the conclusion of a TE is then justified insofar as that conclusion is justified by the reconstructed argument.

The principal exponent of the extreme inflationary response is James Brown (1991, 1992), who offers an account of what he terms "platonic TE's". He maintains that we find such TE's in mathematics, where "we can sometimes prove things with pictures". In such cases, he argues, "we grasp an abstract pattern" via a kind of "intellectual perception". TE's in the natural sciences can also function platonically, according to Brown. He points, here, to Galileo's "Tower" TE. The move to the final conclusion is "immediate" because it involves an exercise of intellectual intuition which reveals a law of nature. Such laws, for Brown, are relations between universals, and sometimes a TE can lead us to grasp such laws.

Since I shall later appeal to a moderate inflationist strategy in countering certain objections to literary cognitivism, I shall discuss this strategy in more detail. The moderate inflationist stresses how TE's allow the scientist to mobilize cognitive resources not available in the kinds of scientific reasoning celebrated by deflationist accounts such as Norton's. She argues that TE's are epistemically singular. They cannot be reconstructed as deductive or inductive arguments without epistemic loss, because of *the way in which* they mobilize cognitive resources available prior to the formulation of the TE. Ernst Mach (1905) is widely acknowledged as the progenitor of this approach. He argued that we have "instinctive knowledge", derived from experience but never articulated and perhaps even incapable of being articulated or made explicit. This knowledge is activated when we imagine ourselves in a hypothetical experimental situation. It is only in virtue of the mobilization of such instinctive knowledge that we are able to "immediately" draw the required conclusion from the TE narrative.

A number of more recent commentators have echoed Mach's strategy. Tamar Gendler (1998), for example, responds to Norton's criticisms of Galileo's "Tower" TE. She argues that attempts to reconstruct the latter as a deductive argument (see Norton 1996) fail to capture how the representation of the phenomena in the TE invokes experientially grounded "tacit knowledge". The demonstrative force of Galileo's TE, she argues, depends crucially upon such "tacit knowledge". Richard Arthur (1999), agreeing with Gendler, draws comparisons between the role accorded here to "tacit knowledge" and the role accorded to "natural interpretations" by Paul Feyerabend (1975) in the latter's reconstruction of Galileo's method of argumentation. And David Gooding (1993) argues that the narratives whereby both real and thought experiments are presented are persuasive only if they manage to convey the "relevant experimental know-how", much of it what Gooding terms "an experimenter's embodied familiarity with the world", upon which the experimenter herself necessarily draws in her experimental practice.

This kind of moderate inflationist strategy has also been developed by Nancy Nersessian (1993) and Nenad Miscevic (1992). Each draws on work by cognitive scientists on the construction and manipulation of mental models. Nersessian, in particular, bases her account on Johnson-Laird's work (1983) on the use of mental models in narrative comprehension. TE narratives, it is claimed, are used by the receiver to construct a quasi-spatial "mental model" of the hypothetical situation. In running the TE, the receiver operates directly upon the model, deriving the experimental conclusion by manipulating the latter rather than operating upon the linguistic representations comprised by the narrative used in constructing the model. Crucially, in constructing and manipulating the model, the receiver mobilizes a number of other cognitive resources. These include her everyday understandings of the world, based on practical experience; other forms of tacit knowledge, such as individual expertise, practical know-how, and the "embodied familiarity" with the world discussed by Gooding; and geometrical intuitions. It is in virtue of the role played by these unarticulated (and often inarticulable) cognitive resources in the mental modeling of TE's that the latter yield determinate conclusions and have a bearing on the real world.

Miscevic sets out clearly (1992, 24) how the "mental modeling" approach allows us to solve the original puzzle about TE's. TE's enable us to produce new data by manipulating old data, by providing us with the means to generate a manipulable representation of a problem. In constructing and manipulating this model, we can mobilize various kinds of cognitive resources in ways not possible if we were to work directly on a regimented propositional account of that problem. Because of the role played, here, by tacit, unarticulated, and often inarticulable, forms of knowledge, we cannot reconstruct a TE as an argument without epistemic loss. Thus the final conclusion of Galileo's "Tower" TE strikes us as "immediate" because we run the TE in a mental model that, in virtue of the tacit experientially-based knowledge that guides its very construction, rules out other possible reasons for differential rates of fall, such as the colors of the falling bodies.

II

An epistemological challenge parallel to the one posed by Kuhn for scientific TE's confronts the literary cognitivist. How is it that works of fiction can teach us about the real world if fictional narratives are (by definition) fictitious, or, at least, written without the over-riding concern with mapping reality that is supposed to guide non-fictional narratives? In this section, I shall elaborate further on the kinds of cognitive functions that have been ascribed to works of literary fiction and the kinds of challenges to which literary cognitivism is open.

There are at least four ways in which literary fiction has been represented as a source of knowledge or understanding of the real world[9]:

[9]For an overview of these issues, see Novitz (1987, Chapter 6).

First, fictional narratives can be seen as sources of *factual information* about the world. Authors often incorporate true statements about the real world into their narratives, perhaps in order to set the fiction in a real context, as the Sherlock Holmes stories are set in turn of the century London and its environs. Readers may then come to believe these statements as a result of reading a fictional work. Thus a reader might acquire certain true beliefs about Victorian London through reading the fictional works of Dickens or Conan Doyle.

Most literary cognitivists, however, have larger cognitive fish to fry. They claim that literary fictions can provide readers with an *understanding of general principles* operative in the real world. The narrated events may explicitly or implicitly exemplify and make salient to the reader such principles—moral, metaphysical, or psychological, for example. While these principles are sometimes explicitly stated by a narrator or a character, those who claim that an understanding of general principles can be gained by reading fictional narratives usually have in mind principles taken to be implicit in the narrative. It might be said, for example, that implicit moral insights are to be found in the novels of Henry James[10], or implicit psychological insights in the works of Jane Austen.

Third, a number of writers have praised literary fictions as a source of *categorical understanding*. In presenting a fictional world, a narrative may furnish the reader with new categories or kinds whose application to the real world illuminates certain matters of fact. For example, works like *1984* or *The Trial* provide us with conceptual frameworks in terms of which to critically examine the ways in which socio-political structures can exercise control over the life of the individual. What we can thereby acquire, it is claimed, are new and insightful ways of *classifying and categorizing* things and situations. Nelson Goodman[11] talks, here, of fiction as a "way of worldmaking" which remakes our world by providing us with new classifications like "Quixotic", "Catch 22", and "Kafkaesque".

Finally, literary fictions have been viewed as a source of a kind of *affective knowledge*—knowledge of "what it would be like" to be in a particular set of circumstances. This can be viewed as an ethically valuable feature of fictions in so far as it bears upon our ability to comprehend, and respond appropriately to, morally complex situations that we encounter in the actual world. Effective moral agency, some have claimed (see e.g., Putnam 1976), presupposes an ability to grasp how others are affected by our actions and by their circumstances, and, more generally, the ability to understand the moral complexity of a situation—the way in which it impacts upon the welfare or the legitimate expectations of the individuals concerned.

It might be questioned, however, whether literature can provide *knowledge*, or even *warranted belief*, of any of these kinds, if the latter requires true beliefs or right classifications to which we are in some way entitled. In physical science and history, which might stand as paradigm examples of knowledge-yielding practices,

---

[10]This is argued in Nussbaum (1985).

[11]Goodman (1978, Chapter IV); see also Goodman (1976).

various assertions are made about the world upon whose truth or warranted acceptability we can frequently obtain consensus by appeal to shared epistemic norms. We can also engage in reasoned debate, by appeal to such norms, where consensus is not forthcoming. But literary works do not, it would seem, have cognitive value through making explicit assertions about the real world assessable in such ways. Even when a literary work contains sentences that express factual truths, the fiction does not work through the assertion of these sentences, but rather by inviting the reader to make-believe what the sentences affirm. Furthermore, at least in the case of purported "factual knowledge" derivable from our reading of fictions, it seems we must avail ourselves of other cognitive resources to verify that certain sentences in a fiction are indeed true of the actual world. For novelists may insert false details into their narratives in order to give them an air of authenticity, and may also, in good faith, insert details concerning which they have false beliefs.

Some critics, generalizing from the foregoing objection, argue that the most we can get from reading fiction are *hypotheses* about the general ordering of things in the world or about the affective dimensions of a particular kind of situation, *beliefs* about specific aspects of the world, and *potentially insightful* ways of categorizing things in our experience. The claim, here, is that talk of "learning" from fiction is justified only to the extent that what we derive from our reading is subjected to further testing. Only if those hypotheses, beliefs, and categorizations pass further tests can we talk of cognitive value arising out of our engagement with standard fictional narratives. We find this sentiment expressed even by philosophers who are sympathetic to some of the cognitive claims of fictions. Hilary Putnam, for example, while extolling the contribution of literature to moral learning, is quick to stress that this contribution must be an indirect one:

> No matter how profound the psychological insights of a novelist may seem to be, they cannot be called *knowledge* if they have not been tested. To say that the perceptive reader can just *see* that the psychological insights of a novelist are not just plausible, but that they have some kind of universal truth, is to return to the idea of knowledge by intuition of matters of empirical fact . . . If I read Celine's *Journey to the End of Night* I do not *learn* that love does not exist, that all human beings are hateful and hating . . . What I learn is to see the world as it looks to someone who is sure that hypothesis is correct . . . It is knowledge of a possibility. It is *conceptual* knowledge . . . (Putnam 1976, 89–90).

Putnam's reservations are developed in a more systematic and forthright manner by Jerome Stolnitz, who has argued (1992) for the "cognitive triviality" of literature. Stolnitz raises a number of distinct challenges to literary cognitivism, but we need only concern ourselves here with one of these, which is a variant on what Noel Carroll (2002) terms the "no-evidence" argument: even if there are truths, particular or general, contained in literary fictions, the fictions themselves provide us with no good reasons to accept those truths.[12] Art, Stolnitz maintains, never "confirms"

[12]Stolnitz also charges that the "profound truths" supposedly obtainable from reading literary works are, once we succeed in spelling them out, banal, and that the "truths" supposedly embedded in different fictions may contradict one another without any established method for resolving the conflict. I think, however, that these challenges are easily answered given the kind of response

its "truths". In the case of general principles that might be extracted, as "thematic meanings", from the fictional narratives of literary works, the supposed "evidence" for the "reality" of these principles is flawed in three ways: (a) the work cites no actual cases, (b) it relies on a single example, and (c) it is gerrymandered to support such principles, having been carefully designed to exemplify them.

Some defenders of literary cognitivism seem to grant this objection without seeing how serious it is for their views. David Novitz, for example, argues that we can validate the conceptual and cognitive resources that we derive from literary works, and the beliefs about the non-fictional world that are generated in our reading, by "projecting" what is gleaned from a literary work onto the world.

> Readers can only acquire conceptual or cognitive skills from fiction by tentatively projecting the factual beliefs gleaned from the work on to the world about them. They try to see specific objects, events, and relationships in terms of these new beliefs, and they attempt to rethink, perhaps to explain, what was previously baffling or bewildering. If their application of these beliefs is met with obvious rewards, if it helps them to dispel puzzles and doubts, to make sense of or come to terms with enigmas of one sort or another, they are likely to adopt these ways of thinking and observing (1987, 138).

Literature and science, Novitz maintains, are analogous enterprises in that, in each case, we are offered hypotheses that must be tested against the experienced world before any claims to knowledge can be justified.

But the comparison with science, which might seem to be an answer to the epistemological challenge to literary cognitivism, in fact concedes the very point at issue. The cognitive credentials of science rest upon its being a practice that encompasses not merely the formulation of hypotheses but also their comparative assessment in light of experimental testing.[13] Indeed, the vast majority of the hypotheses proposed by scientists prove to be unacceptable as measured by the norms of scientific practice. But Novitz seems to grant that literature merely furnishes us with hypotheses, and is thus a valid source of knowledge only if taken together with an *independent* practice of subjecting such hypotheses to empirical scrutiny.

The charge that the best we can hope to get from fictions are interesting hypotheses that are acceptable only if subjected to independent empirical testing obviously resonates with the literature disputing the cognitive credentials of scientific TE's examined in the previous section—the claims made by Duhem (1914) and Hempel (1965), for example. But the full significance of the "no-evidence" objection emerges only when it is combined with what Carroll terms the "no argument" objection against literary cognitivism. This is most fully developed by Peter Lamarque

---

developed in section "IV" to the "no-evidence" and "no-argument" argument (for the latter, see below).

[13] This is not to subscribe to a discredited atomistic conception of theory assessment in science, according to which, in Quine's famous metaphor, theories meet the tribunal of experience singly rather than collectively. But it is to insist that, even on the most holistic conception of science, bringing experimental or other empirical evidence to bear in the assessment of theories plays a central role, albeit a role that does not rule unequivocally on the status of a "tested" theory. I am grateful to Catherine Elgin for pointing out the need to clarify this point.

and Stein Olsen (1994) in their attack on what they term the "Propositional Theory of Literary Truth".

The Propositional Theory holds that, while works of literary fiction "at the literal level" have only fictional content, at a different "thematic" level they imply or suggest general propositions about human life whose truth we must assess if we are to properly appreciate the works. It is these propositions that make literature valuable. Such "thematic statements" may occur explicitly in the literary work, but are more often implicit yet accessible to readers through interpretation. Lamarque and Olsen argue, against the Propositional Theory, that it is not part of the ordinary activity of readers or critics to assess, or inquire into the truth or falsity of, general thematic statements expressed in or by literary fictions, and that this indicates that determining such truth or falsity is not a proper part of literary appreciation. Those who advocate the Propositional Theory misunderstand the function of such general thematic statements in literary fictions, according to Lamarque and Olsen. They are not properly viewed as conclusions which we are invited to accept as true on the basis of our reading of the work. Rather, they are devices for organizing and producing aesthetically interesting structure in the story's narrative content.

As may be apparent, the "no-evidence" and "no-argument" objections complement one another. It is important, however, to spell this out more fully to counter a possible objection to the significance I am ascribing to the "no-evidence" argument against literary cognitivism.[14] It might be claimed that the latter fails to recognize the cognitive legitimacy of both "experimental" and "theoretical" science. The latter seeks to develop and refine hypotheses and theories, rendering them more precise and teasing out their empirical consequences. Only given the work of the theoretical scientist do we have hypotheses that are both testable and worth testing, which is where the work of the experimental scientist comes in. It would be absurd to claim that theoretical science is not cognitively valuable because it doesn't involve experimental testing. In fact, it plays a crucial and indispensable part in the generation of scientific knowledge. So why can't the literary cognitivist maintain that, even if literary fictions do not provide evidence for the "hypotheses" that they embody, the generation of literary fictions is itself cognitively valuable in just the way that theoretical science is?

This kind of cognitivist response to the "no-evidence" argument falls foul of the "no-argument" argument, however. For the work of the theoretical scientist has a bearing on the acquisition of scientific knowledge only because it is complemented by the work of her experimental colleagues. It is only in virtue of their collective endeavors that *science* is properly viewed as a source of knowledge of the world. In the literary case, however, it isn't clear that, insofar as we can identify "hypotheses" somehow embodied in literary fictions, the empirical "testing" of those hypotheses against the world is itself part of our *literary* practice. Indeed, the "no-argument" objection against cognitivism explicitly denies that "testing" the thematic contents of literary fictions plays any legitimate part in that practice. If so, then this seems to

[14] I am grateful to Roman Frigg for raising this objection.

undermine any talk of distinctively *artistic* knowledge parallel to our talk of scientific knowledge. Any "projecting" of the thematic meanings of literary works onto the world is quite extraneous to the proper engagement with literary works *as literature*, it will be claimed. It is upon this distinction between literature as a source of hypotheses and literature as a source of knowledge that the "no-evidence" argument, buttressed by the "no-argument" argument, insists. But this distinction has no force if applied to science, for the reasons above.

In the remaining sections of this paper, I shall examine how it might be argued that the "testing" of literary hypotheses by readers of literary fictions is no less intrinsic to our literary practice than the refining and expression of such hypotheses by the authors who craft those fictions. The literary cognitivists whose views I examine in section "III"—Carroll, Elgin, and Young—agree that some kind of "testing" of literary hypotheses is integral to a proper readerly engagement with works of literary fiction. But, while their individual accounts engage implicitly or explicitly with the anti-cognitivist arguments, they fail to provide satisfactory responses to one or other of these arguments. In section "IV", I suggest and critically assess a novel resource—drawn from the debates over the cognitive virtues of scientific TE's—that might remedy this deficiency in cognitivist accounts.

III

I want first to examine two recent attempts, by Noel Carroll (2002) and Catherine Elgin (2007), to defend literary cognitivism against its critics by taking literary fictions to be thought experiments. Carroll explicitly addresses the foregoing anti-cognitivist arguments, and it is clear that Elgin also intends the association of literary fictions with TE's as an answer to criticisms of literary cognitivism. Their approaches are in other respects very different, but neither, I shall suggest, fully comes to terms with the anti-cognitivist arguments. This also applies to James Young's more detailed defense of literary cognitivism. While Young doesn't explicitly draw an analogy between literary fictions and thought experiments, his account closely resembles Elgin's in certain key respects and, I argue, is open to the same objections.

As Elgin notes, she and Carroll differ as to the epistemological function of TE's. Carroll takes as his model the use of TE's in philosophy, noting that the very philosophers who bring anti-cognitivist arguments against literature seem perfectly happy to employ fictional narratives for cognitive purposes in their philosophical use of TE's. Philosophical TE's, for Carroll, function as arguments by "excavating conceptual refinements and relationships". The imaginary situations canvassed in the narratives are designed to mobilize conceptual knowledge we already possess and elicit from us intuitions grounded in that knowledge. Philosophical TE's are not open to the "no-evidence" argument leveled at literary fictions, since they do not require empirical evidence, being aimed at unearthing conceptual knowledge. And they are not open to the "no-argument" argument since they function as arguments in virtue of the reflective processes that go on in the reader when she entertains the TE. Carroll talks here of the mobilization of "implicit knowledge" of concepts which provides the necessary dialectical supplementation to the philosophical text in our reflective reading of it. At least some literary fictions, according to Carroll, function in the same way as philosophical TE's. In support of this claim, he cites the philosophical use of some literary fictions as TE's—for example, the use of Borges' short story, *Pierre Menard, Author of the Quixote*, in support of particular views in the ontology of art. He also cites literary works such as Graham Greene's *The Third Man* and E. M. Forster's *Howards End*. These, he maintains, were designed to function as extended TE's in the interests of conceptual refinement and discrimination. Literary fictions, he further maintains, are particularly adept at aiding the clarification of concepts, and at refining our ability to apply concepts, in the moral sphere.

We should, I think, grant that there are literary fictions of the sort that Carroll describes, although Lamarque and Olsen might argue that, insofar as they are intended or made to serve as devices for conceptual clarification, they are no longer functioning as literary works. And we can perhaps agree that the kinds of philosophical TE's cited by Carroll play a legitimate cognitive role in getting clearer on our concepts, although some might view this as less a matter of unearthing conceptual knowledge already possessed and more a matter of providing resources for arriving at a rational equilibrium between our concepts and the practices they are intended to serve. But we might be concerned that such a defense of the cognitive virtues of literature hardly does justice to the claims of the literary cognitivist. The latter standardly maintains that literary fictions can help us to better understand the world, and not merely serve as a means of tuning up our conceptual apparatus. More significantly, while Carroll explicitly addresses the "no-argument" argument, his response to the "no-evidence" argument only applies if the cognitivist's claims are restricted to conceptual knowledge.

Kuhn (1964), in his seminal article on scientific TE's, rejects the idea that the role of TE's in science is simply to clarify our concepts. TE's, he maintains, are like real experiments in that they also change our beliefs about the world. They reveal that the world is such that it doesn't comport with certain assumptions built into our concepts. To take this point on board, we might look for a more "Kuhnian" view of the way in which literary fictions can function cognitively as TE's. This is in fact what we find in Elgin's article. Elgin argues that the idea that we can make cognitive progress through reading fictions appears puzzling only if we operate with what she terms the "information transfer" view of cognitive progress, where we progress by amassing information about the world. Elgin responds that the main obstacle to cognitive progress is not a lack of information, but a lack of "right" ways of organizing, classifying, and properly orienting ourselves towards the information we already possess. This can be viewed as a matter of determining salience, but it is only salience for one possessing the necessary cognitive resources that counts. Cognitive progress, for Elgin, is made by creative "reconfiguration" which allows us to arrive at new and valuable ways of configuring our experience and thereby the world. Both real experiments and TE's in science are ways of doing this. In each case, we set up a constrained situation and determine the consequences of particular ways of configuring things in those circumstances. If the choice of constraints in setting up the situation is felicitous, then a TE functions as a "laboratory of the mind" in which we can not only control, elaborate, and test various ways of configuring things in the fiction, but also expect that those configurations that work in the fiction will also work in the world. The general picture, then, is as follows: "An experiment or a TE brings it about that certain features are exemplified and manifests why they matter in the (artificial, carefully contrived) experimental setting. It thereby affords reason to suspect that they matter elsewhere. So it indicates that we would do well to consider such factors salient in related real-life situations".

Literary fictions, she argues, perform the same cognitive functions as scientific TE's so conceived. A fiction serves to advance understanding by presenting us with a fictional world that exemplifies certain features. It draws out consequences in this world of those features, and thereby affords us reason to think that these same features are a fruitful way of configuring things outside the fiction. Literary fictions, as thought experiments, differ from TE's in philosophy in being much more detailed, and from TE's in science in lacking an established background theory that provides a thick context within which agreement is achieved on their import. The narratives of literary fictions, she claims, have certain of Goodman's proclaimed "symptoms of the aesthetic" in symbolic functioning (Goodman 1976, 1978)—they are replete and semantically dense. For this reason, the appropriate determiner of the import of a literary fiction is what she terms the widely read, aesthetically sensitive reader: "Her experience equips her to know what to look for, what to focus on, what characters are important. Approaching fiction thoughtfully and sensitively, she reflects on a work and her reactions to it. She reads a work in light of her understanding of the world and understands the world in light of the works she has read". Such a reader "tests her insights to see whether they make sense of the text and whether they ring true when projected beyond the text, thus heightening her awareness of patterns, perspectives, and possibilities both in the work and the world".

There is much to admire in Elgin's account—not least its attempt to provide a systematic defense of both literary fictions and TE's in science in terms of the cognitive value of "reconfiguration". But the real teeth of the anti-cognitivist challenge have not been drawn. For, while the widely read, aesthetically sensitive reader may be more adept at extracting its thematic content from the dense and replete literary symbol, it is not clear why this also makes her a good judge of the cognitive value of this content. Indeed, the test of truth or correctness of the cognitive content of a literary fiction seems to be, as with Novitz, the results of projecting what we extract from the fiction onto the world to test it out. Thus, in spite of the sophistication of Elgin's analysis, it might be thought that she has not really moved us much beyond the skeptical view of Stolnitz which mirrors the extreme deflationism of Duhem vis à vis TE's in science. While testing through "projection" may indeed provide evidence for the thematic claims of a literary work, it is not clear why such testing is properly viewed as integral to our engagement with literary fictions as literature, and thus why we are entitled to view literature as a source of knowledge rather than as a source of hypotheses.

Similar remarks apply to James Young's (2001) defense of literary cognitivism. Young neither draws a comparison between literary fictions and TE's, nor provides the richer contextualization that Elgin offers in her account of "reconfiguration". But the notion of "illustrative demonstration" which is central to his account seems closely analogous to Elgin's notion of "exemplification". He claims that works of fiction can represent reality even though they contain fictional narratives, because what they characteristically represent are not concrete individuals or situations, but, rather, *types* of individuals or situations. Fictions function as what he terms "illustrative representations", which represent in virtue of various kinds of resemblance between the experience elicited by the work and the experience elicited by the thing it represents. Illustrative representations, he claims, provide knowledge not through making statements and rationally demonstrating conclusions, after the manner of the sciences, but through providing "illustrative demonstrations" which place the receiver in a position where she can acquire knowledge. A literary work gives the reader a new way of interpreting—a new "perspective" on—the type of object or situation it represents.

Such perspectives, according to Young, can be assessed as right or wrong, depending upon whether they assist us in acquiring knowledge about, and making sense of, the type of entity on which they are perspectives. We can determine the *rightness* of a perspective if we bring to our engagement with a literary work other knowledge we possess, and apply that perspective in our attempts to acquire further knowledge of the object, and to make sense of other features of our experience. As with Elgin, the cognitive claims of literary fictions are defended in terms of two factors: (a) knowledge that the reader brings to her encounter with the literary work, and (b) the "projection" of the thematic content of the literary work onto the real world, as a "test" of its rightness. Again, as with Elgin, this doesn't seem to address the "no-argument" argument. For, while the knowledge that a reader brings to her encounter with a literary work can plausibly be claimed to enter into the activity of reading itself, and thus to be intrinsic to a proper literary engagement with the work, the "projection" that seems to act as the test of the literary work's claims to furnish us with knowledge, and thus to answer the "no-evidence" argument, might be thought to be extrinsic to such an engagement. To counter this sort of anti-cognitivist move, we need to bring the two "factors" together, to show how the process of "testing" can rightly be viewed as intrinsic to the activity of reading, where, as we read, we bring to bear in novel ways knowledge we already possess.

IV

But we already have insights into how this might be done. First, we may recall Carroll's talk of the mobilization of implicit knowledge of concepts in philosophical TE's, where this provides the necessary dialectical supplementation to the text. Second, we may recall, from section "I", the moderate inflationist's claim that at least some scientific TE's draw essentially upon cognitive resources that, while possessed in some form prior to the TE, are only mobilized through the details of the TE narrative. In the final section of this paper, I shall critically assess the idea that we might develop a more comprehensive response to the anti-cognitivist arguments by appealing to the role that implicit understandings mobilized by a literary fiction play in our reading of that fiction. Such implicit understandings might then serve to justify our coming to believe the thematic contents we extract from the fiction, just as the mobilization of implicit cognitive resources has been held by some to play an essential role in justifying the conclusions at which readers arrive when entertaining scientific TE's.[15]

The "no evidence" objection, we may recall, is that the most we can get from reading standard fictional narratives are hypotheses about the general ordering of things in the world, or beliefs about specific aspects of the world, or *potentially insightful* ways of categorizing things in our experience. Only if those hypotheses or beliefs pass further tests can they acquire the status of knowledge, it is claimed. To evade the "no-argument" argument, any "tests" to which the cognitivist points in attempting to answer the "no-evidence" argument must be intrinsic to the proper engagement with literary works *as literature*.

Take, for example, the claim that we learn about the dynamics of complex human relationships through reading Henry James, or about the rhythms of lived experience through reading Virginia Woolf. The challenge is to provide some reason why we should accept such claims, without further "extrinsic" empirical test, or why our responses to such works are to be trusted. The suggestion is that, as with scientific TE's on the moderate inflationist account, our responses to such fictional narratives mobilize unarticulated cognitive resources based in experience. The fiction is able to elicit such responses because it makes manifest to us patterns underlying the complexity of prior and present actual experience—this is reflected in our feeling that the novel has indeed revealed such patterns to us. And this feeling is to be trusted because it reflects the operation of such unarticulated cognitive resources in our reading. This answers the "no-argument" argument because it makes the process

[15]It should be clear why neither of the deflationary responses to Kuhn's "epistemological puzzle" can help us to defend literary cognitivism, since they are effectively arguments against a cognitivist view of scientific TE's. But what of the "extreme inflationist" response? We can certainly envision a defense of literary cognitivism that parallels that response—think, for example, of the "Romantic" view of the literary artist, as one whose words transmit to others her intuitive insight into the inner nature of things. But extreme inflationism seems unpromising as a defense of the epistemic virtues of scientific TE's, for it cannot prevail over moderate inflationism if the principal arguments offered in its favor are intended to show that it is preferable to some form of deflationism. And this indeed seems to be an objection to Brown's form of extreme inflationism (1991, 1992), which is defended largely by pointing to aspects of the functioning of TE's for which the moderate deflationist cannot provide a plausible account. Miscevic (1992) argues that the "mental modelling" account can explain all of these aspects of the functioning of TE's. If Miscevic is right, the greater explanatory burden that must be shouldered by the extreme inflationist—who must tell a convincing story about our capacity to grasp relations between universals in our engagement with TE's—tells in favor of the moderate position. I think analogous difficulties would plague an "extreme inflationist" defense of literary cognitivism.

of "empirical testing" of what is exemplified in the fiction *internal* to the process of reading, rather than something we have to do after we have read the fiction.

But caution is needed in a number of respects if we are to endorse such a defence of literary cognitivism. In the first place, it is implausible to think that an account that appeals to the mobilization of unarticulated cognitive resources can justify each of the different kinds of claims to knowledge made on behalf of literary cognitivism. In the case of knowledge of particular matters of fact, justified belief would seem to require that the fictional narrative be, or be rightly believed to be, a reliable source of knowledge of facts of this kind. I cannot, in such cases, put reasonable trust in my personal conviction that what is presented as part of the fiction is factually true of the actual world, since (save in the case where the fiction reminds me of something that I already knew) no amount of unarticulated knowledge can serve to validate such a conviction. The same, I think, applies to claims about affective knowledge derivable from fictions. We may find ourselves convinced, in reading a fictional narrative, that the narrator has correctly characterized the affective nature of an experienced situation of a kind that we have not actually encountered. But this may reflect the narrative skill of the author rather than the correctness of our intuitions. Only if the author is, or is rightly believed to be, a reliable source of knowledge as to the affective nature of such an experienced situation—most obviously because she has herself experienced the situation in question and is sincere in her communicative intentions—can the claim to derive affective knowledge from the reading of fiction be supported.

This leaves us with two kinds of claim made on behalf of literary cognitivism that might be defended by appeal to the moderate inflationist account of scientific TE's: first, the claim that literary fictions can yield knowledge of general principles operative in real events, and, second, the claim that, in our reading of such fictions, we can acquire new ways of classifying real entities or events that illuminate the nature of those entities or events. Even here, the claim that such cognitive benefits can accrue in the act of reading a fictional narrative, without the need to carry out some additional empirical verification, requires clarification of what is to be included in "the act of reading". In the case of fictional narratives presented cinematically, for example, our sense that the author has furnished us with insights into the structures of reality is unlikely to solidify in the very act of watching the film, but only in our subsequent reflections upon it. This can still be cashed out in terms of the bringing of unarticulated cognitive resources to bear upon what is presented in the fiction, rather than in terms of carrying out some kind of empirical inquiry to "test" the latter's applicability to real entities or events, but it renders less sharp the distinction between *answering* the epistemological challenge to literary cognitivism and *conceding* to that challenge.

The same point can be made for *literary* fictions, although the extended nature of the process of reading such fictions may often permit the mobilization of the relevant unarticulated cognitive resources during the process of reading itself. Peter Kivy (1997) has written very illuminatingly about what he terms the "gappiness" of our reading of fiction and the "reflective afterlife" of the literary artwork, and it seems that an adequate account of learning in the process of "reading" a literary fiction must take "reading" in this broad sense. What must nonetheless be excluded, if we are to answer the anti-cognitivist arguments, is any recourse to further investigation and inquiry in a process of "projecting" the thematic or categorical content of the novel onto the world.

A final worry needs to be addressed.[16] It is undeniable that many "popular" fictional narratives, such as those to be found in mainstream American war films and romantic comedies, exemplify, in the narrated events, categorizations and explanatory principles that, if applied to real entities and events, would provide a hopelessly simplistic or ideologically distorted classification or explanation of those entities and events. For example, films that present geopolitical events as clashes between forces of good and forces of evil do not furnish us with cognitively useful resources if we are to understand and negotiate the nuanced nature of geopolitical realities. But it is arguable that some of those making decisions as to how the latter should be understood and negotiated have indeed trusted their feeling that the fictional narratives are genuine sources of understanding of reality. Is there not, then, a danger in any suggestion that we can trust our sense that a given fictional narrative illuminates reality, and should we not always require that purported "insights" in fictions be subjected to independent test?

The objection suggests that, to avoid such possible cognitive misuses of fiction, we should resist the literary cognitivist idea that our engagements with fictional narratives can themselves yield knowledge, and grant that all that such engagements yield are *possible* cognitive benefits that stand in need of independent testing. But the literary cognitivist should resist such a line of argument, and can do so by clarifying what it is that she is claiming. The claim is not that, because our sense that we are learning something about the real world—either about general principles operative therein or about how certain ways of categorizing things bring illumination—draws upon unarticulated cognitive resources, we can trust this sense and rest content, without further exploration, that we are indeed learning what we believe ourselves to be. For our sense that we are learning is trustworthy only in proportion to the adequacy of the unarticulated cognitive resources upon which we draw. This is no different from the situation in respect of scientific TE's. If the cogency of the latter sometimes draws upon the receiver's embodied knowledge of how things work in a real experimental context, for example, as Gooding suggests, then only the responses of one who possesses such knowledge can serve to test the TE in question.

So, in the case of fictional narratives, we should admit that there is genuine learning through the reading of such narratives only to the extent that we also allow that the unarticulated knowledge of the world upon which the reader's intuitions of rightness are based is itself adequate to validate those intuitions. One who enters the reading or viewing process of a narrative that over-simplistically represents geopolitical events with a naive or over-simplistic, although unarticulated, general sense of such events will indeed find the narrative to be illuminating, and may indeed

---

[16] I am grateful to Catherine Elgin for raising this worry.

(wrongly) trust her sense of being illuminated in her further dealings with reality. But the claim, again, is not that, in the case of the relevant kinds of cognitive resources, a reader's sense of having learned from fiction itself justifies her belief that she has so learned. The claim is only that, when the sense of having learned from a fiction is *in fact* grounded in the right kind of unarticulated knowledge, the reader can indeed be said to have learned what she believes herself to have learned. The claim is, in this respect, an externalist rather than an internalist one, depending upon how the agent in fact stands in relation to the knowledge claim, rather than how she sees herself as standing in relation to it.

If qualified in the ways suggested in the preceding paragraphs, therefore, I think the claim that we can answer the epistemological challenge to literary cognitivism by appeal to the moderate inflationary account of scientific TE's is plausible. But the qualifications are as important as the claim itself, if we are to better understand what is at issue in assessing the cognitive claims of literary fictions.

Acknowledgments Earlier versions of this paper were presented at the London School of Economics/Courtauld Institute of Art conference, "Beyond Mimesis and Nominalism: Representation in Art and Science" in June 2006, at a workshop on "Thought Experiments" held at the University of Toronto in May 2007, and at a colloquium at London School of Economics in December 2007. I am grateful to all those who offered helpful comments and criticisms on those occasions. I am also grateful to students in the graduate pro-seminar on "Thought Experiments" that I co-taught at McGill in Fall 2006 for their feedback on some of the ideas in this paper. I am especially grateful to the editors of this volume, and to two anonymous referees, for critical comments and suggestions on an earlier draft of this paper, which helped me to clarify and refocus the paper in ways that (I hope) considerably improved it. Finally, I wish to thank the Social Sciences and Humanities Research Council of Canada, a research grant from whom facilitated the research for this paper.

References

Arthur, R. (1999), "On Thought Experiments as a Priori Science", *International Studies in the Philosophy of Science* 13, 3: 215–229.

Brown, J. R. (1991), *The Laboratory of the Mind: Thought Experiments in the Natural Sciences.* London: Routledge.

Brown, J. R. (1992), "Why Empiricism Won't Work", in *PSA 1992*, vol. 2. East Lansing, MI: Philosophy of Science Association, 271–279.

Carroll, N. (2002), "The Wheel of Virtue: Art, Literature, and Moral Knowledge", *Journal of Aesthetics and Art Criticism* 60, 1: 3–26.

Currie, G. (1990), *The Nature of Fiction*. Cambridge: Cambridge University Press.

Davies, D. (2005), "Fiction", in B. Gaut and D. Lopes (eds.), *Routledge Companion to Aesthetics*, 2nd ed. London: Routledge, 347–358.

Davies, D. (2007a), *Aesthetics and Literature*. London: Continuum.

Davies, D. (2007b), "Thought Experiments and Fictional Narratives", *Croatian Journal of Philosophy* VII, 19: 29–46.

Duhem, P. (1914), *The Aim and Structure of Physical Theory*, trans. P. Weiner. Princeton NJ: Princeton University Press (1954).

Elgin, C. Z. (2007), "The Laboratory of the Mind", in W. Huerner, J. Gibson, and L. Pocci (eds.), *A Sense of the World: Essays on Fiction, Narrative, and Knowledge*. London: Routledge, 43-54.

Feyerabend, P. (1975), *Against Method*. London: New Left Books.

Gendler, T. (1998), "Galileo and the Indispensability of Thought Experiments", *British Journal for the Philosophy of Science* 49: 397–424.
Gooding, D. (1973), "What is *Experimental* about Thought Experiments?" in F. Hull and K. Okruhlik (eds.), (1993), in *PSA 1992*, vol. 2. East Lansing, MI: Philosophy of Science Association, 280–290.
Goodman, N. (1976), *Languages of Art*, 2nd ed. Indianapolis: Hackett.
Goodman, N. (1978), *Ways of Worldmaking*. Indianapolis: Hackett.
Häggqvist, S. (1996), *Thought Experiments in Philosophy*. Stockholm: Almqvist & Wiksell International.
Hempel, C. (1965), *Aspects of Scientific Explanation*. New York: Free Press.
Hull, D., Forbes, M. and Okruhlik, K. (eds.) (1993), in *PSA 1992*, vol. 2. East Lansing, MI: Philosophy of Science Association.
Johnson-Laird, P. N. (1983), *Mental Models*. Cambridge, MA: Harvard University Press.
Kivy, P. (1997), "The Laboratory of Fictional Truth", in P. Kivy, *Philosophies of Arts: An Essay in Differences*. Cambridge: Cambridge University Press, 120–139.
Kuhn, T. (1964), "A Function for Thought Experiments", reprinted in T. Kuhn, *The Essential Tension* (1977). Chicago: University of Chicago Press, 240–265.
Lamarque, P. and Olsen, S. H. (1994), *Truth, Fiction, and Literature*. Oxford: Clarendon Press.
Mach, E. (1905), "On Thought Experiments", reprinted in E. Mach, *Knowledge and Error* (1975). Dordrecht: Reidel, 134–147.
McAllister, J. (1996), "The Evidential Significance of Thought Experiments in Science", *Studies in History and Philosophy of Science* 27, 2: 233–250.
Miscevic, N. (1992), "Mental Models and Thought Experiments", *International Studies in the Philosophy of Science* 6, 3: 215–226.
Nersessian, N. (1993), "In the Theoretician's Laboratory: Thought Experimenting as Mental Modeling", in F. Hull and K. Okruhlik (eds.), in *PSA 1992*, vol. 2. East Lansing, MI: Philosophy of Science Association, 291–301.
Norton, J. (1996), "Are Thought Experiments Just What You Always Thought?" *Canadian Journal of Philosophy* 26, 3: 333–366.
Novitz, D. (1987), *Knowledge, Fiction, and Imagination*. Philadelphia PA: Temple University Press.
Nussbaum, M. (1990), "Finely Aware and Richly Responsible: Literature and the Moral Imagination", in M. Nussbaum (ed.), *Love's Knowledge*. Oxford: Oxford University Press, 148–167.
Putnam, H. (1975), "The Meaning of 'Meaning'", in H. Putnam (ed.), *Mind, Language and Reality*. Cambridge: Cambridge University Press, 215–271.
Putnam, H. (1976), "Literature, Science, and Reflection", in *Meaning and the Moral Sciences*. London: Routledge & Kegan Paul, 83–94.
Searle, J. (1980), "Minds, Brains, and Programmes", *Behavioural and Brain Sciences* III-3. Reprinted in D. Rosenthal (ed.), *The Nature of Mind* (1991). Oxford: OUP, 509–519.
Sorensen, R. (1992), *Thought Experiments*. Oxford: Oxford University Press.
Stolnitz, J. (1992), "On the Cognitive Triviality of Art", *British Journal of Aesthetics* 32, 3: 191–200.
Thompson, J. J. (1971), "A Defense of Abortion", *Philosophy and Public Affairs* 1, 1. Reprinted in G. Sher (ed.), *Moral Philosophy* (1987). New York: Harcourt Brace Jovanovich, 631–645.
Young, J. (2001), *Art and Knowledge*. London: Routledge.

Models as Make-Believe

Adam Toon

This paper proposes an account of representation for scientific models. Section "Representation in Modeling" sets out the problem of scientific representation and engages with Craig Callender and Jonathan Cohen recent dismissal of the problem. Section "Models as Make-Believe" offers a solution based on Kendall Walton's "make-believe" theory of representation in art. Finally, section "Models Without Actual Objects" demonstrates one advantage this account has over existing accounts of scientific representation. Existing accounts analyze scientific representation in terms of relations, such as similarity or denotation. By contrast, the account proposed in this paper does not take representation in modeling to be essentially relational. For this reason, it can accommodate models which do not represent an actual object.

Representation in Modeling

The Problem of Scientific Representation

When we think of scientific models, perhaps the first things that come to mind are "ball-and-stick" models of molecules or astronomical models of the solar system. Let us refer to such models as *physical models*, to indicate that they are actual, physical objects. Most philosophical work focuses not on physical models but on what I shall call *theoretical modeling*. Suppose we want to predict the behavior of a bob bouncing on the end of a spring. To do so we might use Hooke's law to formulate the equation of motion for a simple harmonic oscillator, $md^2x/dt^2 = -kx$, where m is the mass of the bob, k is the "spring constant" and x is the displacement from the equilibrium position. In using this equation we make a number of assumptions: we take the bob to be a point mass m subject only to a uniform gravitational field and a

A. Toon (✉)
University of Bielefeld, Bielefeld, Germany
e-mail: adam.toon@uni-bielefeld.de

R. Frigg, M.C. Hunter (eds.), *Beyond Mimesis and Convention*, Boston Studies in the Philosophy of Science 262, DOI 10.1007/978-90-481-3851-7_5,

linear restoring force exerted by a massless frictionless spring with spring constant *k* attached to a rigid surface. This is what Nancy Cartwright (1983) calls a "prepared description" of the bouncing spring system. We realize that this description is false, but using it allows us to apply our equation of motion and calculate predictions for the bob's behavior. This is an example of theoretical modeling: we *model* the bob as a simple harmonic oscillator.[1]

Many physical models *represent* some object or event in the world. Crick and Watson's famous model represents the DNA molecule. The astronomical models represent the solar system. An engineer's scale model might represent a bridge. We also represent the world through theoretical modeling. Of course, despite Cartwright's terminology, we cannot regard our "prepared description" or equation of motion as straightforward descriptions of the bouncing spring; we realize that the bob is not a point mass and do not claim that it is. And yet we do represent the spring when we model it. Intuitively, we might say that we represent it *as* a simple harmonic oscillator. Put simply, the problem of representation for scientific models is to understand how such cases of representation work. In the case of theoretical modeling, this problem takes different forms depending on which view we adopt of the ontology of theoretical models.[2] For example, according to Ronald Giere, a theoretical model like our model of the bouncing spring is not the prepared description and equation of motion that we write down, but some form of abstract entity that they define. Giere offers an *indirect*, two-stage, view of theoretical modeling: First, prepared descriptions and theoretical laws define abstract objects. Second, these objects represent (or, as Giere would put it, are used to represent) the system being modeled. If we adopt this view, then, understanding how we represent the bouncing spring is a matter of understanding the relation between the spring and the abstract simple harmonic oscillator defined by our prepared description and equation of motion.

To understand the problem of representation for models, it is helpful to look to another representational device: pictures. Like models, many pictures are representational, and some represent actual objects or events. Jacques-Louis David's *Napoleon Crossing the Saint Bernard* represents Napoleon. Constable's *Salisbury Cathedral from the Meadows* represents Salisbury Cathedral. The problem for theories of pictorial representation is to understand how they do this. In itself, Constable's painting is merely a collection of brushstrokes on a piece of canvas. And yet it depicts horses pulling a cart through a stream and the Cathedral beneath a rainbow. How does it do this? In virtue of what does Constable's painting represent the Cathedral? The problem of representation for scientific models may be presented in the same way. The reconstruction of Crick and Watson's original DNA model in the Science Museum is simply a collection of metal rods and plates held in place by

[1]Note that I use the term "theoretical" only to indicate that scientists do not construct a physical model of the system modeled, and not to imply that the model is derived from some existing theory, like Newtonian mechanics. Recent case studies suggest scientists must often go beyond existing theory to model a system; for example, see Morgan and Morrison (1999).

[2]For my own view of the ontology of theoretical modeling, see Toon (2010).

clamps. And yet it represents the complex helical structure of the DNA molecule. How does it do this? In virtue of what does the model represent the molecule?

We have a name for the sort of representation pictures provide. We say that David's painting *pictures* or *depicts* Napoleon, and that Constable's landscape *depicts* Salisbury Cathedral. Of course, pictures represent in other ways too, apart from depiction. David's painting might be said to represent the glory of France, or Constable's "the culmination of his numerous treatments of Salisbury Cathedral".[3] Such is the vagueness of the term "represent". But it is one particular form of representation that pictures offer, namely depiction, which theories of pictorial representation seek to explain.

We lack a name for the way that models represent. If we say merely that models represent their objects then we are likely to be misled, for the word "representation" is used in so many different ways. Crick and Watson's model might also be said to represent the greatest achievement of British science or Bohr's model a belief in the simplicity of the atomic realm. In analogy to pictorial representation, then, we might label the form of representation we are interested in *model-representation*. Crick and Watson's model, we shall say, *model-represents* the DNA molecule and Bohr's model *model-represents* the atom.

We must be careful here, however. The variety of scientific models is remarkable. What reason do we have for thinking that all of these models represent in the same way? Does Crick and Watson's model represent the DNA molecule in the same way as Bohr's model represents the atom, for example, or our model represents the bouncing spring? Might there not be many forms of model-representation? Here the contrast with depiction is telling. The variety of things we call "pictures" is also remarkable. It includes figurative paintings, Impressionist landscapes, political cartoons, children's drawings, stick figures and more. And yet despite their obvious differences, it is often thought that there is one form of representation that is common to all of these pictures, namely depiction. We lack the same intuitions for scientific models. Whether or not there is a form of representation common to both Crick and Watson's and Bohr's models, for example, would seem to be an open question that a theory of scientific representation must address.

We should not assume, then, that there is one form of representation common to all scientific models: there may be many different forms of model-representation.[4] And we should also be careful not to assume that any of these forms of representation are unique to scientific models. Any, or even all, of the forms of model-representation that we identify may turn out to be employed by other representational devices, used either within or outside of science. Our theory of representation does not need to go on to say how, if at all, scientific models differ from these other representational devices, although this may be an interesting question in its own right.

---

[3] *Salisbury Cathedral from the Meadows* on www.nationalgallery.org.uk

[4] Versions of this point may be found in Frigg (2006), Hughes (1997) and Suárez (2003), although each draw rather different lessons from it.

The task of explaining how models represent is usually taken to be that of providing an account of a *relation*, between a model and that part of the world that it represents. For example, according to Craig Callender and Jonathan Cohen, the central question concerning representation for scientific models is "what constitutes the representation relation between a model and the world?"[5] The task for theories of depiction is often presented in the same way. A theory of depiction, it is often said, must tell us what the relation is between a picture and its subject, in virtue of which it depicts that subject. The difficulty with presenting the task in this way, of course, is that many pictures have no actual subject. And yet it seems that a picture of a unicorn is still depictive, even though there is no unicorn that it depicts. In the final part of this paper, I will argue that the same problem arises for scientific models: many models are representational, even though they represent no actual object.[6] If we want to allow that such models are representational then we are faced with a dilemma: either we postulate some entity that they represent or we cease to think of model-representation as essentially relational. We ought, therefore, to refrain from presenting our task as that of giving an account of representation as a relation if we do not want to commit ourselves prematurely to the first route out of this dilemma.

Misrepresentation

Most models are inaccurate (or incorrect or unrealistic) in some way. Often this is deliberate. When we model the bouncing spring, for instance, we neglect the effects of air resistance. Sometimes, inaccuracy is unintentional: before building their famous double-helical model, for example, Crick and Watson constructed and rejected a number of different models of the DNA molecule. For our present purposes, the important point to notice is that inaccurate (or incorrect or unrealistic) models are still representations, and so must be accommodated by our theory of model-representation. Our simple harmonic oscillator model and a model that accounts for air resistance both represent the bouncing spring, and Crick and Watson's early efforts, like their final double-helical model, all represent the DNA molecule.

Many people share the intuition that an account of scientific representation should accommodate inaccurate, as well as accurate, models. For example, Mauricio Suárez (2003, 226) writes that:

> we shall not require a theory of representation to mark or explain the distinction between accurate and inaccurate representation, or between a reliable and unreliable one, but merely between something that is a representation and something that is not.[7]

---

[5]Callender and Cohen (2006, 68). See also Frigg (2006a) and Hughes (1997).

[6]This problem is also raised by Suárez (2003) and Callender and Cohen (2006). As we shall see in section "Models Without Actual Objects", however, neither provide a solution.

[7]See also Callender and Cohen (2006) and Frigg (2006a).

Some will disagree, however. Use of terms such as "representation" or "depiction" is often vague and subject to dispute. Are doodles really depictions? What about stick men? For some, "representation" carries implications of realism, or at least empirical adequacy, when used with regard to scientific models. Although I do not understand the term in this way, I do not, of course, deny that the question of what makes one model more accurate or realistic than another is an important one. I claim only that this question need not be addressed by our theory of representation for models, which is concerned with the prior question of what makes something a model-representation. Once we understand how models represent, we will want to make further distinctions among models, distinguishing good from bad along various different dimensions. If someone wishes to reserve the term "representation" for those models that fall only on the good side of one or more of these divides then we needn't quibble too much. The more important point is that our account of representation should provide us with the resources to make these distinctions amongst models.

The situation is similar with pictures. We often judge the realism of pictures, counting a Rembrandt more realistic than a cave painting or a Picasso.[8] This raises the longstanding question of what makes one picture more realistic than another. Many theories of depiction suggest a natural answer to this question. For example, the view that pictures depict in virtue of similarity suggests a straightforward account of realism: the more a picture resembles its object, the more realistic it is. However, the question of what constitutes realism need not be addressed by a theory of depiction. Unrealistic pictures are still pictures; they still depict their objects. It is this which a theory of depiction must try to understand.

Does the Problem Exist?

The problem of representation in scientific modeling is now the focus of a burgeoning literature in the philosophy of science. Indeed, it is often referred to simply as "the problem of scientific representation". However, in a recent paper, Craig Callender and Jonathan Cohen have argued that this attention is unwarranted. In fact, they claim, "there is no special problem about scientific representation" (Callender and Cohen 2006). In this section, and the one that follows, I shall attempt to show why Callender and Cohen are wrong. In doing so, I hope to clarify further the nature of the problem that faces us.

Callender and Cohen argue that we should approach the problem of scientific representation from a stance which they label as "General Griceanism". According to this view:

> among the many sorts of representational entities (cars, cakes, equations, etc.), the representational status of most of them is derivative from the representational status of a privileged core of representations. ... artistic, linguistic, representation and culinary representation

---

[8]Of course, this is not to claim that realism in modeling is the same as realism in painting.

> ... can be explained (in a unified way) as deriving from some more fundamental sorts of representations, which are typically taken to be mental states (2006, 70).

A General Gricean account of representation therefore consists of two stages:

> First, it explains the representational powers of derivative representations in terms of those of fundamental representations; second, it offers some other story to explain representation for the fundamental bearers of content (2006, 71).

It is by adopting this General Gricean position that Callender and Cohen believe (2006, 67) we may "solve or dissolve the so-called 'problem of scientific representation'":

> Our proposal ... is that scientific representation is just another species of derivative representation to which the General Gricean account is straightforwardly applicable. This means that, while there may be outstanding issues about *representation*, there is no special problem about *scientific* representation (2006, 77; emphasis in original).

Callender and Cohen offer little argument in support of a General Gricean position. But let us, for the moment, suppose that we were to accept their proposal. What would this mean for our enquiry into model-representation? Presumably, the first stage would be to provide an account of how models represent in terms of some other, more fundamental form of representation such as mental or linguistic representation. Let us call such an account a *derivative* account of model-representation. A derivative account would attempt to show how the representational power of models derives from some other form of representation. By contrast, a *non-derivative* account would attempt to explain how models represent in non-representational terms. According to Callender and Cohen, providing a derivative account of model-representation "amounts to a relatively trivial trade of one philosophical problem for another" (2006, 73). But if we could take this first step then we would have at least reduced the problem of explaining how models represent to the problem of explaining some other form of representation. And we might even feel at this stage that our work as philosophers of science was complete. The second step, of providing an account of the more fundamental form of representation, might be left to those working in the philosophy of mind or language.

However, immediately after they propose that we adopt the General Gricean approach to explain how models represent, Callender and Cohen expand on their claim in the following way:

> [W]e propose that the varied representational vehicles used in scientific settings (models, equations, toothpick constructions, drawings, etc.) represent their targets (the behavior of ideal gases, quantum state evolutions, bridges) by virtue of the mental states of their makers/users. For example, the drawing represents the bridge because the maker of the drawing stipulates that it does, and intends to activate in his audience (consumers of the representational vehicle, including possibly himself) the belief that it does (2006, 75).

This further claim comes as a surprise. Rather than being asked to take the first step in our General Gricean account of scientific representation, we are told that this step has already been taken. We do not need to provide a derivative account of representation for models. In fact, this account has a very simple form: all that is required for a model to represent its target is that the user of the model stipulate that it does, and that he intend to bring about the belief that it does. Consequently, Callender and Cohen claim, "scientific representation ... is constituted in terms of a stipulation, together with an underlying theory of representation for mental states" (2006, 78). The representational relation between a drawing and a bridge, for example, is "the product of mere stipulative fiat" (2006, 75). If this is correct, then there is indeed no special problem about scientific representation. Philosophers of science need no longer occupy themselves with finding even a derivative account of how models represent. It requires only an act of stipulation to bring about an instance of scientific representation. The remaining puzzles may be left to philosophers of mind.

What are we to make of this claim? Callender and Cohen seem to think that it follows directly from the General Gricean position. But it is difficult to see why this should be the case. It is one thing to claim that the representation relation between model and target exists only in virtue of some other, more fundamental, form of representation, such as mental representation; it is quite another thing to claim that an act of stipulation is sufficient to bring about this representational relation. In the first case, we claim merely that *some* form of derivative account of scientific representation may be found; in the second we commit ourselves to one particular, very simple, form that this account might take.

The parallel with depiction is helpful here. Suppose that we were to adopt the General Gricean position with regard to depiction. This would commit us to offering a derivative account that explained depiction in terms of some other, more fundamental form of representation. In fact, there are rather a lot of existing accounts that we could draw upon here. Consider the reconstruction of Plato's account of depiction offered by Alan Goldman (2003, 194):

> a picture represents an object if and only if (a) its artist successfully intends by marking a surface to create a visual experience that resembles that of the object, (b) such that the intention can be recovered from the experience, perhaps together with certain supplementary information, and (c) the object can be seen in the picture.

This account attempts to explain depiction in terms of, amongst other things, the intention of the artist to create a certain visual experience of an object. If successful, it will reduce the problem of depiction to some other problem (or problems) concerning mental representation. Or, to take another example, consider Kendall Walton's make-believe theory of depiction (1990, Chapter 8). I will be discussing Walton's views more fully later on. For now, all that is important is that for Walton, depiction is explained in terms of particular acts of imaginings engaged in by the viewer of the picture: she imagines of her looking at the picture that it is an instance of looking at the object. Again, Walton's is a derivative account: it aims to explain depiction in terms of the representational capacities of mental states (in this case, imagination).

Either Goldman's or Walton's theories might constitute the first step in a General Gricean account of depiction. Yet the two accounts are very different, and the continuing debate over depiction suggests that taking either step would be far from trivial. Moreover, neither Goldman nor Walton's account parallels the derivative account of scientific representation offered by Callender and Cohen. Presumably, such an account would claim that a picture depicts its subject if the painter stipulates that it does and intends to bring about the belief in the viewer that it does. Neither Goldman nor Walton take such an act of stipulation to be sufficient for depiction. And it is clear why they do not, for stipulation is plainly not sufficient for depiction. Suppose we took a blank canvas and stipulated that it represented Napoleon, and that we intended to bring about the belief in others that this canvas represented Napoleon. And suppose further that this intention was recognized and our audience did believe that the canvas represented Napoleon. The blank canvas might, then, be said to represent Napoleon, in some sense, but it would not *depict* him.

Adopting the General Gricean position with regard to pictures, then, does not commit us to the view that stipulation is sufficient for depiction, but instead leaves open many different ways of explaining depiction in terms of other, more fundamental forms of representation. Similarly, we might adopt a General Gricean approach to models without taking stipulation to be sufficient for model-representation. Just as there are many different candidates for a derivative account of depiction, so there might be many different derivative accounts of model-representation. Nevertheless, we might still ask whether the account that Callender and Cohen propose is successful.

Stipulation and Salt Shakers

Is an act of stipulation sufficient for model-representation? To support their claim that it is, Callender and Cohen ask us to suppose that we were to pick up a salt shaker and stipulate to our dinner partner that it represents Madagascar. As long as our stipulation is understood, they point out:

> when your dinner partner asks you what is your favorite geographical land mass, you can make the salt shaker salient with the reasonable intention that your doing so will activate in your audience the belief that Madagascar is your favorite geographical land mass (2006, 74).

According to Callender and Cohen, this shows that an act of stipulation, if properly recognized, is sufficient to establish an instance of scientific representation. Is this correct? Would we say that the salt shaker represents Madagascar? In some sense of the term "represents" no doubt we would; again, the term is loose enough to support many different uses. But would we say that the salt shaker is a *model-representation* of Madagascar? Would it represent Madagascar in the same way that Crick and Watson's model represents the DNA molecule, for example, or Bohr's model represents the atom?

Let us again look to depiction. Perhaps the account of depiction that comes closest to claiming that stipulation is sufficient for depiction is Nelson Goodman's conventionalist account. According to Goodman, the relation between a picture and what it depicts is like that between a name and its referent; both refer to, stand for, or denote, their objects. Resemblance or similarity are neither necessary nor sufficient conditions for a picture to denote its object. In fact, "almost anything may stand for almost anything else" (Goodman 1976, 5). One way to establish denotation, it seems, is by stipulation. If we stipulate that the blank canvas represents Napoleon then the canvas may be said to denote Napoleon. However, even Goodman does not take denotation to be sufficient for depiction. Instead, he recognizes that his theory must account for the considerable intuitive differences between pictorial and non-pictorial representations. And he attempts to do so by presenting a number of formal criteria that are intended to distinguish pictorial symbol systems from non-pictorial ones, such as linguistic or diagrammatic symbol systems.

Both David's portrait and the name "Napoleon" may be said to represent Napoleon. Perhaps both "refer to" or "denote" him. Similarly, both Crick and Watson's model and "DNA molecule" might be said to represent or refer to or denote the DNA molecule. But any theory of depiction which counted the name "Napoleon" a *depiction* of Napoleon would have failed to capture something important about the way that David's portrait represents Napoleon. Similarly, it seems that any theory of model-representation that counted "DNA molecule" a model-representation of the DNA molecule would have failed to capture something important about the way Crick and Watson's model represents the DNA molecule. It would have failed to characterize the particular form of representation that Crick and Watson's model provides. Our intuitions regarding scientific models are perhaps less clear-cut than our intuitions regarding pictures. But there still seem to be many differences between Crick and Watson's model and the name "DNA molecule" that our theory must explain. The form of the name "DNA molecule" is ultimately arbitrary, for example; any combination of letters could have done the job just as well. But the form of Crick and Watson's model was the subject of years of research and careful adjustment. Unlike the name "DNA molecule", Crick and Watson's model seems to "tell us" something about the DNA molecule, and we feel that in some way what it tells us can be right or wrong, accurate or inaccurate. Our theory of how models represent must account for these intuitions.

In the next section I will propose an account of scientific representation. First, however, let us sum up the problem that faces us. Many scientific models are representational. Some represent actual objects or events. The problem of scientific representation asks how they do this. Why does Crick and Watson's model represent the DNA molecule, or our model represent the bouncing spring? There may turn out to be many different forms of model-representation. Any, or even all, of these forms of representation may be employed by other representational devices, apart from scientific models. We want an account of each of these forms of model-representation. Theories of depiction aim to state conditions that are necessary and sufficient for something to be a depiction. Similarly, if possible, we want to provide a set of conditions that are both individually necessary and jointly sufficient to establish an instance of each form of model-representation that we identify. Our intuitions are less clear-cut in the case of models than for pictures. But our account should be able to distinguish models from merely denoting entities, like names or Callender and Cohen's salt shaker, as well as excluding non-representational entities, like ordinary chairs, tables or trees.[9] And it should also accommodate inaccurate or incorrect models, as well as accurate or correct ones.

Models as Make-Believe

Walton's Theory: Props and Games

According to Walton, representations are props in games of make-believe. Suppose that some children play a game in the woods in which they imagine tree stumps to be bears. In Walton's terminology, in this game the tree stumps are *props* and the convention that the children establish by their agreement that stumps "count as" bears is a *principle of generation*. Together, props and principles of generation make propositions *fictional*. To say that a proposition is fictional, on Walton's theory, is to say that there is a prescription to imagine it. (A *fictional truth* is simply the fact that a certain proposition is fictional.) Thus, given the rule that stumps "count as" bears, if a participant in the game comes across a stump in a thicket, they are to imagine that there is a bear there; it is fictional that there is a bear there.[10]

What is fictional in a game of make-believe need not be the same as what is imagined. A stump which remains hidden under a pile of leaves still makes it fictional that a bear lurks there, even if this is never imagined by anyone playing the game. An oddly shaped stump might prompt one of the participants to imagine a wolf and not a bear, but the proposition that there is a wolf before them is only imagined, not fictional. Fictional truths therefore possess a certain kind of "objectivity"; participants can be unaware of fictional truths and mistaken about them.

The stumps in the children's game are not representations, however. A *representation*, in Walton's sense, is not something that merely happens to be used as a prop; it is something of which it is the function to serve as such (see Walton 1990, Section 1.7). Whether it is the function of a given object to serve as a prop depends upon social context. Walton's theory does not aim to analyze our ordinary use of the term "representation", but to "carve out a new category" that may be applied to what we might call works of *fiction*, including novels, paintings, sculptures, plays, films and musical works (1990, 2). Many other entities that we might normally call "representations", such as most history books, newspaper articles, biographies or textbooks, Walton thinks, do not count as representations in his sense (see Walton 1990, Chapter 2). The function of a biography of Napoleon, it seems, is not to prescribe imaginings about Napoleon, but to make certain claims about him. The biography

[9]Of course, in certain cases such objects may be representational. A chair might be used in a work of abstract art, for example, or a table used to represent a shelter in a play.

[10]These central features of the account are introduced in Section 1.5 of Walton (1990).

does ask us to believe certain things of Napoleon, and it is arguable that believing something requires us to imagine it. But there is no rule that we ought to believe what the biography says about Napoleon simply because it says it. On the other hand, Walton claims, there is a rule that we ought to imagine certain things of Napoleon when we read *War and Peace*, simply because the novel is written as it is. For this reason, the novel counts as a representation in Walton's sense.

Something is an *object* of a representation on Walton's theory if there are propositions about it which the representation makes fictional (see Walton 1990, Chapter 3). Napoleon is an object of *War and Peace*, as is St. Petersburg. Salisbury Cathedral is an object of Constable's *Salisbury Cathedral from the Meadows*. *Representation-as* is a matter of what propositions about an object a representation makes fictional. *War and Peace* makes it fictional that Napoleon invaded Russia in 1812; it represents Napoleon *as* invading Russia in 1812. Sometimes, when we call something "fictional" we do so to imply that it is false or even deceitful. To say a proposition is fictional in Walton's sense, however, is simply to say that there is a prescription to imagine it. This is perfectly compatible with truth. If a child screams when he comes across a stump in the woods, it will probably be fictional that he screams; it is both fictional and true that the child screams. Similarly, of course, it is true, as well as fictional in *War and Peace*, that Napoleon invaded Russia in 1812. In this respect, the novel corresponds to Napoleon. If a representation corresponds completely with its object then it *matches* it. But a work may represent something it does not match and match something it does not represent. It is fictional in *The War of the Worlds* that Martians attack London in the late nineteenth century. The novel represents London, but does not match it. Conversely, a portrait of John may match his twin brother David, but it represents John and not David.

As well as prescribing imaginings, the stumps in the children's game are also objects of those imaginings: the children imagine of the stumps that they are bears. This is not a necessary condition for something to count as a prop. The text of *War and Peace* may prescribe us to imagine many things of Napoleon, but we do not imagine the text of the novel itself to be Napoleon. Some works of fiction do prescribe imaginings about themselves, however. For example, we are to imagine that the first chapter of the novel *Dracula* is an excerpt from a journal. Walton calls these *reflexive* representations (1990, 117).

The principle that wherever there is a stump, fictionally, there is a bear, was established by participants in the game by explicit stipulation. But Walton's theory does not demand that principles of generation be established in this way, nor that they be explicitly formulated. And indeed, many implicit rules are likely to operate in the children's game: it may well be that if the stump in the thicket is taller than the stump under the leaves, then, fictionally, the bear in the thicket is taller than the bear hiding in the leaves. In the case of novels or paintings, principles of generation are difficult to specify explicitly, complex, and vary from case to case. The principles that apply to novels are conditional upon the text of the novel; those that apply to paintings or statues depend upon the distribution of paint on the canvas or on the form of the sculpted marble.

Make-Believe and Model-Representation

With this outline of Walton's theory in place, let us now begin to apply it to scientific models.[11] First, consider a physical model, such as a 1:1,000 scale model of the Forth Road Bridge. I think we may regard this model as a representation in Walton's sense: the model functions as a prop in games of make-believe. These games are governed by certain principles of generation, appropriate for such models. One principle is that, if part of the model has a certain length, then we should imagine the corresponding part of the bridge to be a thousand times longer. Together, the model and principles of generation determine what users of the model are supposed to imagine; in Walton's terminology, they generate fictional truths. Some of these fictional truths are about the bridge itself. For example, if the model is a meter long, it will be fictional that the bridge is a thousand meters long. The bridge is therefore an *object* of the model; the model represents it *as* a thousand meters long. Since the Forth Road Bridge is in fact 1,006 m long, the model represents the bridge but does not match it.

I propose that we regard all physical models in this way, as props in games of make-believe, which represent their objects by prescribing imaginings about them. The principles of generation by which models prescribe imaginings will vary from case to case. Were the bridge model built to carry out structural tests, for example, one principle of generation in effect may be that if the model is built from a certain material then it is fictional that the bridge is also built from that material. If, instead, the model were built for a museum display, however, this principle may not hold. Furthermore, not all physical models are scale models. The famous Phillips machine represents the workings of the macro-economy by the ebb and flow of colored water in a hydraulic system. The principles guiding our imaginings when we use the Phillips machine will be very different from those that apply to the bridge model. One principle may be that if water is flowing through a certain pipe then, fictionally, taxes are being paid. Many physical models are reflexive representations, in Walton's sense: they prescribe imaginings about themselves. When we use the bridge model, for example, we not only imagine things of the bridge; we also imagine that the model itself is the bridge. Similarly, we imagine the balls of a ball-and-stick chemical model to be atoms, and the sticks to be bonds between them. Physical models need not be reflexive, however. When we use the Phillips machine, perhaps we do not imagine the flow of water itself to be the payment of taxes, but only that taxes are being paid.

Let us now turn to consider theoretical modeling. When we model the bouncing spring we write down an equation of motion $md^2x/dt^2 = -kx$, and a prepared description, which takes the bob to be a point mass m subject to a linear restoring force, and so on. I believe these may be understood in the same way that Walton

[11] The suggestion that Walton's theory may be applied in the context of scientific modeling is also found in Barberousse (2006), Barberousse and Ludwig (2000) and Frigg (2010). See below for a discussion of Frigg's views.

understands literary works of fiction. Consider the following passage from *The War of the Worlds*: "The dome of St. Paul's was dark against the sunrise, and injured, I saw for the first time, by a huge gaping cavity on its western side" (Wells 2005, 170). Clearly, this is not a description of St Paul's Cathedral: when Wells wrote this he was not claiming that there really was a hole in the side of St Paul's. Nevertheless, on Walton's view, the passage still represents St Paul's; St Paul's is an *object* of *The War of the Worlds*. Usually, Walton thinks, when we read a linguistic work of fiction that uses proper names, we take ourselves to be prescribed to imagine things of the normal referents of those names. On this view, the above passage represents (the actual) St. Paul's, because it requires readers to imagine certain things of St Paul's, namely that it has a large hole in its dome. In Walton's terminology, the passage makes it fictional that St Paul's has a large hole in its dome.

I think we may use Walton's analysis to provide an account of our prepared description and equation of motion. We have seen that these are not straightforward descriptions of the bouncing spring. Nevertheless, I believe, they do *represent* the spring, in Walton's sense: they represent the spring by prescribing imaginings about it. When we put forward our prepared description and equation of motion, I think, those who are familiar with the process of theoretical modeling understand that they are to imagine certain things about the bouncing spring. Specifically, they are required to imagine that the bob is a point mass, that the spring exerts a linear restoring force, and so on. Unlike some physical models, our theoretical model is not a reflexive representation: we do not imagine that our description or equation are themselves a point mass or subject to a linear restoring force. Instead, our description and equation prescribe imaginings about the bouncing spring system. The bouncing spring is an *object* of our model; our model *represents* it *as* a point mass, subject to a linear restoring force and a uniform gravitational field. Using Walton's terminology, we may say that our prepared description and equation of motion make it *fictional* that the bob is a point mass, that it is subject to a linear restoring force and so on.

My suggestion, then, is that models function as props in games of make-believe; model-representation is an instance of representation in Walton's sense. Tentatively, I claim that this notion of model-representation applies to all physical and theoretical modeling. In physical modeling, the prop is a physical object, while in theoretical modeling, it is usually a prepared description and equation of motion. In some cases, the prop might be a diagram or picture. Just as for novels or paintings, the principles of generation governing the games in which these props function are complex and vary from case to case. In each case, however, the model represents in virtue of prescribing us to imagine things. We may formulate this account as follows:

M is a model-representation if and only if *M* functions as a prop in a game of make-believe (MM)

As we have seen, something is an object of a representation, on Walton's theory, if there are propositions about it which the model makes fictional. Taking this criterion together with the account (MM) allows us to state the conditions under which a model will represent some actual system:

M model-represents *T* if and only if *M* functions as a prop in a game of make-believe in which propositions about *T* are made fictional (MM_1)

On the account I propose, then, where a model represents an actual system, it does so by prescribing imaginings about that system; in Walton's terminology, it makes propositions about the system fictional. However, the primary statement of the account remains that given in MM: a model *M* is a model-representation, if and only if, if functions as a prop in a game of make-believe; it need not prescribe imaginings about any actual system. We shall see the importance of this feature of the account when we come to consider models without actual objects.

In the remainder of this paper, I will try to demonstrate the advantages of the account of representation I have proposed. First, however, it is important that this account is distinguished from another recent application of Walton's theory to scientific models. Frigg (2010; see also "Fiction and Scientific Representation" in this volume) also suggests that the descriptions presented in theoretical modeling should be understood as props in Walton's sense. On his view, however, these descriptions prescribe us to imagine what he calls "model systems", where these are to be understood as imagined physical systems, i.e. as hypothetical entities that, as a matter of fact, do not exist spatio-temporally but are nevertheless not purely mathematical or structural in that they would be physical things if they were real (Frigg 2010, 253). Frigg calls the means by which prepared descriptions and equations of motion give rise to these model systems "p-representation". For him, it is only this p-representation that is to be understood using Walton's theory. There is then a second representation relation, which he calls "t-representation", that exists between model systems and the world.

Frigg's account is therefore very different from that proposed in this paper. Like Giere, Frigg offers an indirect, two-stage, view of scientific modeling: prepared descriptions and equations of motion first give rise to model systems (p-representation) and these in turn represent the system being modeled (t-representation). By contrast, I do not take prepared descriptions and equations of motion to give rise to model systems. On my account, the prepared description and equation of motion that we write down when we model the bouncing spring, for example, do not prescribe us to imagine an "imagined" or "hypothetical" ideal oscillator. Rather, they prescribe us to imagine propositions *about the actual bouncing spring*: we imagine of the actual bob that it is a point mass and of the actual spring that it is massless, and so on.[12] On my account, then, there are not two forms of representation relation, but only one, given by MM_1: the prepared description and equation of motion represent the bouncing spring directly, by prescribing imaginings about it.

Frigg's proposal is interesting, and its relationship to the account I have put forward merits further investigation. One reason I have for preferring my own application of Walton's theory has to do with questions regarding the nature of model systems. What exactly are "imagined physical systems" or "hypothetical entities"? Like a number of other authors, Frigg compares model systems to fictional

[12]For more on the ontology of theoretical modeling, see Toon (2010).

entities, like unicorns or Count Dracula.[13] Of course, the nature of fictional entities, and in particular the question of whether such entities exist at all, is itself the subject of considerable controversy. Frigg acknowledges this, but claims that his account incurs no "ontological commitments" since Walton's theory is antirealist with regard to fictional entities (2010, 264). And yet it is difficult to see how Frigg can take an antirealist stance on fictional entities, and thereby model systems, if these model systems are central to his account of theoretical modeling. If there are no model systems, what will do the t-representing?

Make-Believe and Stipulation

In our discussion of Callender and Cohen's views, we saw that we may accept their arguments in favor of adopting a derivative account of scientific representation, while rejecting their claim that stipulation is sufficient for scientific representation. The account I have proposed offers a derivative account: it explains the representational power of models in terms of the representational power of certain mental states, namely those of the imagination. For example, the bridge model represents the bridge in virtue of prescribing users to imagine that the bridge is a certain shape, length and so on. Unlike Callender and Cohen's stipulation view, however, my account is able to distinguish model-representation from cases of mere denotation or reference.

According to MM_1, in order to be a model-representation of some object, a model must not only refer to that object; there must be an understanding amongst those who use the model that various imaginings are prescribed that depend upon the features of the model. This is absent in the case of Callender and Cohen's salt shaker. The act of stipulation they describe may establish that the salt shaker refers to Madagascar, but there is no understanding among the diners that they are to imagining anything about Madagascar, given the properties of the salt shaker. For the same reason, my account is also able to exclude proper names: no convention exists such that we are to imagine certain things of the DNA molecule depending upon the properties of the name "DNA molecule", such as the number of letters it has or whether it is written in English or French.

Earlier, we observed that the form of a name like "DNA molecule" is ultimately arbitrary, while that of a scientific model is often crucial to its representational function. Furthermore, we noted that scientific models seem to "tell us" something about their objects, while names do not, and that what the model tells us can be right or wrong. We are now in a position to explain these differences. The reason that the properties of a model are important to its representational function, while those of

[13]The suggestion that models might be understood as fictional entities is found in Godfrey-Smith (2006) and Frigg (2006b). Contessa (2010) follows this approach by developing his own "dualist" account of fictional entities, while Thomson-Jones (2007) also explores versions of this view.

names or Callender and Cohen's salt shaker are not, is that the imaginings the model prescribes about its object are conditional on those properties. What a model "tells us" about its object is dependent on the content of those imaginings, and what it tells us is right or wrong depending on whether the propositions it asks us to imagine are true or false of that object.

Under certain circumstances, the salt shaker could become a model-representation of Madagascar. For example, we might imagine the shaker being used to indicate the location of Madagascar with respect to Africa (the dinner plate). In this case, the salt shaker (together with the dinner plate) would constitute a model-representation on my account: the salt shaker's properties prescribe us to imagine something about Madagascar, according to rules such as "if the shaker is to the right of the plate, you are to imagine that Madagascar is to the east of Africa". One way to establish this rule would be to declare it explicitly. As we have seen, however, principles of generation need not be stated explicitly. Many suggest themselves to us almost "automatically". Once we have explicitly specified that the salt shaker denote Madagascar and the plate denote Africa, it is almost inevitable that we will associate the relative positions of the salt shaker and the plate with the relative positions of Madagascar and Africa. The ease with which we understand such conventions, however, should not mislead us into neglecting their importance. No familiar principles of generation come to mind when we are told that the salt shaker represents Madagascar. (Its shape does not readily suggest taking it to be a scale model of Madagascar, for example.) In the absence of such principles, the salt shaker fails to model-represent Madagascar and merely refers to it; its properties are irrelevant to its representational function, and it can tell us nothing about Madagascar.

Make-Believe, Misrepresentation and Realism

When introducing the problem of scientific representation, I argued that our theory of model-representation should be able to accommodate inaccurate (or incorrect or unrealistic) models as well as accurate ones. The account I have offered meets this criterion. According to MM_1, a model represents an object if it makes propositions about that object fictional. Once again, recall that propositions can be fictional in Walton's sense and still be true. For example, our model of the bouncing spring makes it fictional that the bob has mass *m* and that it is attached to a spring. However, it is not a condition for model-representation on my account that all, or even any, of the propositions a model makes fictional be true. For this reason, my account is able to accommodate inaccurate (or incorrect or unrealistic) models. Our model still represents the spring, even though much of what it asks us to imagine about it is false: the bob is not a point mass, the spring is not massless, and so on. Or again, like their final double-helical model, Crick and Watson's early models represent the DNA molecule because they prescribe us to imagine things about the molecule. It is simply that some, or even all, of what the early models ask us to imagine is false.

However, the accuracy, or realism, with which a model represents a system is often of considerable importance, of course. There are many questions that we might ask in this regard. Can we say anything general about the accuracy or realism of scientific models? If we can, how realistic are scientific models in general and in what respects? Are we justified in believing that scientific models are realistic representations of their objects? In this paper I am concerned not with these questions, but with the prior question of how scientific represent their target systems. As I have already noted, however, it would be desirable if our theory of model-representation provided us with a framework in which to address these questions about realism. The theory of model-representation I have proposed does provide such a framework, but this framework differs from that commonly employed.

On most accounts of scientific modeling, accuracy is judged in terms of some form of similarity or fit between a model and the world. For example, as we have seen, Giere takes theoretical models to be abstract objects defined by the prepared descriptions and equations of motion scientists write down when they model a system. The accuracy of a theoretical model is then a matter of the similarity between this abstract object and the system in certain respects and to certain degrees. In contrast to this indirect view of theoretical modeling, I (2010) propose a direct view. On my account, there is no abstract object (or fictional entity or any other kind of object) that satisfies scientists' prepared description and equation of motion; instead, the prepared description and equation represent the system directly, by prescribing imaginings about it. However, this account still provides us with a simple way of understanding the accuracy or realism of a theoretical model: put simply, a model is accurate in a certain respect if and only if what it prescribes us to imagine in that respect is *true* of the object it represents.

For example, consider our model of the bouncing spring. Whether this model is accurate is not a matter of whether some abstract ideal oscillator is similar to the bouncing spring. The model is accurate if what it prescribes us to imagine of the spring is true. For instance, the model prescribes us to imagine that the bob oscillates with a time period of $T = 2\pi\sqrt{m/k}$. The model is accurate in its prediction if, and only if, the bob does in fact oscillate with period $T = 2\pi\sqrt{m/k}$. On my account, then, the accuracy of a model is dependent upon the truth (or perhaps the approximate truth) of the propositions it prescribes us to imagine about the system it represents. This view may be applied to physical, as well as theoretical models; as we have seen, on my account, even physical models prescribe us to imagine propositions about their objects.

Models and Works of Fiction

Many of entities to which Walton applies his theory, such as novels, painting and films, are central examples of works of fiction. If Walton does indeed offer the correct analysis of these works then, on my account, model-representation turns out to be an instance of a wider form of representation also instantiated by such works. Some will object to this comparison. Surely there are many differences between our model of the bouncing spring and works of fiction such as *War of the Worlds*, or between an architect's scale model and a statue of Napoleon? Although I claim that models employ the same form of representation that Walton ascribes to works of fiction, I do not deny that there are many important differences between the two, as there are amongst works of fiction themselves. Similarly, to claim that some scientific drawings employ the same mode of representation as cartoons and Surrealist paintings, namely depiction, would not prevent us recognizing the enormous differences between these different representations.

And, although there clearly *are* important differences between models and some works of fiction, I think it is less clear where to draw a line between them, if one can be drawn at all. It is clearly not correct to say that the imaginings models prescribe are generally true, or even approximately true, whereas those prescribed by works of fiction are not. As we have seen, even good models prescribe many false imaginings about their objects. Conversely, works of historical fiction often prescribe many true imaginings about actual characters and events, as do many portraits. Moreover, given that we know that something is a work of historical fiction or a portrait, it is arguable that we are entitled to expect the work to be accurate in these ways. The same considerations also show that we cannot draw the distinction in terms of whether or not the works *aim* at truth. One important function of many scientific models is that of providing us with predictions. But, again, this does not give a clear criterion for distinguishing models from works of fiction. On the one hand, it seems that some models are not used to provide predictions. Obvious examples here are the models we will consider in the following section, which do not represent an actual object. And on the other hand, it is arguable that some works of fiction offer predictions. One example here might be Orwell's *Nineteen Eighty-Four*.[14]

Models Without Actual Objects

The Variety of Models Without Actual Objects

As we noted earlier, the problem of scientific representation is usually presented as that of giving an account of a *relation* between a model and some actual system, just as the problem of depiction is often said to be that of identifying a relation between a picture and its subject. We also observed, however, that many pictures

[14]Note also that the position I advocate is distinct from what Arthur Fine (1998) calls *fictionalism*. As Fine characterizes it, fictionalism is an anti-realist position which argues that a scientific theory may be reliable without being true and without the entities it invokes existing. To classify a model as a representation in Walton's sense is to say nothing about the truth of the propositions the model prescribes or about the existence of the entities it invokes.

seem to be depictive, even though they depict no actual subject. An illustrated edition of *Dracula* might contain a picture of Count Dracula, for example, his fangs dripping with blood. It seems that the picture represents or depicts Dracula, in a similar way to that in which a portrait like David's *Napoleon Crossing the Saint Bernard* depicts Napoleon. Of course, Count Dracula does not exist in the same way as Napoleon did. But if the painting represents Dracula, must he not exist in *some* sense? These problems also arise for discourse about fiction. If we say "Dracula sucks blood" it seems we assert something true. And yet if Dracula does not exist, to what does the name "Dracula" refer? Solutions to these problems fall into two camps. *Accommodationist* theories grant fictional entities like Count Dracula some place in our ontology. *Eliminativist* theories attempt to show how fiction, and our discourse about it, may be understood without granting the existence of fictional entities.[15]

Many scientific models pose parallel problems. Obvious examples are models of entities we once thought to exist but now know not to. Nineteenth century physicists constructed mechanical models of the ether. Even if, as we now believe, the ether does not exist, these models still seem to be representational. Intuitively, we want to say that ether models represent something, even though we know there is no ether. Just as we seem to need Dracula to understand pictures of the Count, so we seem to need the ether to understand the physicists' models. The problem also arises for our discourse: just as we make statements that seem to refer to Dracula, so we might make statements that appear to refer to the ether, like "the ether is at rest".

The problems posed by models without actual objects is rarely recognized. Where it is recognized, it is always models of discredited entities like the ether or phlogiston that are offered as examples. In fact, however, problems with fictional entities arise for a much wider range of cases. Many of these are rather mundane. For example, suppose that engineers constructing a bridge invite architects to submit models of their proposed designs. Like the ether models, a model proposing an unsuccessful design would still seem to be representational, even if there is no actual bridge that it represents. Many scientific experiments create events which may never otherwise occur; a scientist might formulate a theoretical model of such an event even if funding runs out and the experiment never takes place. Or again, while using a ball-and-stick chemical model we might construct any number of models that represent configurations of atoms that do not exist.

In addition to examples such as these, there are clearly many cases of models that represent no *particular* actual object or event. We say that Bohr's model "represents the hydrogen atom", for example, but presumably it does not represent any particular hydrogen atom (although it might be used to do so). In fact, it is arguable that most scientific models are of this form. In some cases, such as that of the Bohr model, we might think that the model represents a type of entity or event. R.I.G. Hughes elects to "assume without argument that our concept of denotation allows us to denote a

[15]The terms "accommodationist" and "eliminativist" are taken from Lamarque (2003).

type" and offers Bohr's model as an example (Hughes 1997, S330–S331). However, even allowing that we may make sense of the notion of a model representing a type, there are many models, or uses of models, that cannot be thought of in this way.

A comparison with pictures might once again be helpful. Many pictures would also appear to represent types. Examples might include encyclopaedia illustrations representing certain species of plants or the famous diagrams of man and woman on the plaque of the Pioneer spacecraft. But clearly not every picture that fails to represent a particular actual object may be thought of in this way. For example, Vermeer's *The Milkmaid* shows a woman pouring milk from a jug by a window. Even if Vermeer used a model when painting the work, there is no actual woman that the painting represents, nor does it represent a type of woman. Instead, the painting simply represents a particular fictional or "imaginary" woman. There are numerous pictures of this sort. As Goodman puts it, "the world of pictures teems with anonymous fictional persons, places, and things" (1976, 26).

Analogous cases exist in scientific modeling. Consider the Phillips machine. The machine could be used to represent some actual economy, such as that of Britain. Alternatively, perhaps it could be used to represent a type of economy. But we could also use the machine simply to represent a particular "imaginary" or fictional economy. (We might begin by saying "suppose there were an economy like this ...".) Or, to take another example, suppose that the "prepared description" and equation of motion that we write down when we model the bouncing spring system were to appear instead in a textbook, written to instruct students on how to model a bouncing spring like ours. In this case, it seems there will be no actual system that the model represents, nor type of system. Instead, it represents an "imaginary" or fictional bouncing spring that the student is to imagine encountering.[16]

Need an account of representation for scientific models accommodate those without actual objects? Callender and Cohen suggest we might "bite the bullet and hold that, in cases where *x* doesn't exist, agents don't succeed in representing *x* but merely believe they are representing *x*" (2006, 81, n11). As we have seen, this would be to exclude a considerable number of models from our account of representation. Moreover, in many of the cases we have considered, agents do not even *believe* that they are representing an actual object. Most importantly, however, I think it is simply wrong to deny that models without actual objects are representational.

A comparison with pictures is helpful. We take for granted that pictures without actual objects are representational. Of course, we recognize that when we say that *The Milkmaid* is a "picture of a milkmaid" this does not license the inference that the milkmaid exists. However, even if she does not, the picture is undoubtedly still depictive. Indeed, our experience of the picture depends very little upon whether or not the milkmaid exists. We can still stand before the painting and admire her care and concentration in her task, just as we might look at David's portrait and admire

[16]This example reminds us that the same prepared description and equation of motion may serve very different representational functions.

Napoleon's bravery and determination. The same is true of models. Consider the architects' models discussed earlier, each showing proposals for a bridge design. Suppose that these models were all put on display after the bridge is built. If we were to inspect the models without knowing which was chosen, our experience of the unsuccessful models would be very similar to that of the successful one. Looking at these models, which might be built from balsa wood or paper or construction kit, and might be a meter or ten meters high, we could still recognize each as representing a bridge to be built across the river, and discuss whether that bridge is ugly or beautiful, flimsy or strong. Similarly, we realize that when we say a model "represents the ether" we cannot conclude that there is an actual object that it represents. But the model is still *representational.* Indeed, the representational properties of ether models may have played an important role in allowing scientists to determine whether or not the ether exists.

In the next section, I will consider whether existing accounts of scientific representation can accommodate models without actual objects. First, however, it is important that our present problem is distinguished from another way in which scientific modeling is sometimes thought to give rise to fictional entities. This route to fictional entities arises from theoretical modeling of actual objects, like our model of the bouncing spring. When we model the spring we make assumptions that are true of no actual system: no actual pendulum is a point mass, no actual spring is massless, and so on. Recently, as we have already seen, a number of authors have suggested that our model of the bouncing spring is itself a fictional entity that satisfies these modeling assumptions. On this indirect view of theoretical modeling, our prepared description and equation of motion define a *fictional* idealized oscillator, and this, in turn, represents the bouncing spring. Theoretical models are *themselves* taken to be fictional entities.

I have argued against this view elsewhere (Toon 2010). For now, we may simply note that the ontology of theoretical models themselves is not the problem that concerns us here. We want to know how our account of representation can accommodate models without actual objects. It is possible for these two problems to become confused. Speaking loosely, we might say that our model of the bouncing spring "represents" a point mass or a massless spring. Point masses and massless springs do not exist, of course, and it is tempting to label them as "fictional entities". Speaking more carefully, however, we should say that our model represents an *actual* pendulum bob *as* a point mass and it represents an *actual* spring *as* massless and frictionless. For this reason, it does not present the same problem as models like the ether model.

Moreover, even if we take theoretical models to be fictional entities, rather than linguistic entities or abstract objects, this does not solve the problem posed by models that represent no actual object. To see this, consider the theoretical model mentioned above, which represents an experimental event that never occurs. The problem we are faced with is that of explaining how it is that this model is representational, given that there is no actual object that it represents. Taking the model *itself* to be a fictional entity, rather than, say, a linguistic entity or abstract object, does not solve this problem. Or again, suppose that, before it was discovered not to exist, someone had produced a theoretical model of the ether that was thought to offer a highly simplified account of its behavior. Even if we were to take the scientists' ether model to be a fictional entity, defined by whatever assumptions and equations they wrote down, we would still be left with the problem that this model, like a mechanical ether model, seems intuitively to represent the ether, even though there is no ether.[17]

Existing Accounts of Scientific Representation and Models Without Actual Objects

Most existing accounts conceive of representation in modeling as a relation. This includes the similarity and isomorphism accounts criticized by Mauricio Suárez (1999, 2003) and Roman Frigg (2006a). It also includes Ronald Giere's (2004) more sophisticated similarity account, on which scientists use models to represent systems by forming "theoretical hypotheses" detailing their similarities.[18] Although Hughes' "D.D.I. account" is not intended to provide necessary and sufficient conditions for representation, Hughes does endorse the maxim "no representation without denotation" (1997, S331).[19] Finally, as we have seen, on Callender and Cohen's view, representation in modeling is a relation established by an act of stipulation connecting a model and its object. As they stand, none of these accounts can explain why models without actual objects are representational. An ether model cannot represent in virtue of its similarity or isomorphism to the ether if the ether does not exist, nor could a scientist list the model and ether's similarities in a theoretical hypothesis. The model also cannot denote or stand for the ether, and we cannot establish a representation relation between the model and the ether by stipulation.

If accounts that take representation in modeling to be a relation are to be applied to models without actual objects, then their proponents must posit some object for these models to represent. That is, they must adopt an accommodationist stance on fictional entities. Whether this is thought to be problematic would depend upon

[17]Similarly, Frigg (2010) suggests that the problem of models without actual objects can be avoided simply by adopting his distinction between p-representation and t-representation. This alone does not seem sufficient to solve the problem, however: we still require an account of t-representation that can explain how some model systems (like the simplified ether model system) can be representational, without representing any actual object.

[18]Giere allows there may be other ways in which models are used to represent, although does not specify any; see also Giere (1988, 1999).

[19]"D.D.I." stands for "denotation, demonstration, and interpretation". According to Hughes, these combine in the following way: elements of the physical world are denoted by elements of the model; the model possesses an internal dynamic that allows us to demonstrate theoretical conclusions; these in turn need to be interpreted if we are to make predictions (Hughes 1997, S325).

which accommodationist view was adopted and how palatable its ontological commitments were taken to be. However, it would be a mistake to assume that the problem disappears once we posit fictional entities. In fact, questions would remain for each of the accounts. For example, would the objects posited to serve as fictional entities have the right properties to enter into relations of similarity or isomorphism with models? The claim that models may *denote* fictional entities, just as they denote actual entities, would also be open to debate. Fictional objects are dependent on representations for their existence in a way that actual objects are not; the relation between a representation and a fictional object, if there are any, would therefore appear very different from that between a representation and an actual object (Walton 1990, 127). Finally, could we say that the ether model is representational because it was stipulated that it represent a fictional ether? If any stipulation occurred it was surely that the model represent the real ether.

The only account of scientific representation that attempts to accommodate models that represent no actual object is Mauricio Suárez's "inferential conception". On this view, a representational source A represents some target B "only if (i) the representational force of A points towards B and (ii) A allows competent and informed agents to draw specific inferences regarding B" (Suárez 2004, 773). At first sight, then, it seems that the inferential conception also regards representation as a relation. However, Suárez argues that this account can accommodate what he calls "fictional representation, that is, representations of nonexisting entities", and in fact, he claims that on his account "there is absolutely no difference in kind between fictional and real-object representation—other than the existence or otherwise of the target" (2004, 770).

How is this supposed to work? Consider an ether model. Even though the ether does not exist, perhaps there is a sense in which we might say that the model possesses a representational force "towards the ether" just as, for example, a model of the Forth Road Bridge possesses a representational force towards the bridge. The ether model is rather like a description such as "the only inhabitant of London": both purportedly pick out an object; they simply fail to do so because that object does not exist. However, it is not clear that we may say this in all cases. For example, consider the case discussed above, in which the Phillips machine is used to represent an "imaginary" economy. Unlike the creator of an ether model, the user of the Phillips machine does not attempt, but fail, to represent an actual object. When used in this way, the Phillips machine does not purport to represent any actual object. As a result, it is difficult to make sense of the idea that this model possesses a representation force, even a thwarted one, presuming that representational force always points towards actual objects or events. Of course, we might attempt to get round this problem by granting the existence of fictional entities and allowing that representational force may point to them too. But then the claim that there is no difference between representation of fictional entities and of actual ones would require further argument for, as mentioned already, the relation between a representation and a fictional object and the relation between a representation and an actual object would appear to be rather different.

Models as Make-Believe and Models Without Actual Objects

Unlike similarity and isomorphism accounts, Hughes' D.D.I. account or Callender and Cohen's stipulation view, the account of models I set out above (MM) does not take representation in modeling to be a relation. It is therefore able to accommodate models that represent no actual object without postulating some object for them to represent. Something is a model-representation if it has the function of serving as a prop in games of make-believe; it is not a necessary condition for model-representation that there be any object that the model prescribe imaginings about. Ether models, or the textbook model of a bouncing spring or the model for an experiment that does not take place, are all representational because they function as a prop in a game of make-believe; all of them are taken to prescribe imaginings. They can still fulfil this role even if there is no object that they prescribe imaginings about.

Of course, other accounts of representation in art and elsewhere, apart from Walton's, acknowledge that it may fail to be a relation in certain cases. For example, although Hughes bases his D.D.I. account on Goodman's theory of representation, Goodman does not endorse Hughes' maxim "no representation without denotation". Instead, Goodman allows that denotation may fail in certain cases and considers it necessary only that representations are "*ostensibly* provided with denotata" (1976, 228; emphasis in original). For example, we might say that "the only inhabitant of London" is ostensibly provided with a denotatum; it simply fails to denote because no one happens to satisfy the description. Walton's position is more radical. On his account, the notion of having an object is not central to representation in any way. He asks us to imagine a society of people who make pictures, say, "of people" or "of trees", but never pictures that depict actual people or trees. Drawing or painting a person is thought of as creating or making an imaginary person and not representing any real person. And yet, Walton argues, these pictures would still be representational. Unlike Suárez's inferential conception, then, my account can accommodate cases in which there is not even attempted reference to any object, like Phillips machine being used to represent an "imaginary" economy.

Problems concerning fictional entities have not been entirely dispelled, however. For in addition to prescribing many unproblematic imaginings, such as that the speed of light is constant or that electromagnetic waves are transverse, intuitively it seems that an ether model will also prescribe imaginings "about the ether". For example, it may ask us to imagine that the ether is at rest. Once again, then, we meet the problem of fictional entities, this time for imagination: how are we to understand the contents of imaginings that appear directed towards fictional entities like the ether? This is certainly a problem, but it is not one that a theory of representation for scientific models need address. Instead, it is a general problem that faces all theories of intentionality. And it is a problem that will exist whatever account of representation we adopt for scientific models; even those who hold similarity or isomorphism accounts will concede that we often appear to imagine things of the ether.[20] The same is true of the problem posed by discourse apparently referring to fictional entities, like "the ether is at rest" or "the bridge is stable" (said in reference to a failed bridge model). This too is a general problem that exists whatever our account of scientific representation and is the subject of longstanding debate.

Conclusion

Scientific models are props in games of make-believe, which represent their objects by prescribing imaginings about them. Analyzing models in this way allows us to accommodate models which represent their objects inaccurately, while showing how models differ from merely denoting entities, like Callender and Cohen's salt shaker. Since this account does not take representation in modeling to be essentially relational, it is also able to accommodate an important group of models that have been largely ignored by recent philosophical work on modeling, namely those that are representational, but represent no actual object.

Acknowledgments This paper is based on a talk delivered at the "Beyond Mimesis and Nominalism: Representation in Art and Science" conference held at London School of Economics and the Courtauld Institute of Art in June 2006. Parts of the paper were also presented at the Philosophy Workshop in Cambridge in June 2006, the CMM Graduate Conference held in Leeds in June 2007, and the "Scientific Models: Semantics and Ontology" workshop, held in Barcelona in July 2007. I would like to thank participants at all of these events. Thanks also to Nancy Cartwright, Stacie Friend, Manuel Garcia-Carpintero, Ronald Giere, Mary Leng, Mauricio Suárez, Paul Teller, Martin Thomson-Jones and Kendall Walton for helpful discussion and correspondence, and to Roman Frigg and two anonymous referees for comments on drafts of this paper. Finally, I would like to thank my PhD supervisor Martin Kusch, and my advisor, the late Peter Lipton. Research for this paper was supported by The Arts and Humanties Research Council, The Darwin Trust of Edinburgh and The Rausing Fund for History and Philosophy of Science. I am very grateful to all of these institutions for their support.

[20]Callender and Cohen also attempt to defer the problem posed by models without actual objects, observing that "the worry arises for all species of representation—not just scientific representation—and there is no reason to suspect that whatever ultimately explains representations of unicorns and golden mountains won't work for representation of phlogiston and the ether" (2006, 81). There is an important difference between Callender and Cohen's deferral strategy and my own, however. Callender and Cohen simply express a hope that a solution to the problem for other forms of representation may be applied to scientific models. They tentatively suggest a "Humean strategy", which provides a relational theory for "atomic" representations and explains representations without actual objects as constructed as "compounds" of other representations. But they do not show whether this can be applied to scientific models, nor whether their account would remain intact if it were. This amounts simply to deferring the problem for scientific representation itself. By contrast, my own account reduces the problem of understanding models without actual objects to the more general problem of understanding imaginings about fictional entities.

References

Barberousse, A. (2006), "Images of Theoretical Models", unpublished paper delivered at A.P.A. Pacific Division Mini-Conference on Scientific Images, March 2006.

Barberousse, A. and Ludwig, P. (2000), "Les modèles comme fictions", *Philosophie* 68: 16–43.

Callender, C. and Cohen, J. (2006), "There Is No Special Problem About Scientific Representation", *Theoria* 55: 7–25.

Cartwright, N. (1983), *How the Laws of Physics Lie.* Oxford: Oxford University Press.

Contessa, G. (2010), "Scientific Models as Fictional Objects", *Synthese* 172: 215–229.

Fine, A. (1998), "Fictionalism", in E. Craig (ed.), *Routledge Encyclopedia of Philosophy*, URL = <http://www.rep.routledge.com/article/Q035>. Retrieved December 05, 2007.

Frigg, R. (2006a), "Scientific Representation and the Semantic View of Theories", *Theoria* 55: 49–65.

Frigg, R. (2006b), "Scientific Models", with Stephan Hartmann, in S. Sarkar and J. Pfeifer (eds.), *The Philosophy of Science: An Encyclopedia.* New York: Routledge, 740–749.

Frigg, R. (2010) "Models and Fiction", *Synthese* 172: 251–268.

Giere, R. (1988), *Explaining Science.* Chicago: Chicago University Press.

Giere, R. (1999), *Science Without Laws.* Chicago: Chicago University Press.

Giere, R. (2004), "How Models are Used to Represent Reality", *Philosophy of Science* 71: S742–S752.

Godfrey-Smith, P. (2006), "The Strategy of Model-Based Science", *Biology and Philosophy* 21: 725–740.

Goldman, A. (2003), "Representation in Art", in J. Levinson (ed.), *The Oxford Handbook of Aesthetics*, Oxford: Oxford University Press, 192–210.

Goodman, N. (1976), *Languages of Art.* Indianapolis: Hackett.

Hughes, R. I. G. (1997), "Models and Representation", in *PSA 1996*, vol. 2. East Lansing, MI: Philosophy of Science Association, S325–S336.

Lamarque, P. (2003), "Fiction", in J. Levinson (ed.), *The Oxford Handbook of Aesthetics.* Oxford: Oxford University Press, 377–391.

Morgan, M. and Morrison, M. (eds.) (1999), *Models as Mediators. Perspectives on Natural and Social Science.* Cambridge: Cambridge University Press.

Suárez, M. (1999), "Theories, Models, and Representations", in L. Magnani, N. J. Nersessiand and P. Thagard (eds.), *Model-Based Reasoning and Scientific Discovery.* New York: Kluwer/Plenum, 75–83.

Suárez, M. (2003), "Scientific representation: against similarity and isomorphism", *International Studies in the Philosophy of Science* 17: 225–244.

Suárez, M. (2004), "An Inferential Conception of Scientific Representation", *Philosophy of Science* 71: S767–S779.

Thomson-Jones, M. (2007), "Missing Systems and the Face Value Practice", URL = < http://philsci-archive.pitt.edu/archive/00003519 >.

Thomson-Jones, M. (2010), "Missing Systems and the Face Value Practice", *Synthese* 172: 283–299.

Toon, A. (2010), "The Ontology of Theoretical Modelling: Models as Make-Believe", *Synthese* 172: 301–315.

Walton, K. (1990), *Mimesis as Make-Believe: on the Foundations of the Representational Arts.* Cambridge, MA: Harvard University Press.

Wells, H. G. (2005), *The War of the Worlds.* London: Penguin (1st ed., 1898).

Fiction and Scientific Representation

Roman Frigg

Introduction

Scientific discourse is rife with passages that appear to be ordinary descriptions of systems of interest in a particular discipline. Equally, the pages of textbooks and journals are filled with discussions of the properties and the behavior of those systems. Students of mechanics investigate at length the dynamical properties of a system consisting of two or three spinning spheres with homogenous mass distributions gravitationally interacting only with each other. Population biologists study the evolution of one species procreating at a constant rate in an isolated ecosystem. And when studying the exchange of goods, economists consider a situation in which there are only two goods, two perfectly rational agents, no restrictions on available information, no transaction costs, no money, and dealings are done immediately. Their surface structure notwithstanding, no competent scientist would mistake descriptions of such systems as descriptions of an *actual* system: we know very well that there are no such systems. These descriptions are descriptions of a *model-system*, and scientists use model-systems to *represent* parts or aspects of the world they are interested in. Following common practice, I refer to those parts or aspects as *target-systems*.

What are we to make of this? Is discourse about such models merely a picturesque and ultimately dispensable *façon de parler*? This was the view of some early twentieth century philosophers. Duhem (1906) famously guarded against confusing model building with scientific theorizing and argued that model building has no real place in science, beyond a minor heuristic role. The aim of science was, instead, to construct theories, with theories understood as classificatory or representative structures systematically presented and formulated in precise symbolic

R. Frigg (✉)
London School of Economics, London, UK
e-mail: r.p.frigg@lse.ac.uk

R. Frigg, M.C. Hunter (eds.), *Beyond Mimesis and Convention*, Boston Studies in the Philosophy of Science 262, DOI 10.1007/978-90-481-3851-7_6,

language. With some modifications this view also become dominant among the logical positivists of the Vienna Circle and the Berlin Group; see, for instance, Carnap (1938) and Hempel (1965).

Early resistance against this understanding of science came from Campbell (1920) and Hesse (1963), who emphasized the importance of models to scientific theorizing. The tides changed in the 1970s and 1980s. On the one hand the positivist view that theories were partially interpreted logical calculi (now referred to as the "syntactic view of theories") was replaced by the so-called semantic view of theories, according to which a theory simply *is* a collection of models; see Suppe (1977). Parallel, but by and large unrelated to the rise of the semantic view, a tradition of philosophy of science arose that emphasized the importance of scientific practice to philosophical analysis, and so places models again at the heart of a philosophical account of science; see the essays collected in Morgan and Morrison (1999). Hence, current philosophies of science of all stripes agree with a characterization of science as an activity aiming at representing parts of the world with the aid of scientific models.

For this reason the questions of what scientific modes are and how they represent have become central to the concerns of philosophers of science. This chapter proposes a novel approach to the issue of models and representation, one that draws essentially on the analogy between models and literary fiction. But before we can sketch the outlines of this account, some setting up is needed.

As the above examples show, when presenting a model scientists offer us the description of a hypothetical system, one that does not actually exist in nature, which they proffer as an object of study.[1] Scientists sometimes express this fact by saying that they talk about "model-land"; see for instance Smith (2007, 135). The rationale for doing so is that this hypothetical system has two desirable properties. First, it is chosen such that it is easier to study than the target-system and therefore allows us to derive results. Second, it is assumed to represent its target system, and representation is something like a "licence to draw inferences". Representation allows us to "carry over" results obtained in the model to the target-system and hence it enables us to learn something about that system by studying the model.

Thus, scientists actually perform two acts when they propose a model: they introduce a hypothetical system as the object of study, and they claim that this system is a representation of a target-system of interest. This is reflected in the ambiguous usage of the term "model" in the sciences. On the one hand "model" is often used to denote the hypothetical system we study (e.g., when we say that the model consists of two spheres). On the other hand it is employed to indicate that a certain system represents, or stands for, another system (e.g., when we observe that the Newtonian

[1] Some scientific models are material objects (for instance the wood models of a car that we put into a wind tunnel), but most models are not of this kind. I here focus on models that are, in Hacking's (1983, 216) words, "something you hold in your head rather than your hands".

model of the solar system misrepresents its target in various ways). In practice, however, these two acts are often carried out in tandem and scientists therefore rarely, if ever, clearly distinguish the two.

While this may well be a legitimate way of proceeding efficiently in the heat of battle, it is detrimental to philosophical analysis where it is germane that these two acts be kept separate. In this chapter I endeavor to clearly separate these two acts and to present an analysis of each. To this end, let me first introduce some terminology. I use the term "model-system" to denote the hypothetical system proffered as an object of study. I call those descriptions that are used to introduce the model-system as "model-descriptions". Representation then is the relation between a model-system and its target-system. The term "model" could refer to either the model-system or representation, or the combination of the two, or yet other things; I will therefore avoid it in what follows. I use the term "modeling" to refer to the practice of devising, describing and using a model-system. In this more regimented language, the two acts performed in utterances of the kind mentioned above are, first, presenting a model-system and specifying some of its essential properties, and, second, endowing this model-system with representational power.

This separation may do some violence to common sense, which regards representational power as an intrinsic property of things that are models and sees this dissociation of model-systems from representation as artificial at best. Common sense is wrong. It has been pointed out variously—and in my view correctly—that, in principle, anything can be a representation of anything else.[2] Representations are not a distinctive *ontological* category and it is wrong to believe that some objects are, *intrinsically*, representations and other are not. It is one question to ask what an object is in itself; but it is quite a different one to ask what, if anything, an object represents and in what way. Taking model-systems to be intrinsically representational is a fundamental mistake. Model-systems, first and foremost are objects of sorts, which can, and de facto often are, used as representations of a target-system. But the intrinsic nature of a model-system does not depend on whether or not it is so used: representation is extrinsic to the medium doing the representing.

Hence, understanding scientific modeling can be divided into two sub-projects: analyzing what model-systems are, and understanding how they are used to represent something beyond themselves. The first is a prerequisite for the second: we can only start analyzing how representation works once we understand the intrinsic character of the vehicle that does the representing. Coming to terms with this issue is the project of the first half of this chapter. My central contention is that models are akin to places and characters of literary fictions, and that therefore theories of fiction play an essential role in explaining the nature of model-systems. This sets the agenda. Section "Model-Systems and Fiction" provides a statement of this view, which I label the *fiction view of model-systems*, and argues for its *prima facie* plausibility. Section "Strictures on Structures" presents a defense of this view against its

---

[2]The point is Goodman's (1976); in recent years Teller (2001), Giere (2004) and Callender and Cohen (2006) have discussed it with special focus on scientific representation.

main rival, the structuralist conception of models. In section "Model-Systems and Imagination" I develop an account of model-systems as imagined objects on the basis of the so-called pretense theory of fiction. This theory needs to be discussed in some detail for two reasons. First, developing an acceptable account of imagined objects is mandatory to make the fiction view acceptable, and I will show that the pretense theory has the resources to achieve this goal. Second, the term "representation" is ambiguous; in fact, there are two very different relations that are commonly called "representation" and a conflation between the two is the root of some of the problems that (allegedly) beset scientific representation. Pretense theory provides us with the conceptual resources to articulate these two different forms of representation, which I call p-representation and t-representation respectively. Putting these elements together provides us with a coherent overall picture of scientific modeling, which I develop in section "The Anatomy of Scientific Modeling".

While p-representation turns out to be internal to pretense theory (and hence is explained by pretense theory itself), an analysis of t-representation has to draw on different resources. This resource is maps. In section "A First Stab at T-Representation" I present an analysis of how maps represent their target systems and claim that the general structure of this account doubles as the general structure of t-representation. In other words, the view that I am proposing is that one can think of the model-system as a kind of a "generalized map" and explain how it represents (t-represents) its target along the lines of how maps represent their targets. In section "Re-reading the Newtonian Model of the Sun–Earth System" I use this view to analyze the Newtonian model of the solar system and show that it not only gives a plausible understanding of what happens in this model, but even makes important features of it visible that are usually concealed. Far from being an idle philosophical pastime, the fiction view of models, I claim in conclusion, can actually help us to better understand what is involved in the representational activities essential to scientific models.

Model-Systems and Fiction

What kind of things are model-systems? Referring to them as "model-systems" has a homely ring to it which obscures the fact that we don't know what they are. As we have seen, the descriptions in question are not descriptions of any actual system. So what, if anything, are they descriptions of? What sense can we make of the common practice to qualify claims about such systems as true or false? And how do we find out about the truth and falsity of such claims?

My answers to these questions take as their starting point the realization that model-systems share important aspects in common with literary fiction. This is more than just an interesting but eventually inconsequential observation. My claim is that thinking about model-systems as being akin to characters and places in literary fiction provides essential clues to solving pressing problems in the philosophy of science. In other words, drawing an analogy between scientific modeling and literary fiction is not idle musing; it is the driving force behind an approach to scientific modeling that aims to provide an understanding of a central aspect of scientific practice.

The core of the fiction view of model-systems is the claim that model-systems are akin to places and characters in literary fiction. When modeling the solar system as consisting of ten perfectly spherical spinning tops physicists describe (and *take themselves* to be describing) an imaginary physical system; when considering an ecosystem with only one species biologists describe an imaginary population; and when investigating an economy without money and transaction costs economists describe an imaginary economy. These imaginary scenarios are tellingly like the places and characters in works of fiction like Madame Bovary and Sherlock Holmes. These are scenarios we can talk about and make claims about, yet they don't exist.

Although hardly at the center of attention, the parallels between certain aspects of science and literary fiction have not gone unnoticed. It has been mentioned by Maxwell, and occupied center stage in Vaihinger's (1911) philosophy of the "as if". In more recent years, the parallel has also been drawn specifically between models and fiction. Cartwright observes that "a model is a work of fiction" (1983, 153) and later suggests an analysis of models as fables (1999, Chapter 2). McCloskey (1990) regards economists as "tellers of stories and makers of poems". Fine notes that modeling natural phenomena in every area of science involves fictions in Vaihinger's sense (1993, 16), and Sklar highlights that describing system "as if" they were systems of some other kind is a royal route to success (2000, 71). Elgin (1996, Chapter 6) argues that science shares important epistemic practices with artistic fiction. Hartmann (1999) and Morgan (2001) emphasize that stories and narratives play an important role in models, and Morgan (2004) stresses the importance of imagination in model building. Sugden (2000) points out that economic models describe "counterfactual worlds" constructed by the modeler. I have defended the view that models are imaginary objects in my (2003) and my (2009), and Grüne-Yanoff and Schweinzer (2008) emphasize the importance of stories in the application of game theory.[3] Moreover, Godfrey-Smith (2006) has recently set out what amounts to the most explicit and forceful statement of the fiction view of model-systems now available.

What we have to recognize, though, is that the analogy between model-systems and fiction is only a starting point. If put forward without further qualifications, explaining model-systems in terms of fictional characters amounts to explaining the unclear by the obscure. In fact, fictional entities are beset with philosophical problems so severe that avoiding fictional entities altogether would appear to be a better strategy. Fictional entities do not exist: there is no woman called Emma Bovary and there is no detective Sherlock Holmes. Yet they have some kind of reality: we think about them, we talk about them,

[3] Giere (1988, Chapter 3) argues that models are "abstract entities", which could be also interpreted as a fiction based view of models. However, in personal communication he pointed out to me that this is not his intended view.

and they are objects of our emotions. Fictional entities are the subject matter of discussions, and claims about them can be true or false: we say that it is true that Holmes is a detective but false that he is a ballet dancer. How can this be if there is no Holmes? And how can sentences containing the name "Holmes" even be meaningful if Holmes does not exist? It seems that the sentence would then be about nothing, and yet we qualify such sentences as true or false. On what grounds do we do this?

These and other related concerns have led many philosophers to dismiss fictional entities. So how is appeal to something as problematic and obscure as fictional entities going to help us work through the thorny problem of scientific representation? Before turning to the details of the account that I favor (section "Model-Systems and Imagination"), I want to mention four reasons for believing that thinking about modeling in this way is helpful.

First, works of fiction characteristically do not portray actual states of affairs. The names of persons and objects in literary fiction characteristically do not denote real persons or objects, and there is nothing in the world of which the text of a novel is a true description.[4] Nevertheless, fictional discourse is genuinely meaningful: readers neither make a mistake, nor are they under an illusion when they believe that they understand the content of a novel. Yet, at the same time they are fully aware that the sentences they read when engaging with a work of fiction do not describe anything in the actual world. The same is true of modeling discourse in science. As we have seen above, scientific discourse abounds with descriptions that are meaningful yet fail to be plain descriptions of physical systems from the domain of enquiry of the scientific discipline in question.

Second, we can truly say that in David Lodge's *Changing Places* Morris Zapp is a professor of English literature at the State University of Euphoria. We can also truly say that in the novel he has a heart and a liver, but we cannot truly say that he is a ballet dancer or a violin player. Only the first of these claims is part of the explicit content of the novel, yet there is a matter of the fact about what is the case "in the world of the story" even when claims go beyond what is explicitly stated. Whether or not claims about a story's content are correct is—somehow—determined by the text without being part of its explicit content. Such determinations are not merely decided by each reader on a whim. The situation with model-systems is the same. Model-descriptions usually only specify a handful of essential properties, but it is understood that the model-system has properties other than the ones mentioned in the description. Model-systems are interesting exactly because more is true of them than what the initial description specifies; no one would spend time studying model-systems if all there was to know about them was the explicit content of the initial description. It is, for instance, true that the Newtonian model-system representing the solar system is stable and that the model-earth moves in an elliptic orbit; but none of this is part of the explicit content of the model-system's original specification.

---

[4]This is not meant to be a *definition* of fiction. A failure of reference, although typical for fiction, is neither necessary nor sufficient for a text to qualify as fiction. I come back to this point later on.

Third, a fictional story not only has content that goes beyond what is explicitly stated, we also have the means to learn about this "extra content" by using certain (usually implicit) rules of inference. It is an integral part of our response to fiction that we supplement the explicit content and fill in facts about the plot even where the text is silent. In fact, a good part of the intellectual pleasure we get from reading a novel derives from this imaginative "filling in" of the "missing content". The same goes for model-systems. Finding out what is true in a model-system beyond what is explicitly specified in the relevant description is a crucial aspect of our engagement with the system. In fact the bulk of the work that is done with a model-system is usually expended on establishing whether or not certain claims about it hold true. Is the solar system stable? Do the populations of predators and prey reach some equilibrium? Do prices stabilize? These are questions we want to answer given what we know about the model and certain other rules we regard as valid in the context in which the model-system is discussed.

Fourth, sometimes we read just for pleasure, but in particular when we read serious literature we often engage in comparisons between the characters and situations in the fiction and real situations and characters with which we are familiar. We recognize aspects of the protagonist's behavior in someone we know and suddenly begin to understand some of his behavioral patterns: we learn about the world by reading fiction. Again, this has parallels in the context of modeling, where we learn from models about the world. Once we think about models as fictions this parallel becomes salient and urges us to think about how "knowledge transfer" from a fictional scenario to the real world takes place.

Needless to say, this list of communalities between scientific modeling and literary fiction is neither complete, nor should it be understood as suggesting that there are no important differences between the two. The purpose of this list is to make it plausible that thinking about models as alike to literary fiction is a fruitful point of departure.

In the next section I defend this conception of model-systems against its structuralist rival. Those already convinced by the fiction view can skip this section without loss and continue with section "Model-Systems and Imagination" where I present a detailed formulation of the fiction view of models.

Strictures on Structures

Stop and rewind. Many will think that this discussion has taken a wrong turn right at the beginning and has gotten onto a path leading straight into a thicket of confusions. The wrong turn is to take talk about nonexistent systems seriously. Worse, trying to make good on this idea by working out a theory of fiction is a pilgrimage to the devil. Those whom I expect to issue such a verdict are those who hold the view that models are set-theoretic structures. This view originates with Suppes (1960) and is now held by many, among them van Fraassen (1980, 1997, 2002), Da Costa and French (1990), and French and Ladyman (1997).

At the core of this approach to models lies the notion that models are structures. A structure (sometimes "mathematical structure" or "set-theoretic structure") S is a composite entity consisting of a non-empty set U of individuals called the domain (or universe) of the structure S and a non-empty indexed set R of relations on U. Often it is convenient to write these as an ordered triple: $S=[U, R]$.[5]

For what follows it is important to be clear on what we mean by "individual" and "relation" in this context. To define the domain of a structure it does not matter what the individuals are—they may be whatever. The only thing that matters from a structural point of view is that there are so and so many of them. Or to put it another way, all we need is dummies or placeholders. Relations are understood in a similarly "deflationary" way. It is not important what the relation "in itself" is; all that matters is between which objects it holds. For this reason, a relation is specified purely extensionally, that is, as class of ordered n-tuples and the relation is assumed to be nothing over and above this class of ordered tuples. Thus understood, relations have no properties other than those that derive from this extensional characterization, such as transitivity, reflexivity, symmetry, etc. This leaves us with a notion of structure containing dummy-objects between which purely extensionally defined relations hold.[6]

Let us illustrate this with a simple example. Consider $S_t = [U = (a, b, c), R = (\langle a, b\rangle, \langle b, c\rangle, \langle a, c\rangle)]$, a structure consisting of a three object domain (with the objects a, b, and c) endowed with a transitive relation R, (where "$\langle a, b\rangle$" is an ordered tuple expressing that R holds between a and b).[7] In fact, the formula in the previous sentence is all we need in order to completely define the structure. It does not matter what they objects are: their materiality is immaterial. It doesn't matter whether they are books, railway bridges, or supernovae—all that is needed is that they are objects. In the same way it does not matter whether the relation R is "greater than" or "older than" or "more appreciated than"—all that matters is that R holds between a and b, and b and c, and a and c, no matter what R "in itself" is.

A view that takes model-systems in science to be structures in this sense is too austere to serve as a basis for an account of scientific modeling. Although structures do play an important role in scientific modeling, model-systems cannot be *identified* with structures. What is missing in the structuralist conception is an analysis of the "material" character of model-systems: even perfectly spherical planets are taken to have mass, populations are taken to consist of rabbits and foxes, etc. The view of model-systems that I advocate regards model-systems as imagined physical

---

[5]Sometimes structures are defined so that they also include operations. Although convenient in some contexts, this is unnecessary because ultimately operations reduce to relations (Boolos and Jeffrey 1989, 98–99).

[6]See Russell (1919, 60) for clear account of this feature of structures.

[7]A relation is transitive iff it is true that whenever the relation holds between objects a and b, and between b and c, then it also holds between a and c. Examples for transitive relations are *more expensive than* and *taller than*; and example for a non-transitive relation is *liking* (since it may well be that a likes b, and b likes c, but a does not like c at all).

systems, i.e., as hypothetical entities that, as a matter of fact, do not exist spatio-temporally but nevertheless have non-structural properties in the same way in which literary characters do. I will explain below in detail how to understand this claim and address the problems that it faces. The aim of this section is to argue that this is the right way of thinking about model-systems.

There are several reasons to prefer this take on model-systems over the structuralist account. The first is the evidence from scientific practice: scientists often talk about model-systems as if they were physical things. Newton, when introducing his model of the planetary system, did not present a mathematical structure. Rather he described a hypothetical situation in which one sphere orbits around another sphere in the absence of confounding factors. This way of thinking about model-systems is typical in mechanics as well as many branches of physics. And the same is true in biology. Godfrey-Smith (2006, 736–738) points out that Levins' work on population biology—as well as the models of Maynard Smith and Szathmáry's in evolutionary theory, and hence most of the work in their respective fields—is best understood as describing imagined concrete populations. Further, Godfrey-Smith adds that this way of looking at model-systems in these fields is integral to the discovery of novel phenomena and to making sense of the treatment of certain issues (e.g., the discussion of robustness in Levins), as well as to the communication of the results in books and papers, even where the models make essential use of mathematical techniques.

Closely related to this point is the fact that the fictional scenario plays a crucial role in understanding how a model relates to reality. This is best illustrated with a simple example from population dynamics.[8] Imagine you have a newborn pair of rabbits, one male the other female, and you also have a large garden which is their habitat. You then want to know how many pairs of rabbits you will have at some later time, and so you turn to a text on population dynamics where you find a simple model (going back to Leonardo of Pisa, also known by his nickname "Fibonacci"). The model tells you that the population at time t_n equals the population at time t_{n-1} plus the population at time t_{n-2}. According to the model, then, we have $P(t_n) = P(t_{n-1}) + P(t_{n-2})$, where $P(t_n)$ is the population at time t_n and where the distance between two instants of time is the time rabbits need to mature and breed (the numbers $P(t_n)$ are known as "Fibonacci numbers").[9] Let us assume this time is 1 month. Thus, the model tells us that if we start with one young pair, we have five pairs after 5 months, eight pairs after 6 months, thirteen pairs after 7 months, and so on.

If you are now getting excited because you figure that your rabbit population will grow really fast (after 10 months you already have 55 pairs according to the model), you will be disappointed. Quite soon the real number of rabbit pairs will start diverging dramatically from the value the model predicts. This may take you by surprise,

[8]For a discussion of this example see Smith (2007, 24–29).

[9]Strictly speaking this is not a structural formulation of the model, but a structural version could easily be constructed from the equation defining the Fibonacci numbers. However, since such a construction requires some setting up and nothing in my conclusion depends on having such a formulation, I will not dwell on this point here.

but it should not if you understand the *entire* model. The above equation is not about rabbits per se; it is about rabbits that never die, a garden that is infinitely large and contains enough food for any number of rabbits, and rabbits that procreate at a constant rate at constant speed. This is not by any standards an accurate description of the real situation; it is a fictional scenario and $P(t_n) = P(t_{n-1}) + P(t_{n-2})$ is true of this scenario. It is crucial to appreciate this fact if we want to know under what circumstances and to what extent conclusions derived in the model can be expected to bear out in the real system. Real rabbits don't live forever, but they live for some years; the garden is not infinite but large enough to provide food and shelter for about one hundred pairs; etc. So we come to the conclusion that model is probably good for about the first 9 or 10 months and then starts breaking down. This is important to know when using the model, but—and this is the crucial point—there is nothing in the mathematics that tells you any of this! What makes you understand the how the model relates to the world and when and where you can reasonably use it is a comparison between the fictional scenario and the real world. So the fictional scenario is an integral component of the model, and one that cannot be eliminated and replaced by structures.

Some might now reply that the fictional scenario merely plays a pragmatic role in our use of the model (whatever that means) and can therefore be eliminated in a final formulation of the model. I disagree because, as I have just outlined, the fictional scenario is essential to the functioning of the model. But irrespective of how this issue is resolved, the structuralist conception of models faces further difficulties when we think about how a model comes to be a representation of a target-system.

A structure per se is not about anything at all, let alone about a particular target-system; they are pieces of pure mathematics, devoid of empirical content. But a representation must posses "semantic content" or "aboutness"; that is, it must stand for something else. Those who take model-systems to be structures suggest connecting structures to target-systems by setting up an isomorphism between model-system and target.[10] Two structures $S = [U, R]$ and $S_T = [U_T, R_T]$ are isomorphic iff there exists an isomorphism between them. An isomorphism is a mapping $f: U_T \rightarrow U$ such that f is one-to-one (bijective) and it preserves the system of relations in the following sense: the elements $a_1, \ldots, a_n$ of S_T satisfy the relation R^T iff the corresponding elements $b_1 = f(a_1), \ldots, b_n = f(a_n)$ in S satisfy R, where R is the relation in S corresponding to R^T.

This definition of isomorphism brings a predicament to the fore: an isomorphism holds between two structures and not between a structure and a part of the world per se. In order to make sense of the notion that there is an isomorphism between a model-system and its target-system, we have to assume that the target exemplifies a particular structure. The problem is that this cannot be had without bringing non-structural features into play.

---

[10]Other suggestions include partial isomorphism, homomorphism, and embedding—nothing in what follows depends on which one of these one chooses.

The argument for this claim proceeds in two steps (Frigg 2006, 55–56). The first is to realize that *possessing structure S* (where *S* is some particular structure) is a concept that does not apply unless some more concrete concepts apply as well. Hence we cannot say that a target-system has structure *S* unless we also say that it has certain more concrete properties as well. Let us make this more precise with the notion of one concept being more abstract than another concept.

Concept a is more abstract than concept b iff b belongs to a class B of concepts (and $a \notin B$) such that[11]

(i) for a to apply it is necessary that at least one $b' \in B$ applies, and,
(ii) on any given occasion, the fact that $b' \in B$ applies is what the applying of a on that occasion consists in.

In other words, the concepts in B are used to "fit out" the abstract concept a on any given occasion. *Working*, for instance, is more abstract in this sense than *writing a letter* or *attending a meeting*. Condition (i) says that for it to be the case that I am working, I either have to write a letter, attend a meeting, or ...; if I don't do any of these, then I am not working. Condition (ii) says that my working on a given occasion consists in, say, writing a letter. If I complain to someone that I have been writing letters all day, and he then replies "OK, but when did you work?" he is either making a joke or does not get the point (namely that writing letters is working). In other words, the two conditions say that there is no such thing as working and only working.

Having structure S is like *working* in that it needs fitting out on every occasion in which it applies. It follows from the definition of a structure that for something to have structure *S* it has to be the case that *being an object* must apply to some of its parts, and *standing in a relation R* (where *R* is one of the relations of *S*) must apply to these. These concepts are abstract relative to more concrete concepts. Let us take relations first. Recall that relations are defined purely extensionally and hence have nothing but logico-mathematical properties such as transitivity. Consider, then, *standing in a transitive relation*. There are many transitive relations: *taller than*, *older than*, *hotter than*, *heavier than*, *stronger than*, *more expensive than*, *more recent than* (and their respective converses: *smaller than*, *younger than*, etc.), and with a little ingenuity one can extend this list ad libitum. By itself, there is nothing worrying about that. However, what we have to realize is that *standing in a transitive relation* applies to two objects only if either *greater than*, or *older than*, or ... applies to them as well. We cannot have the former without the latter: something cannot be a transitive relation without also being one of the above listed relations. *Being taller than*, say, is what *being a transitive relation* consists in on a particular occasion. So *standing in a transitive relation* is abstract relative to more concrete concepts like *being hotter than* and hence there simply is no such thing in the physical world as a relation that is nothing but transitive.

---

[11] This definition is adapted from Cartwright (1999, 39).

Similarly for objects. What is needed for something to be an object is not an easy question, and an answer depends on the relevant context as well as the kinds of things we are dealing with (medium size physical objects like tables, social entities such as families, etc.). But nothing in the world is such that the only property it possesses is "objectness"; whatever the circumstances, some other concepts must apply to it for it to be the case that it is an object. For instance, a medium size physical object has an identifiable shape which sets it off from the environment, which implies that it is colored, has a certain texture, etc. If none of this was the case, we just would not have a medium size physical object.

The crucial point in all this is that the more concrete concepts that are needed to ground structural claims are not structural themselves. *Being a transitive relation* is structural, *being taller than* is not, as becomes clear from has been said about structures above. In other words, structural claims ride on the back of non-structural claims.

This by itself would not have to worry the structuralist who claims that model-systems are structures. He could point out that although, as the above argument shows, structures are grounded in something else (which is non-structural), it is the structural features of reality that models relate to and that therefore models are structures. The problem with this response—and this is the second step of the argument—becomes apparent when we realize that the descriptions we choose to fit out abstract structural claims almost never are true descriptions of the target systems. The above examples make this sufficiently clear. The structure on which the formal treatment of the solar system is based is not fitted out by a realistic description of the solar system, but by a description that takes planets to be ideal spheres with homogenous mass distributions gravitationally interacting only with each other and nothing else. Similarly, the structure on which the calculations of the population sizes is based does not attach to a realistic description of animal life and so on. So the structural claims that give rise to the equations that we study when dealing with a problem at hand (at least in the overwhelming majority of cases) are not true descriptions of the target system, and hence the target does not have the structure at stake.[12]

Hence, taken literally, descriptions that ground structural claims (almost always) fail to be descriptions of the intended target system. Instead, they describe a hypothetical system which is distinct from the target system. This has unfortunate consequences for the structuralist. If the descriptions employed to attribute a structure to a target system were just plain descriptions of that system, then the claim that model-systems are just structures would appear at least prima facie plausible. But once we acknowledge that these descriptions describe hypothetical systems rather

[12]This is what Downes has in mind when he says that there is no empirical system corresponding to the equation of the ideal pendulum (1992, 145), and what Thomson-Jones (2007) emphasizes when he points out that science is full of "descriptions of missing systems"; in a different way the same point is also made by Cartwright (1983, Chapter 7) who emphasizes that we have to come up with a "prepared description" of the system in order to make it amenable to mathematical treatment.

than real target systems, we also have to acknowledge that hypothetical systems are an important part of the theoretical apparatus we employ, and that they therefore have to be included in our analysis of how scientific modeling works. This can, of course, be done in different ways. My suggestion is that these hypothetical systems in fact *are* the models-systems. I therefore I reserve the term "model-system" for the hypothetical physical entities described by the descriptions we use to ground structural claims; I refer to the relevant structures as "model structures". This facilitates the analysis in what follows, but ultimately nothing hangs on this choice; one could just as well say that model-systems are composite entities consisting of a hypothetical and a structural system. What does matter, however, is that we acknowledge that scientific modeling indeed involves such hypothetical systems.[13]

At least some proponents of structuralist conception will reject this argument.[14] The bone of contention is what model-systems represent. So far I have assumed that a model-system represents a piece of the real world, for instance the solar system or a population of rabbits. This, so the objection goes, is the wrong point of departure since models don't represent systems in this sense. What a model-system ultimately represents is a *data model*, not an object of some sort. Data are what we gather in experiments. When observing the motion of the moon, we choose a coordinate system and observe the position of the moon in this coordinate system at consecutive instants of time. We then write down these observations. This can be done in different ways. We can simply write a list with the coordinates of the moon at certain instants of time; we can draw a graph consisting of various points standing for the position of the moon at different times; or we can choose yet another form of taking down the data. The data thus gathered are called the *raw* data. The raw data then undergo a process of cleansing, rectification and regimentation: we throw away data points that are obviously faulty, take into consideration what the measurement errors are, take averages, etc. Often (but not always) the aim of this process is to fit a smooth curve through the various data points so that the curve satisfies certain theoretical desiderata (such as having minimal least-square-distance from the actual data points). The end result of this process is a so-called *data model*.

[13]One could try to avoid the commitment to hypothetical systems by renouncing a literal understanding of the relevant descriptions and arguing that it does not follow from the fact that descriptions are poor or highly idealized that they are not descriptions of the target at all; it just means that they are idealized descriptions. This move is of no avail. Being an idealized description is not a primitive concept and it calls for analysis. On the most plausible analysis, D is an approximate description of object O iff what D literally describes is in some relevant sense an idealization of O. But what D literally describes is a hypothetical system, and so we find ourselves back where we started.

[14]The German structuralists explicitly acknowledge the need for a concrete description of the target-system (Balzer et al. 1987, 37–38). Moreover, they consider these "informal descriptions" to be "internal" to the theory. Unfortunately they do not say more about this issue. Nevertheless, it is important to emphasize that there is no conflict between structuralism thus construed and the view developed in this chapter; in fact they can be seen as complementary.

The claim then is that model-systems do not represent parts of the world (like the earth and the sun), but rather data-models that have been constructed from observations made on these parts of the world. So what a model of the motion of planet earth is about is not the earth itself, but the smooth curve that we have fitted through the data gained when observing the motion of the earth. In this vein van Fraassen declares that "... the theoretical models (proffered ... as candidates for the representation of the phenomena) are confronted by the data models. ... to fit those data models is ultimately the bottom line" (2002, 164).[15] In brief, the suggestion is that representation be explicated in terms of setting up an isomorphism between the model-system (on this view a structure) and the data model. This move indeed renders the above argument obsolete since data models are mathematical entities and as such can be considered to have a well-defined structure.[16]

This suggestion is wrong because it is descriptively inadequate: it is not the case that models represent data. This point is not new. It has been argued by Bogen and Woodward (1988) and Woodward (1989), and has recently been reiterated in different guise by Teller (2001).[17] In essence I agree with these authors; however, my focus differs slightly from theirs and I present the subject matter in a way that suits my needs.

In nuce, Bogen's and Woodward's point is that science is not about data; it is about phenomena. A theory about the melting point of lead is not about the data we gather when we find out at what temperature lead melts; it is about the melting of lead itself. This carries over to models: models do not represent data. In fact, most models do not per se contain anything that could be directly compared to data we gather; or more specifically, they do not involve structures that could plausibly be thought of as being isomorphic to a data model.

Let me illustrate this with an example from Bogen and Woodward: the discovery of weak neutral currents (1988, 315–318). What the model at stake consists of is particles: neutrinos, nucleons, the Z^0, and so on, along with the reactions that take place between them.[18] Nothing of that, however, shows in the relevant data. What was produced at CERN in Geneva were 290,000 bubble chamber photographs of which roughly one hundred were considered to provide evidence for the existence of neutral currents. The notable point in this story is that there is no part of the

[15] See also van Fraassen (1980, 64, 1989, 229, 1997, 524) and French (1999, 191–192).

[16] There is an exegetic question here. Although structuralists certainly suggest that representation *is* data matching, they never explicitly say so. I here explore the stronger version of the view on which representation indeed *consists in* data matching since the weaker version, on which data matching is distinct from representation, does not provide a viable criticism of the above argument from abstractness.

[17] McAllister (1997) presents an antirealist critique of Bogen and Woodward. But his concern is orthogonal to mine: even if one construes phenomena in an antirealist way they turn out to be more than just data.

[18] The model I am talking about here is not the so-called standard model of elementary particles as a whole. Rather, what I have in mind is one specific model about the interaction of certain particles of the kind one would find in a theoretical paper on this experiment.

model (which quantum field theory provides us with) that could be claimed to be isomorphic to these photographs (or any data model one might want to construct on the basis of these). It is weak neutral currents that occur in the model, not any sort of data we gather in an experiment.

This is not to say that these data have nothing to do with the model. The model posits a certain number of particles and informs us about the way in which they interact both with each other and with their environment. Using this we can place them in a certain experimental context. The data we then gather in an experiment are the product of the elements of the model and of the way in which they operate in a given context. Characteristically this context is one which we are able to control and about which we have reliable knowledge (e.g., knowledge about detectors, accelerators, photographic plates and so on). Using this and the model we can derive predictions about what the outcomes of an experiment will be. But, and this is the salient point, these predictions involve the entire experimental set-up and not only the model and there is nothing in the model itself with which one could compare the data. Hence, data are highly contextual and there is a gap between observable outcomes of experiments and anything one might call a substructure of a model of neutral currents.[19]

But what, then, is the significance of data, if they are not the kind of things that models represent? The answer to this question is that data perform an evidential function. That is, data play the role of evidence for the presence of certain phenomena. The fact that we find a certain pattern in a bubble chamber photograph is evidence for the existence of neutral currents, and for the fact that the model is a (more or less) faithful representation of what is happening in the world. Thus construed, we do not denigrate the importance of data to science, but we do not have to require that data have to be isomorphically embeddable into the model at stake.

In sum, understanding the fictional scenario of which the formal apparatus of a model is literally true is essential to understanding and using a model. Furthermore, one has to recognize that structures cannot be connected to anything in the world without the mediation of non-structural concepts, and attempts to bypass this conclusion by appeal to data models fail.

---

[19]To underwrite this claim consider the following example. Parallel to the research at CERN, the NAL in Chicago also performed an experiment to detect weak neutral currents. The data obtained in this experiment were quite different, however. They consisted of records of patterns of discharge in electronic particle detectors. Though the experiments at CERN and at NAL were totally different and the data gathered had nothing in common, they were meant to provide evidence for the same theoretical model. But the model does not contain any of these contextual factors. It posits certain particles and their interaction with other particles, not how detectors work or what readings they show. The model is not idiosyncratic to a special experimental context in the way the data are, and therefore it is not surprising that the model does not contain a substructure that could plausibly be claimed to be isomorphic to the data. The model represents an entity—weak neutral currents—and not data used in its discovery.

Model-Systems and Imagination

So far, I have proposed that model-systems are best understood as akin to characters and objects of literary fiction. However, as I have indicated above, fictional entities are beset with philosophical problems (see Friend (2007) for a discussion of these) and hence explaining models in terms of fiction hardly seems to be progress. Hence the burden of proof is on the side of the proponent of the fiction view, who has to show that there is a workable conception of fiction that serves the needs of a theory of scientific modeling. Developing such a view is the aim of this section.[20] This involves a lengthy discussion of philosophical subtleties that at first may seem peripheral to the concerns of scientific modeling. I appeal to the forbearance of the reader and promise that this effort is not in vain. For one, without a tenable conception of fiction the fictions view is without foundation, and the only way to prove that it stands firm is to explicitly formulate a tenable account of fiction. For another, one of the results of this excursion into the philosophical jungles of fiction is the distinction it allows us to draw between two different conceptions of representation, p-representation and t-representation. This distinction, I think, is crucial to understanding how scientific modeling works, and a failure to keep the two separate has led to considerable confusion.

What do we expect from an account of fiction in order for it to be able to serve as the foundation of the fiction view of model-systems? I think it has to provide responses to five questions (Q1–Q5) and to satisfy two meta-theoretical criteria (C1–C2). These questions and criteria are as follows:

(Q1) *Identity conditions*. When are two model-systems identical? This question is pressing because unlike in the context of literature, where we can point to canonical texts and authors' intentions, model-systems in science are often presented by different authors (in different papers or textbooks) in different ways. Nevertheless, many different descriptions are actually meant to describe the same model-system. Under what circumstances is that the case? That is, when are the model-systems specified by different descriptions identical?

(Q2) *Attribution of properties*. In the previous section I have argued that model-systems have "physical", "concrete", or "material" properties. As the scare-quotes indicate, there is something problematic about this claim. In fact, it has even been claimed that such statements are outright contradictory because abstract objects like the ideal pendulum cannot have the same properties as concrete physical systems (Hughes 1997, 330). How is it possible for a model-system to have "material" properties if model-systems do not exist in space and time? What sense can we make of statements like "the ball is charged" or "the population is isolated from its environment" if there are no balls and populations?

(Q3) *Comparative statements*. As we have seen above, comparing a model and its target-system is essential to many aspects of modeling, and it plays a crucial role in the account of representation developed below. We customarily say things like "real

---

[20] This section and the next are based on my (2010).

agents do not behave like the agents in the model" and "the surface of the real sun is unlike the surface of the model sun". How can we compare something that does not exist with something that does? Likewise, how are we to analyze statements that compare features of two model-systems with each other like "the agents in the first model are more rational than the agents in the second model"?

(Q4) *Truth in model-systems.* There is right and wrong in a discourse about model-systems. It is true that the population in Fibonacci's model never decreases and it is wrong that the earth in Newton's model moves on parabolic orbit. But on what basis are claims about a model-system qualified as true or false, in particular if the claims concern issues about which the description of the system remains silent? What we need is an account of truth in model-systems, which, first, explains what it means for a claim about a model-system to be true or false and which, second, draws the line between true and false statements at the right place (for instance, an account on which all statements about a model-systems come out false would be unacceptable).

(Q5) *Epistemology.* We do investigate model-systems and find out about them; truths about the model-system are not forever concealed from us. In fact, we engage with model-systems because we want to explore their properties. How do we do this? How do we find out about these truths and how do we justify our claims?

(C1) *Naturalism.* The account we offer in response to the above issues should be able to make sense of scientific practice. That is, it should be able to explain how scientists build models and how they reason about them.

(C2) *Metaphysical commitments.* The metaphysics of fictional entities is an issue fraught with controversy. For this reason we need to know what kind of commitments we incur when we understand model-systems along the lines of fiction, and how these commitments, if any, can be justified. However, it is not, in my view, a condition of adequacy that the account we propose be metaphysically parsimonious. As a matter of fact, the account I develop below eschews commitment to fictional entities, but this is accidental, as it were. To say it a different way, it just so happens that the theory that provides the most convincing answers to the above questions is also metaphysically parsimonious; but if it had turned out that a metaphysically substantial theory (i.e., one that is committed to fictional entities) had provided the best answers, then we should have chosen that theory. In other words, I think that accounts of fictional entities should not be dismissed merely on the grounds of being metaphysically "thick". That I dismiss such accounts has to do only with their failure to answer other questions in a satisfactory way.[21]

---

[21] For want of space I cannot discuss competing approaches. In a nutshell, their problems seem to be the following. The paraphrase account (Russell 1905) does not offer a workable theory of truth in fiction (Crittenden 1991, Chapter 1). The neo-Meinongean view (Parsons 1980) runs into difficulties with incompleteness (Howell 1979, Section 1) and as a consequence does not offer a satisfactory answer to (Q5). Finally, Lewis' (1978) account is too permissive about what counts as true in a fictional context (Currie 1990, Section 2.3, Lamarque and Olsen 1994, Chapter 4).

That said, it is the contention of this chapter that Kendall Walton's (1990) pretense theory of fiction best fits this bill.[22] In this section I provide a brief introduction to this theory and show how it answers (Q1)–(Q5) and (C1)–(C2). In the next section, "The Anatomy of Scientific Modeling", I formulate a general account of modeling on the basis of this discussion.

The point of departure of Walton's approach is the capacity of humans to imagine things.[23] Sometimes we imagine something without a particular reason. But there are cases in which our imagining something is prompted by the presence of a particular object, in which case this object is referred to as a "prop". "Object" has to be understood in the widest sense possible; anything capable of affecting our senses can serve as a prop. An object becomes a prop due to the imposition of a rule or "principle of generation" (1990, 38), prescribing what is to be imagined as a function of the presence of the object. If someone imagines something because he is encouraged to do so by the presence of a prop he is engaged in a game of make-believe. Someone who is involved in a game of make-believe is pretending; so "pretense" is just a shorthand way of describing participation in such a game (1990, 391) and has (in this context) nothing to do with deception (1990, 392). The simplest examples of games of make-believe are children's games (1990, 11). In one such game, stumps may be regarded as bears and a rope put around the stump may mean that the bear has been lassoed; or pointing the index finger at someone and saying "bang" may mean that the person has been shot.

A prop becomes a prompter if someone notices the prop and as a result starts engaging in a rule-guided imaginative activity. The set of prompters and the set of props overlap, but neither is a subset of the other. For one, a prop that is never perceived by anybody and hence never causes anybody to imagine something is not a prompter (but still a prop). For another, an object can prompt imaginations without being part of a game of make-believe (i.e., in the absence of rules of generation), for instance when we see faces in the clouds and imagine how these faces talk to each other. Even within a game we can make errors (e.g., mistakenly take a mole heap for a stump and then say that it is a bear), in which case the mole heap is a prompter (because it prompts imaginings) but it is not a prop (because there is not a rule).

Pretense theory considers a vast variety of different props ranging from novels to movies, from paintings to plays, and from music to children's games. In the present context I only discuss the case of literature. Works of literary fiction are, on the current account, regarded as props because they prompt the reader to imagine certain things. By doing so a fiction generates its own game of make-believe. This game can be played by a single player when reading the work, or by a group when someone tells the story to the others.

---

[22] Strictly speaking, Walton (1990) restricts the use of "pretense" to verbal (or more generally behavioral) participation, which does not include the activity of someone reading on his own. However, it has become customary to use "pretense" as synonymous with "make-believe" and I stick to this wider use in what follows.

[23] I here discuss pretense theory as it is presented by Walton (1990); Currie (1990) and Evans (1982, Chapter 10) develop different versions. Parenthetical references in the text of this and the following section are to Walton's book.

Some rules of generation are ad hoc, for instance when a group of children spontaneously imposes the rule that stumps are bears and play the game "catch the bear". Other rules are publicly agreed on and hence (at least relatively) stable. Games based on public rules are "authorized"; games involving ad hoc rules are "unauthorized".

By definition, a prop is a representation if it is a prop in an *authorized* game. On this view, then, stumps are not representations of bears because the rule to regard stumps as bears is an ad hoc rule that is neither shared by others in the society nor stable over time (stumps may not be props to other people and even the children playing the game now may regarded them as elephants on the next walk). However, *Hamlet* is a representation because everybody who understands English is invited to imagine its content, and this has been so since the work came into existence. Within pretense theory "representation" is used as a technical term. Representations are not, as is customary, explained in terms of their relation to something beyond themselves (e.g., resemblance or denotation); representations are things that possess the social function of serving as props in authorized games of make-believe (I will come back to this point below).

Props generate fictional truths by virtue of their features and principles of generation. Fictional truths can be generated directly or indirectly; directly generated truths are "primary" and indirectly generated truths are "implied" (1990, 140). Derivatively, one can call the principles of generation responsible for the generation of primary truths "principles of direct generation" and those responsible for implied truths "principles of indirect generation". The leading idea is that primary truths follow immediately from the prop, while implied ones result from the application of some rules of inference. When little Jimmy sees a stump and shouts "here is a bear" this is a direct truth because it follows from fact that there is a stump and the direct rule "stumps are bears", which is constitutive of the game. The boys may then stay away from the bear because they think the bear is dangerous and might hurt them. This fictional truth is inferred because it does not follow from the basic laws of the game that stumps are bears, but from the additional principle that bears in the game have the same properties as real bears.

The distinction between primary and inferred truths is also operative in literary fiction. The reader of *Changing Places* reads that Zapp "embarked ... on an ambitious critical project: a series of commentaries on Jane Austen which would work through the whole canon, one novel at a time, saying absolutely everything that could possibly be said about them". The reader is thereby invited to imagine the direct truth that Morris Zapp is working on such a project. She is also invited to imagine that Zapp is overconfident, arrogant in an amusing way, and pursues a project that is impossible to complete. None of this is explicitly stated in the novel. These are inferred truths, which the reader deduces from common knowledge about academic projects and the psyche of people pursuing them.[24] What rules can legitimately be used to reach conclusions of this sort is a difficult issue fraught with

[24]The distinction between primary and inferred truths is not always easy to draw, in particular when dealing with complex literary fiction. Walton also guards against simply associating primary truth with what is explicitly stated in the text and inferred ones with what follows from them; see

controversy. I will return briefly to it below; for the time being all that matters is that there are such rules, no matter what they are.

This framework has the resources to explain the nature of model-systems. Typically, model-systems are presented to us by way of descriptions, and these descriptions should be understood as props in games of make-believe. These descriptions usually begin with expressions like "consider" or "assume" and thereby make it clear that they are not descriptions of fact, but an invitation to ponder—in the present idiom, imagine—a particular situation. Although it is often understood that this situation is such that it does not occur anywhere in reality, this is not a prerequisite; models, like literary fictions, are not *defined* in contrast to truth. In elementary particle physics, for instance, a scenario is often proposed simply as a suggestion worth considering. Only later, when all the details are worked out, the question is asked whether this scenario bears an interesting relation to what happens in nature, and if so what the relation is.[25]

The "working out" of the details usually consists in deriving conclusions from the primary assumptions of the model and some general principles or laws that are taken for granted. For instance, we derive that the earth moves in an elliptical orbit from the basic assumptions of the Newtonian model and the laws of classical mechanics. This is explained naturally in the idiom of pretense theory. What is explicitly stated in a model description (that the model-earth is spherical, etc.) are the primary truths of the model, and what follows from them via laws or general principles are the implied truths; the principles of direct generation are the linguistic conventions that allow us to understand the relevant description, and the principles of indirect generation are the laws that are used to derive further results from the primary truths.

We can now address the above questions. The attribution of certain concrete properties to models (Q2) is explained as it being fictional that the model-system possesses these properties. To say that the model-population is isolated from its environment is just like saying that Zapp drives a convertible. Both claims follow from a prop together with rules of generation. In other words, saying that a hypothetical entity possesses certain properties involves nothing over and above saying that within a certain game of make-believe we are entitled to imagine the entity as having these properties. For this reason there is nothing mysterious about ascribing concrete properties to nonexistent things, nor is it a category mistake to do so.

Let us now discuss the issue of truth in model-systems (Q4), which will also provide us with solutions to the other open questions. The question is: what exactly do we assert when we qualify "Zapp drives a convertible" as true in the fiction while "Zapp drives a Mini Cooper" as false?[26] To begin with, it is crucial to realize that there are three different kinds of statement in connection with fiction, and that these

---

Walton (1990, Chapter 4) for a discussion. For the purpose of the present discussion these subtleties are inconsequential.

[25]For an accessible account of particle physics that makes this aspect explicit see Smolin (2007), in particular Chapter 5

[26]There is controversy over this issue even within pretense theory. It is beyond the scope of this chapter to discuss the different proposals and compare them to one another. In what follows I

require a different treatment when it comes to the questions of truth; I refer to these as intrafictional, metafictional, and transfictional statements.[27] For someone sitting in an armchair reading *Changing Places* "Morris jumped into the paternoster on the downside" is an intrafictional statement because the reader is involved in playing the game defined by the novel and imagines that the sentence's content is the case. Someone who read the novel a while ago and asserts in discussion with a friend that Zapp jumped into a paternoster makes a metafictional statement because he is talking about the fiction. If he then also asserts that Zapp, his quirks notwithstanding, is more likeable than any literature teacher he ever had or that Zapp is smarter than Candide, he makes transfictional statements as he is comparing Zapp to a real person and a character in another fiction.[28]

Intraficational propositions are made within the fiction and we are not meant to believe them, nor are we meant to take them as reports of fact; we are meant to imagine them. Although some statements are true in the fiction as well as true *tout court* ("1968 was the year of student revolts" is true and true in *Changing Places*), we often qualify false statements as true in the fiction ("Zapp is a literary theorist" is false because there is no Zapp) and true statements as false in the fiction ("white light is composed of light of other colors" is false in Goethe's *Faust*). So truth and truth in fiction are distinct; in fact, truth in fiction is not a species of truth at all (1990, 41). For this reason it has become customary when talking about what is the case in a fiction to replace locutions like "true in the fiction" or "true in a fictional world" by the term of art "being fictional"; henceforth "$F_w(p)$" is used as an abbreviation for "it is fictional in work w that p", where p is a placeholder for an intrafictional proposition like "Zapp pursues an impossible project".[29]

The question now becomes: when is p fictional in w? Let the w-game of make-believe be the game of make-believe based on work w, and similarly for "w-prop" and "w-principles of generation". Then, p is fictional in w iff p is to be imagined in the w-game of make-believe (1990, 39). In more detail:

> p is fictional in work w iff the w-prop together with the w-principles of generation prescribes p to be imagined

---

develop an account of truth in fiction that is based on elements from different theories and that is tailored towards the needs of a theory of model-systems.

[27] All theories of fiction acknowledge this distinction. My terminology is adapted from Currie (1990, Chapter 4) who speaks about the "fictive", "metafictive" and "transfictive" use of fictional names.

[28] Notice that while transfictional statements are recognizable by the presence of terms that are foreign to the work under discussion, intrafictional and metafictional statements are recognizable as such only as a function of the context in which they appear. There are also statements that are difficult to classify. As these typically involve emotional reactions on the part of the reader to the novel (halfway through the book a reader exclaims "I fear the worst for Zapp"), they need not occupy us here.

[29] I here follow Currie (1990, Chapter 2) and assume that sentences like "Zapp drives a convertible" express propositions, something that Walton denies (1990, 391). This assumption greatly simplifies the statement of truth conditions for fictional statements, but nothing in the present paper hangs on it. Essentially the same results can be reached only using sentences and pretense (1990, 400–405).

This analysis alleviates worries about the (alleged) subjectivity of imaginings. In common parlance, "imagination" has subjective overtones, which might suggest that an understanding of models as imagined entities makes them subjective because every person imagines something different. This is not so. In pretense theory, imaginations in an authorized game of make-believe are sanctioned by the prop itself and the rules of generation, both of which are public and shared by the relevant community. Therefore, someone's imaginings are governed by intersubjective rules, which guarantee that, as long as the rules are respected, everybody involved in the game has the same imaginings. So, not only do all participants in the game de facto imagine the same things (which could also be the result of happenstance), but they do so because they participate in a rule-governed activity. What is more, participants *know* that they do; they know that they are participants in an authorized game and as long as they trust that the others play by the rules they can trust that others have the same imaginings.

Furthermore, for a proposition to be fictional in work *w* it is not necessary that it is actually imagined by anyone: fictional propositions are ones for which there is a prescription to the effect that they *have to be imagined* (1990, 39), and whether a proposition is to be imagined is determined by the prop and the rules of generation. Hence, props, via the rules of generation, make propositions fictional independently of people's *actual* imaginings (1990, 38), and for this reason there can be fictional truths that no one knows of. If there is a stump hidden behind a bush, unknown to those playing the game, it is still fictional that there is a bear behind the bush; the prop itself and the rules of generation are sufficient to generate this fictional truth.

With this in place we can now also render concept of a "fictional world" or "world of a fiction" precise: the world of work *w* is the set of all propositions that are fictional in *w*.[30]

This analysis of truth in fiction carries over to model-systems one-to-one simply by replacing *p* by a claim about the model, *w* by the description of the model-system, and *w*-principles of generation by the laws and principles assumed be at work in the model. For instance, "the solar system is stable" is true in the Newtonian model of the solar system systems iff the description of the system together with the laws and principles assumed to hold in the system (the laws of classical mechanics, the law of gravity, and some general assumptions about physical objects) imply that this is the case. This gives us a straightforward answer to the question about identity conditions (Q1): two models are identical iff the worlds of the two models—the set of all propositions that are fictional in the two models—are identical.[31]

[30] Fictional worlds thus defined are rather different from possible worlds as used in modal logic, the most significant difference being that the former are incomplete while the latter are not. See Currie (1990, 53–70) for a discussion of possible worlds and fiction.

[31] An interesting consequence of this identity condition is that not all models with the same prop are identical, because they can operate with different rules of indirect generation. This is the case, for instance, when the "same model" is treated first classically and then quantum mechanically; on the current view, the classical and the quantum model are not identical.

Metafictional propositions make genuine claims that can be true or false in the same way in which claims about chairs and tables can be true or false. But how can such statements be true if the singular terms that occur in them have no referents? A solution emerges when we realize that statements like "Zapp is a professor" are ellipses for "in *Changing Places*, Zapp is a professor". So when we metafictionally assert *p*, what we really assert is "in work *w*, *p*" (1990, 397). Asserting that something is the case in a work of fiction is tantamount to asserting that it is fictional in that work. Hence asserting "in work *w*, *p*" amounts to asserting "*p* is fictional in work *w*", which is the same as "it is fictional in work *w* that *p*". The last sentence is, of course, just $F_w(p)$. Hence metafictionally asserting *p* amounts to asserting $F_w(p)$. The truth condition for this assertion follows from what has been said above:

> $F_w(p)$ is true iff *p* is fictional in *w*, which in turn is the case iff the *w*-prop and together with the *w*-principles of generation prescribes *p* to be imagined.

Derivatively, *p*, when uttered as a metafictional claim, is true iff *p* is fictional when uttered as an intrafictional claim.[32] In sum, once we understand that a metafictional claim has to be prefixed by "In fiction *w*", and hence has the structure $F_w(p)$, the truth of the claim is determined by appeal to the *w*-game of make-believe. Again, this analysis translates to scientific statements without further ado.

Transfictional propositions pose a particular problem because they—apparently —involve comparisons with a nonexistent objects, which does not seem to make sense: we cannot compare someone with Zapp if there is no Zapp. Different authors have offered very different solutions to this problem.[33] Fortunately we need not deal with the problem of transfictional statements in its full generality because the transfictional statements that are relevant in connection with model-systems are of a particular kind: they compare features of the model-system with features of the target-system. For this reason, transfictional statements about model-systems should be read as prefixed with a clause stating what the relevant respects of the comparison are. This allows us to rephrase comparative sentences as comparisons between properties rather than objects, which makes the original puzzle go away.

Crucially, then, truth conditions for transfictional statements in the context of scientific modeling come down to truth conditions for comparative statements between properties, which are unproblematic in the current context (for the problems that attach to them have nothing to do with issues surrounding fictional discourse). For instance, when I say "my friend James is just like Zapp" I am not comparing my

---

[32]In some places Walton ties the truth of such statements to *authorized* games (e.g., 1990, 397–398). This restriction seems unnecessary as the analysis works just as well for unauthorized games.

[33]Lamarque and Olsen (1994, Chapter 4), for instance, solve the problem by introducing characters. Walton, by contrast, renounces the commitment to characters and instead analyzes transfictional statements in terms of unauthorized games (1990, 405–416).

friend to a nonexistent person. What I am asserting is that both James and Zapp possess certain relevant properties (Zapp possesses properties in the sense explained above) and that these properties are similar in relevant ways. Likewise, when I say that the population of rabbits in a certain ecosystem behaves very much like the population in the Fibonacci model, what I assert is that these populations possess certain relevant properties which are similar in relevant respects. What these relevant properties are and what counts as being similar in relevant respects may well depend on the context. But this is not a problem. All that matters from a semantic point of view is that the apparent comparison with a nonexistent objects eventually comes down to the unproblematic comparison of properties. Further, the statement making this comparison is true iff the statement comparing the properties with each other is true. Obviously, statements comparing two nonexistent objects are analyzed in exactly the same way.

These insights provide us with answers to (Q3) and (Q4). And what is more, this take on truth also provides us with an answer to the question about the epistemology of models (Q5): we investigate a model by finding out what follows from the primary truths of the model and the rules of indirect generation. This seems to be both plausible and in line with scientific practice because a good deal of the work that scientists do with models can accurately be described as studying consequences of the basic assumptions of the model—so we can tick off (C1) as well.

What metaphysical commitments do we incur by understanding models in this way? The answer is: none. Walton's theory is antirealist in that it renounces the postulation of fictional or abstract entities, and hence a theory of scientific modeling based on this account is also free of ontological commitments. This, of course, is not a refutation of metaphysically less parsimonious views such as Meinong's, and there may be reasons to eventually prefer such a view over an antirealist one. The point to emphasize here is that whatever these reasons may be, the needs of science are not one among them.

This concludes the discussion of the conditions of adequacy and I hope to have made it plausible that the framework of pretense theory provides convincing responses to the issues that arise in connection with model-systems.

With this in place, we can now distinguish two different kinds of representation, which will be important in understanding scientific representation. As mentioned above, pretense theory *defines* a representation to be a prop in an authorized game of make-believe. On this view, the text of a novel and the description of a model-system are representations. Derivatively one can then say that props represent the imaginings they prescribe. Although this is a common use of "representation", the term is used rather differently in both science and philosophy of science where it is taken to denote a relation between the model-system and its target (and, depending on one's views about representation, also other relata such as users and their intentions). But far from being in conflict with each other, these two notions of representation are actually complementary—I will turn to this point in the next section. For now it is just important not to get them mixed up, and for this reason I call the former "p-representation" ("p" for "prop") and the latter "t-representation" ("t" for target).[34] Using this idiom, pretense theory (as presented in this section) can be understood as an analysis of p-representation. This leaves pending an analysis of t-representation, to which I turn in section "A First Stab at T-Representation" below. I defer this task because I first want to summarize where we stand and formulate a consistent overall picture of scientific modeling, which is the aim of section "The Anatomy of Scientific Modeling".

The Anatomy of Scientific Modeling

We have analyzed model-systems in terms of imagined objects and distinguished two different representational relations, p-representation (which holds between a prop and the imaginings that it mandates) and t-representation (which holds between a thus imagined system and a target-system in the world). Using these notions, the two acts mentioned in the introduction can be described as, first, introducing a p-representation specifying an imagined object and, second, claiming that this imagined object t-represents the relevant target-system.

Putting all this together we obtain a general picture of scientific modeling. This picture is schematically illustrated in Fig. 1

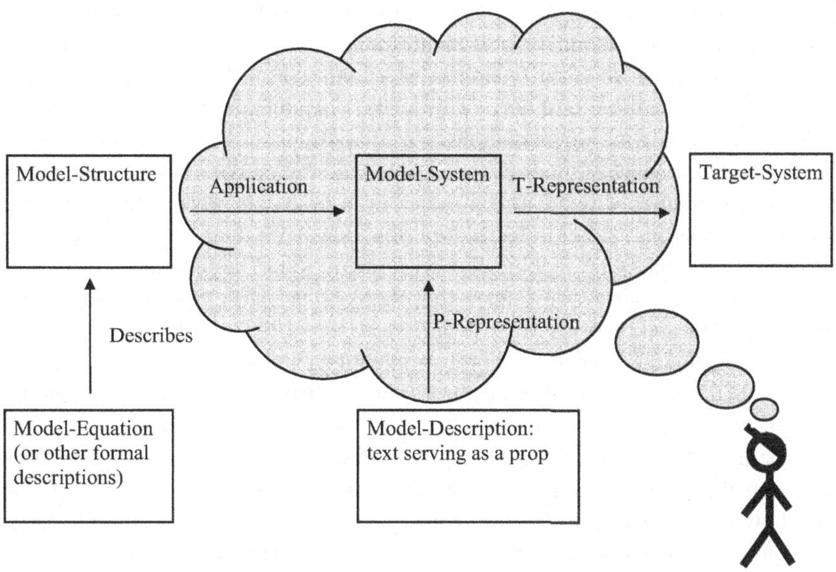

Fig. 1 The elements of scientific modeling

[34]A more intuitive choice of terminology would be to refer the term "representation" for what I here call t-representation, and refer to p-representation as "presentation". However, since this would stand in conflict with the use of "representation" in pretense theory I stick to the somewhat less elegant terminology of p- and t-representation.

The boxes in the middle and on the right emerge from the above discussion and don't need further explanation. Not so the boxes on the left. These account for the use of mathematics. How mathematics applies to something non-mathematical is a time-honored philosophical puzzle, and much has been written about it. However, since this is somewhat peripheral to the concerns of this chapter, I will not discuss this issue further and merely put the relevant boxes into the diagram for the sake of completeness. A discussion of the issue of the applicability of mathematics can be found in Shapiro (2000).

Let me then add some points about this Fig. 1 by way of clarification and explanation. First, there is a temptation to respond to this suggestion by saying: "yes, fine, but where in this scheme is *the model*?" There is no single answer to this question. With the exception of the target-system itself, every part of the above schema (and every combination of parts!) legitimately may be, and sometimes is, referred to as "model", which is why I tried to avoid the term altogether. Once it is acknowledged that scientific modeling involves all the above elements, the determination of which one of these we call "the model" is inconsequential. As long as one is aware of this we can choose terminology as we please.

Second, this picture of scientific modeling is independent of how one understands the relation between models and theories. The model-structure in this diagram is assumed to be a structure used in the treatment of a particular concrete system, and not a general structure. It is, for instance, the structure of the harmonic oscillator, the two-body system, or a conical spinning top on a frictionless plane; it is not Newtonian Mechanics, Quantum Mechanics, Fluid Mechanics, or General Relativity (in a structural rendition). This leaves open the question of how these specific structures relate to overarching theories. In particular, this picture is compatible with the semantic view of theories that would take the model-structure to belong to a family of structures which forms a theory (van Fraassen 1980). This view agrees also with the German structuralist picture that construes model-structures as being the result of a process of specification and restriction of a general theory (Balzer et al. 1987), and a view that denies that there is any straightforward connection between models and theories Morgan and Morrison (1999) and Cartwright (1999).

Third, this diagram has no temporal connotations and there is no view implicit in it about what comes first in the process of the construction of a model. Sometimes we start with a fictional scenario; sometimes we start with an equation we think might be useful; sometimes we have a clear strategy for t-representation in mind right from the start, and sometimes we just "try out" something and worry later about how the model relates to the world. It is not even assumed that all parts of the diagram are belabored by the same scientist. In particular when it comes to large and complex models (such as climate models), different groups may take care of different parts of the model (e.g., one group may develop mathematical tools and another one may take care of their application to the concrete problem at hand). In brief, this picture of modeling is compatible with any view one wants to take on the actual process of model construction and the division of labor therein.

Fourth, there is a time-honored problem about how it is possible that we can represent something that does not exist. How can we represent Santa Claus if there is no Santa Claus? More pertinently, how can we have models representing in great detail mechanical properties of the ether if there is no ether? Thinking about modeling in the way I have proposed makes this problem go away at once, since it becomes clear that equivocating on "representation" is the root of the puzzle. On the one hand, we take representation to be a relation between a picture or model and an item in the real world (which does not exist). On the other hand, it assumes representation to be the infliction of mental content in an observer when she looks at a picture or reads the description of a model (which is, of course, real). This is exactly the distinction between t-representation and p-representation. Santa Claus pictures and ether models do not t-represent because there is no Santa Claus and no ether. But Santa Claus pictures p-represent in that they become props in a game of make believe leading us to imagine all kind of things about a bearded old man in a red outfit bringing gifts, and a description (or graphical representation) of an ether model leads us to imagine a fictional model-system. Once we recognize the distinction between p-representation and t-representation, the problem evaporates.[35]

Fifth, t-representation is not the only element in the above scheme whose absence is as interesting as its presence; structures and equations may similarly be construed. Although formalizations play an important role in modeling, not all scientific reasoning is tied to a formal apparatus. In fact, sometimes conclusions are established by solely considering a fictional scenario and without using formal tools at all. If this happens it is common to speak of a *thought experiment*. Although there does not seem to be a clear distinction between modeling and thought-experimenting in scientific practice, there has been little interaction between the respective philosophical debates.[36] This is lamentable because it seems to be important to understand how models and thought experiments relate to each other. In a recent paper Davies (2007) argues that there are important parallels between fictional narratives and thought experiments, and that exploring these parallels sheds light on many aspects of thought experiments. This take on thought experiments is congenial to the

[35]Model-systems without targets (and hence without t-representation) not only play a role when explaining failures; they are also important as means to explore certain technical tools, in which case they are often referred to as "probing models", "developmental models", "study models", "toy models", or "heuristic models". The purpose of such model-systems is not to represent anything in nature; instead they are used to test and study theoretical tools that are later used to build representational models. In field theory, for instance, the so-called ϕ^4-model has been studied extensively, but not because it represents anything in the world (it was well known right from the beginning that it does not), but because its simplicity allows physicists to study complicated techniques such as renormalization in a simple setting and get acquainted with mechanisms—in this case symmetry breaking—which are important in other contexts (Hartmann 1995). It is an advantage of the proposed view of modeling that it can account for this practice.

[36]Extensive discussions of thought experiments can be found in Brown (1991), Sorensen (1992), and Brown's and Norton's contributions to Hitchcock (2004).

view on models presented in this paper and suggests that modeling and thought-experimenting are intrinsically related: thought experiments (at least in the sciences) are models without formal apparatus.[37]

Sixth, although Walton's general idea of rules of generation is intuitively clear, it turns out to be difficult to give an account of these rules. The two most important rules in the context of literary fiction—the reality principle and the mutual belief principle—suffer from intrinsic problems.[38] Worse, they may also lead to wrong results when put to work in science. So what are the rules of generation in scientific fictions? This is a substantial question that needs to be addressed, but we should not expect a single unified answer. On the contrary, it seems plausible to assume that different disciplines have different rules, and understanding what these rules are will shed light on how modeling in these disciplines works. So we should not expect a ready-made answer, but rather regard the study of rules of generation as part of research programme aiming at understanding the practice of modeling in various branches of science.

Seventh, not all models are introduced by verbal descriptions; sometimes we use drawings, sketches or diagrams to specify the model-system. There are linguistic and non-linguistic props. Although I have discussed pretense theory only in as far as it deals with linguistic props, the scope of the theory is much wider than that. In fact it covers all kinds of props, among them the classical media of visual art (paintings, drawings, etchings, etc.), as well as photography and film. So the current framework is equipped to deal with p-representation that is nonverbal.

Eighth, the fact that the view of modeling advanced here is developed by drawing analogies with literary fiction should not be taken to suggest that there are no

---

[37]As an example consider Galileo's law of equal heights (Sorensen 1992, 8–9). Take a u-shaped cavity, put a ball on the edge of one side, and let the ball roll down into the cavity. Galileo then argued that it would have to reach the same height at the other side—this is the law of equal heights. Of course Galileo realized that the ball's track was not perfectly smooth and that the ball faced air resistance, which is why the ball in an actual experiment does not reach equal height on the other side. So Galileo considered an idealized situation in which there are neither friction nor air resistance and argued that the law was valid in that scenario. This thought experiment fits the above account of model-systems: Galileo considered a fictional scenario specified by a simple description, yet the conclusion he wanted to reach was not part of that description and was reached by using certain general principles that he took to be valid in situations like the one considered. Moreover, had Galileo used a mathematical machinery to derive his conclusion instead of informal arguments, physicists would refer to the product of his endeavor as a model. One would write down a curve specifying the shape of the cavity (for instance a parabola), specify its mechanical properties (frictionlessness), use mechanical laws to calculate the trajectory of the ball, and then find that it ends up at equal height on the other side. This is the sort of thing we find in mechanics textbooks, and which are referred to as mechanical models of a situation.

[38]Roughly, the Reality Principle says that if $p_1 \ldots p_n$ are direct fictional truths, then proposition q is an indirect fictional truth iff: were it the case that $p_1 \ldots p_n$, then it would be the case that q. The Mutual Belief Principle says that that if $p_1 \ldots p_n$ are direct fictional truths, then proposition q is a indirect fictional truth iff: it is mutually believed in the artist's society that were it the case that $p_1 \ldots p_n$, it would be the case that q. See Walton (1990, Chapter 4) for a discussion of these principles.

differences between the two. An in-depth comparison between literature and scientific modeling is beyond the scope of this chapter, but some salient differences are readily stated. Literary plots are often complex and convoluted, while fictional scenarios canvassed in science are extremely simple and it seldom takes more than a few lines to describe the set-up. One of the reasons for this is that they must allow for mathematical treatment. Fictional scenarios in science are also often created with a specific target-system in mind, and the scenario is chosen such that t-representation can be set up—considerations that play only a marginal, if any, role in literature. Aesthetic considerations (style, genre, etc.) are irrelevant for model-descriptions, and so are emotional reactions of the reader to the plot. Finally, authorship is irrelevant in science: we often name models after their progenitors (e.g., the "Bohr model"), but this is merely a sociological fact with no systematic import since ambiguities and open questions are not resolved by appeal to the author's intention or context.

Ninth, needless to say, pretense theory is not without internal problems.[39] Although Walton's account eschews common-sense understandings of imagination (as noted above), more needs to be said about what exactly imagining amounts to in science and about how it differs from imagining in other contexts, as well as how it differs from other activities like considering, pondering, and entertaining. However, I will have to leave this issue for another occasion.

A First Stab at T-Representation

So far I have argued that models are imagined objects and I have shown how this leads to a coherent overall view of scientific modeling (shown in Fig. 1). In particular, I have presented an account of what it means for claims about a model-system to be true, how we learn about model-systems, and how we can meaningfully compare them to either things in the world or other model systems. What is still missing from the analysis is an account of how model-systems represent (i.e., t-represent) something beyond themselves. The structuralist answer (that representation essentially is isomorphism) is not available to the fiction view since only structures can enter into isomorphisms and model-systems, on this view, are not structures. So we have to go back to the drawing board and develop a new account of representation that can explain how a model-system of the kind introduced in section "Model-Systems and Imagination" can represent a target system. This is project for this section.

The first question is what to choose as our source from which we might formulate an account of t-representation. So far I have developed an account of scientific modeling by drawing analogies with literary fiction. Unfortunately this analogy does not seem to be productive when it comes to t-representation. Understanding

[39]For critical discussion see, among others, Lamarque (1991), Budd (1992), and the contributions to the symposium on Walton's book in *Philosophy and Phenomenological Research* 51 (1991). See Currie (2004) for a discussion of different notions of imagination.

t-representation involves establishing and understanding a relation between the fictional scenario and parts (or aspects) of the real world. While we sometimes do this casually (for instance when I compare my friend James with Zapp), there is controversy over whether this is in any way essential to our engagement with fiction, and whether it leads to any interesting insights. Elgin (1996, Chapter 6) argues it does, which is what Kivy (2006, Chapters 24–28) denies. But even if this controversy could be resolved in favor of those who believe in the cognitive value of literature, there is no general method of bringing to bear literary fictions on real-world situations, which could serve as the blue-print for t-representation in science.[40]

The analogy I wish to exploit in what follows is the one between maps and scientific representations. This analogy is of course not new; see Sismondo and Chrisman (2001) for a survey and discussion. But I want to put the analogy to a slightly different use than other writers. While the map analogy has in the past mainly been employed to defend some sort of scientific realism, I wish to remain non-committal about realism and use maps only to explain how representation works at the most general level.[41, 42]

The essence of a map is that it allows us to "read off" properties of the territory from the map: by looking at a map of London we see that Camden lies west of Hackney, Brixton is south of the river, etc. The map is different from a verbal description in that it does not merely state these facts; maps are not long lists with sentences describing a certain area. Facts about the city are inferred from facts about the map itself and a "key of translation", which says how facts about the map translate into facts about the city. This is realization provides us with the elements of the general scheme of representation:

> *X* t-represents *Y* iff:
> (R1) *X* denotes *Y*.
> (R2) *X* comes with a key *K* specifying how facts about *X* are to be translated into claims about *Y*.

---

[40]Elgin's account is based on the notion of exemplification. This account is on the right track, and a worked out version of the account I propose below will draw on many of its insights. However, at least in its basic form, this account does not cover cases in which the representational vehicle and the target do not share the relevant properties. The account suggested below is more permissive in that respect.

[41]Throughout this chapter I use a realistic idiom in the sense that I assume that what is represented, the target system, exists. This is for the ease of formulation and my position could be restated from the point of view of *metaphysical* antirealism. What I want to remain non-committal about is *scientific* realism, roughly the position that theories are more or less truthful mirror images of reality. At a general level representing something does not amount to giving a mirror image, or to make a copy of that item. A representation can be alike to its target, but it does not have to be. There is nothing in the notion of a representation that ties it to imitation or copying. A general account of representation has to make room for non-realistic representations in this sense.

[42]Maps are of course real and not fictional objects. It will become clear as we proceed that representation works in the same way for fictional and real objects. Hence that maps, unlike model-systems, are material objects is no impediment to using them in the current context.

In nutshell, the idea is that the first condition establishes the aboutness of *X*, and the second guarantees the cognitive relevance of *X* for *Y*.[43] Before qualifying these conditions, let me illustrate them in more detail.[44] I have in front of me a map of North London. This is the first condition: the map denotes North London. Now I look at the details. I see a black rectangle on a black line and written next to it is "Camden Road". The explanations that come with the map say that this rectangle stands for an over-ground railway station, the name next to it is the name of the station, and the black line stands for the rail tracks. A bit further up there is a black dot on a black line. The legend say that the dot stands for a tube station, and the name written next to it is the name of the station, in this case "Kentish Town". Between the two there is a thick yellow line, which stand for a main road. Hence, that a black rectangle labelled "Camden Road" is connected with a thick yellow line to a black dot labelled "Kentish Town" (a fact in the map) translates into the fact that Camden Road railway station is connected to Kentish Town tube station by a main road (a fact about North London). Furthermore, from the fact that this yellow line is 4.5 cm long, I can infer that the actual distance between the two is about 1 km since the scale of the map is 4.55 cm to 1 km. Finally, the "Kentish Town" dot lies vertically above the "Camden Road" rectangle, from which I infer that Kentish Town tube station is north of Camden Road railway station.

Our use of a map essentially involves a key, telling us how to translate facts about the map into facts about North London. Some elements of the key are stated at the bottom of the map; for instance, we are instructed that rectangles stand for railway stations and dots for tube stations. Other elements are conventions that are so common that they are assumed without further explanation. The top of the map indicates north, for example, and the distances in the map are proportional to distances in the world (where the "scale" of the map gives the proportionality factor). But these are mere conventions and there is nothing "natural" or "self evident" or even "necessary" about them. We could use rectangles to denote tube stations rather than railway stations. We could draw the map so the south rather than north is on top, and have projection techniques that do not preserve distances.[45] The crucial point,

---

[43]The first condition is Goodman's (1976, Chapter 1) who has argued that denotation lies at the heart of representation.

[44]Common alternatives to the current proposal are isomorphism and similarity accounts of representation; see Frigg (2006) and Suárez (2003) for discussions. Other alternatives have been proposed by Contessa (2007), Hughes (1997), Suárez (2004, 2006) and Toon (2010). For want of space I cannot discuss these here.

[45]Nautical maps, for instance, use the Mercator projection system and do not preserve distances; they preserve angles and one obtains wrong results when translating the distance between two points on a map into the distance between two locations. And this mistake has been made over and over again. As Sismondo and Chrisman (2001, 42–43) point out, about half of a sample of 137 international maritime boundaries are not where they were meant to be. When diplomats met to draw the boundaries between territories they had these charts on the table. They intended to draw the border half way between two territories and so they drew the line on the map mid-point between the territories. This is mistake: even relatively close to the equator the line thus drawn can be over 7 km away from the actual line of equidistance.

though, is that what a map represents depends not only on facts *in* the map, but on the key that is used to translate these facts into claims about the world. And this key does not simply "jump off the page"; they are not "in" the map itself. Instead, one has to know what the key is.

My claim is that model-systems are t-representations in the same way in which maps are: they denote a target system and certain facts obtain in them (in the sense explained in sections "Model-Systems and Imagination" and "The Anatomy of Scientific Modeling") which are then translated into claims about a target system by using a key. As an example, consider the Bohr model of the hydrogen atom. On the current analysis this model consists of a model-system, which is specified by a model description and which is described by a formal apparatus (classical mechanics plus the Bohr-Sommerfeld quantization rule). A number of facts obtain in the model-system, among them that it has discrete energy levels. We then take the model-system to denote real hydrogen atoms, and then use a simple key—here identity (more about this below)—to translate this fact into the claim that hydrogen itself has discrete energy levels.

Let me now add three qualifications. First, (R1) and (R2) provide the *general form* of an account of t-representation, which needs to be concretized in every particular instance of a t-representation. In fact, "denotation" and "key" are abstract in the sense introduced in section "Model-Systems and Fiction" and need fitting out in every particular instance. In order to understand how a *particular* representation works, we need to account for how the particular *X* comes to denote the particular *Y*, and we have to provide a particular key *K*. In the above example, we borrowed denotation from ordinary language by saying "this is a map of North London", and the key was provided to us by cartography. But other cases may work differently since there may be different sources of denotation and there may be any number of keys that can be used to interpret *X*. Moreover, keys are often implicit and determined by context. This is the case with scientific representations, which unlike maps, rarely, if ever, come with something like a legend. It is one of the challenges facing a philosophical analysis of representation to make hidden assumptions explicit, and present a clear statement of them. So there is much more to be said about t-representation than is contained in (R1) and (R2)—they are merely blanks to be filled in every particular instance. Thus, the claim that something is a t-representation amounts to an invitation to spell out how exactly *X* comes to denote *Y* and what *K* is.

Nonetheless, this generality is an advantage. The class of t-representations is large and its members varied. A view that claims that all t-representations work in exactly the same way would be doomed to failure right from the beginning. Maps, graphs, architectural plans, diagrams, photographs, (certain kinds of) paintings and drawings, and of course scientific models, are all t-representations in that they satisfy (R1) and (R2), but they work in very different ways. The differences between them are that these conditions are realized in very different ways: different keys are used and denotation has different sources. The challenge for a complete account of representation is to come up with a taxonomy of different ways in which the two conditions can be realized, and to explain how they differ from each other. Needless to say, this is a Herculean task that I cannot undertake here since there are many different kinds of keys. That said, the value of this account of representation is that it provides us with a framework in which to discuss these questions.[46]

A second qualification I would like to add to the scheme sketched above is to note that there is one important disanalogy between maps and scientific models: where their respective keys come from. In the case of the map we have the target system in front of us, we explore it *directly* (by taking measurements, etc.) and *then* we construct the map. So a map is an elegant summary of what someone already knows, and its sole purpose is to effectively summarize this knowledge and communicate it to those who are not in the business of land surveying. Science is not like this; we do not first survey the hydrogen atom and then construct a model to communicate the findings to those not yet familiar with it. We typically construct models to find out something genuinely new about the target system; something that *no one* yet knows.

This disanalogy does not undermine the saliency of t-representation for our analysis of modeling. Even if the process of constructing a model involves much more than elegantly summarizing observations, once the model-system is constructed (no matter how!) we have to specify how it relates to the world, and this is done by providing a key. However, unlike for maps where we know the key by construction (we have used a certain projection method, certain symbols, etc. when drawing the map), in the case of models the key has the character of a hypothesis.[47] We stipulate that we expect the model to bear this or that relation to its target, and then evaluate this claim against the best available background knowledge and by subjecting it to test using the usual methods of scientific investigation. How exactly this is done depends on the details of the representation. That is, it depends on the key used and the nature of the denotation relation (for instance, an assessment of the accuracy of a key for a model in elementary particle physics will be very different from the assessment of an engineering model of a bridge). Understanding these processes should be part of a future investigation into the nature of different kinds of t-representations (*cf.* the first qualification). For now it is sufficient to point out that keys can be hypothetical, and that this does not undermine the status of models as t-representations.

Third, (R2) states that we need a key specifying how to translate *facts* about *X* into *claims* about *Y*. This is not a slip. An acceptable definition of t-representation has to make room for misrepresentation. A map can contain errors in the sense that even if we use the right key and use it correctly we may obtain wrong results. For instance, it might have happened that the cartographers failed to connect the

---

[46]In passing I would like to point out that this account of representation satisfies the conditions of adequacy that I presented in my (2006). The *ontological puzzle* is addressed by the account of model-systems presented in section "The Anatomy of Scientific Modeling". The *enigma of representation* is met by (R1) and (R2). The *problem of style* now becomes the question of how denotation works and what keys are used.

[47]Although this is reminiscent of Giere's claim that models are connected to their target systems with a "theoretical hypothesis" (1988, 80), the point is a different one. In Giere's account we call a claim to the effect that the model is similar to the target in specific way a theoretical hypothesis; the current view, by contrast, emphasizes the hypothetical—fallible, tentative, and conjectural—character of keys attributed to a model.

black dot and the black rectangle with a yellow line, and so we would have been led to believe that the two stations are not connected by a main road. This would not have turned the map into a non-t-representation; it would still have been a t-representation, but one that misrepresents North London. Saying that we translate facts about the map into *claims* about the target makes room for error because claims can be true or false, while facts cannot. A representation is a *faithful* representation iff all claims about *Y* are true.

There is now also a straightforward way to draw a delineation between cases of misrepresentation and cases of failure of representation. *X* is a misrepresentation if it is not faithful (and notice that misrepresentation comes in degrees!). Something is not a t-representation at all if at least one of the two conditions fails. We have a failure of (R1) if there is no target system; a map of Atlantis fails to be a t-representation of Atlantis because there is no Atlantis, and hence Atlantis cannot be denoted. By contrast, the failure can be put down on condition (R2) if *X* it has no intrinsic properties that are interpreted by using a key. This is why proper names, for instance, are not t-representations: they denote the bearer of the name, but there is no key that translates properties the *name itself* possess into claims about the bearer of the name. If, for some reason, one wants to call proper names "representations" then one can do so, but it is important to realize that they are not t-representations, and being a t-representation is what matters both in the case of maps and in the case of scientific models.

With this in mind we can see what is wrong with Callender and Cohen's (2006) argument that there is no special problem about scientific representation. Because scientific representation comes down to an act of arbitrary stipulation, by their reading, explaining how we make such stipulations lies in the province of philosophy of mind and not in the realm of philosophy of science at all. They ask: "Can the salt shaker on the dinner table represent Madagascar?", and immediately reply "Of course it can, so long as you stipulate that the former represents the latter. ... Can your left hand represent the Platonic form of beauty? Of course, so long as you stipulate that the former represents the latter" (2006, 73–74). If all you mean by representation is denotation, then this is correct. But for something to be a t-representation, more than mere denotation is needed. We would need a key telling us how to translate certain properties of the salt-shaker into claims about Madagascar, or properties about my left hand into properties about the Platonic form of beauty, which, by their own admission, we don't.[48]

Why it is so important for a representation to be a t-representation, and why is simple stipulation not enough? The answer to this question is that maps as well as scientific representations belong to a category of representations that function cognitively: we study *X* to learn something about *Y* that we did not already know. In fact, model-systems are the units on which significant parts of scientific investigation are carried out rather than on the target system itself: we study a model and thereby discover features of the thing it stands for. For instance, we study the nature

[48]For a more extensive discussion of Callender and Cohen's argument see Toon (2010).

of the hydrogen atom, the dynamics of populations, or the behavior of polymers by studying their respective models. We do this by first finding out what is true in the model-system (*cf.* Section "The Anatomy of Scientific Modeling"), and then translating the findings into claims about the target itself. This is possible only if the model-system is a t-representation in the above sense. Denotation is not enough for this to happen. Proper names don't inform us about the properties of things they stand for; we can turn and twist "hydrogen" as long as we wish, but we won't thereby learn anything about hydrogen.

As I mentioned above, I regard the detailed study of different keys as a research programme to be undertaken in the future. However, to get a better idea of what such an investigation involves I now want to discuss two keys often used in science: identity and ideal limits. The simplest of all keys is *identity*, the rule according to which facts in the model (or at least a suitably defined class of facts) are also facts in the world. For example, if X t-represents Y by identity, then it follows from the fact that X has discrete energy levels that Y has discrete energy levels too. Although scientists often talk as if the relation between models and reality was identity, there are actually very few models that work in this way.

A more interesting key is the *ideal limit* key. Many model-systems are idealizations of the target in one way or another. A common kind of idealizations is to "push to the extreme" a property that a system possesses. This happens when we model particles as point masses, strings as massless, planets as spherical, and surfaces as frictionless. Two things are needed to render such idealizations benign: experimental refinements and convergence (Laymon 1991). First, there must be the possibility of *in principle* refining actual systems in a way that they are made to approach the postulated limit (that is, we don't actually have to produce these systems; what matters is that we in principle could produce them). With respect to friction, for instance, one has to find a series of experimental refinements that render a tabletop ever smoother and hence allow real systems to come ever closer to the ideal of a frictionless surface. These experimental refinements together constitute a sequence of systems that come ever closer to the ideal limit. Second, this sequence has to behave "correctly": the closer the properties of a system come to the ideal limit, the closer its behavior has to come to the behavior in the limit. If we take the motion of a spinning top on a frictionless surface to be the ideal limit of the motion of the same spinning top on a non-frictionless surface, then we have to require that the less friction there is, the closer the motion of the real top comes to the one of the idealized model. Or to put it in more instrumental terms, the closer the real situation comes to the ideal limit, the more accurate the predictions of the model. This is the requirement of convergence. If there exists such a sequence of refinements and if the limit is monotonic, then the model is an ideal limit.

If a model is an ideal limit, this implies a key. To see how, let us first briefly recapitulate the mathematical definition of a limit. Consider a function $f(x)$, and then ask the question how $f(x)$ behaves if x approaches a particular value x_0. We say that the number F is the limit of $f(x)$ (in symbols: $\lim_{x \to x_0} f(x) = F$) iff for every positive number ε (no matter how small), there exists another positive number δ such that: if $|x - x_0| < \delta$, then $|f(x) - F| < \varepsilon$. Colloquially, this condition says that the closer x comes to x_0, the closer $f(x)$ comes to F: if we know that x is less than δ way from

x_0, then we also know that $f(x)$ is less than ε away from *F.* This idea can now be used for ideal limits in the above sense. The sequence of experimental refinements plays the role of x, and the ideal limit itself is x_0 (in the example: the ever smoother table tops correspond to different values of x, and the frictionless plane corresponds to x_0). The behavior of the object corresponds to f. If there is a limit we know that if the difference between the friction of the real plane and the ideal frictionless plane is smaller than δ, then difference between the behavior of the real spinning top and the ideal spinning top in the model-system is smaller than ε. So if we are given the friction of the table, we know how to translate facts obtaining in the model-system into claims about the world.[49]

Of course not all model-systems are ideal limits of their target-systems in this sense.[50] For instance, we cannot possibly produce a sequence of systems in which Planck's constant approaches zero. In other cases it may not be clear whether there are such limits. For instance, mathematical knot theory is a branch of topology which deals with one-dimensional strings. But physical strings have finite width. Hence the question arises whether, and if so, in what sense the results of mathematical knot theory carry over to physical situations. So it is an open question how to translate facts in idealized systems into claims about a real-world target if they are not ideal limits—or in the current idiom: there is a question about what the key is—one that should preoccupy us in the future.

Re-reading the Newtonian Model of the Sun–Earth System

Case studies are the touchstone of philosophical analysis, and so it is imperative to show that the account developed in this chapter can shed light on typical cases of scientific modeling. For this reason I now discuss a standard example of a scientific model—the Newtonian model of the sun–earth system—and show that the fiction view not only has the resources to explain what happens in this case, but also makes features of the model visible that are usually overlooked. Hence, the fiction view of models, far from being an idle philosophical pastime, is actually a powerful tool to help us to better understand what is involved in scientific models.

---

[49] I have smuggled in a premise here: that it makes sense to quantify differences in the friction of surfaces and the behavior of spinning tops in terms of numbers. This is not implausible and could be made precise, for instance, by using friction coefficients and a geometrical measure for the closeness of trajectories. The following two questions are more pressing. First, how can we know whether or not a certain model-system is an ideal limit of the target at hand? Second, what is the relation between ε and δ? In real applications on would like to know how close to the limit one would have to come to get a result that is precise to a particular degree. Typical mathematical existence results are of no help here. These are open questions that need to be addressed.

[50] This corresponds to Rohrlich's distinction between factual and counterfactual limits (1989, 1165).

The aim of the Newtonian model is to determine the orbit of the earth moving around the sun.[51] We first posit that the only force relevant to the earth's motion is its gravitational interaction with the sun, and we neglect all other forces, most notably the gravitational interaction with the other planets in the solar system. This force is given by Newton's law of gravity, $F_g = Gm_e m_e / r^2$ where m_e and m_s are the masses of the earth and the sun respectively, r the distance between the two, and G the constant of gravitation. We then make the idealizing assumption that both the sun and the earth are perfect spheres with a homogeneous mass distribution (i.e., the mass is evenly distributed over the sphere), which allows us to treat their gravitational interaction as if the mass of both spheres was concentrated in their center. The sun's mass is vastly larger than the earth's and so we assume that the sun is at rest and the earth orbits around it. Now we turn to classical mechanics and use Newton's equation of motion, $\vec{F} = m\vec{a}$, where $\vec{a}$ is the acceleration of a particle, m its mass and $\vec{F}$ the force acting on it. Placing the sun at the origin of the coordinate system and plugging in the above force law we obtain $\ddot{\vec{x}} = -Gm_s\vec{x}/|\vec{x}|^3$, the differential equation describing the earth's trajectory (where we have, of course, used $\vec{a} = \ddot{\vec{x}}$, i.e., that the acceleration is equal to the second derivative of the position). This equation can be solved and we find that the earth moves on an elliptic orbit around the sun.

When we read the above description, which tells us to regard the earth and the sun as ideal homogeneous spheres gravitationally interacting only with each other, this description serves as a prop and we engage in an authorized game of make believe. We imagine the entity described in the description, where the rules of direct generation are just the rules of ordinary English. We understand the terms occurring in the description and we imagine an entity which has all the properties that the description specifies. The result of this process is the *model-system*, the fictional scenario which is the vehicle of our reasoning: an imagined entity consisting of two spheres, etc. The part of the above description that prescribes us to imagine the model-system is the *model-description*. Now focus on the formal apparatus. $\ddot{\vec{x}} = -Gm_s\vec{x}/|\vec{x}|^3$ is the *model-equation*, which, in this case, is obtained from a general theory—Newtonian mechanics—by specifying the number of particles and their interaction. This equation specifies a *model-structure*, which is instantiated in the model-system (*cf.* section "A First Stab at T-Representation"). A proper analysis of the structure described by this equation would require formal techniques that are beyond this chapter.[52] But for our purposes nothing hinges on giving all the details (since our concern here is not the applicability of mathematics); what matters at this point is only that such an analysis can be given and that its upshot is that the model-equation applies to the model-system (and is literally true of it). The model-equation then is the formal expression of a principle of indirect generation. Using this principle we find that it is true in the model-system that the sphere with mass

---

[51] See, for instance, Feynman, Leighton, and Sands (1963, Sections 9.7 and 13.4) and Young and Freedman (2000, Chapter 12).

[52] Such an analysis can be found in Balzer, Moulines, and Sneed (1987, 29–34, 103–108, 180–191), Frigg (2003, Chapter 8), and Muller (1998, 259–266).

m_e orbits around the sphere with mass m_s on an elliptical orbit. This is an implied truth because it has not been written into the model-description; it is something that we infer from the basic features of the model-system (as given by the model-description) and the rule of generation.

The next step is to connect our model to the target-system. We find clues about how to do this in the above description. Right at the beginning we are told that the model we are constructing is a model of the sun–earth system. This establishes denotation, which is condition (R1). As in the above examples, we borrow denotation from ordinary language by using the expressions "sun" and "earth", which we take to refer to the relevant heavenly bodies. Should these expressions for some reason fail to refer, then t-representation would fail too. Ordinary language also plays a role is specifying the key. The first element of the key is the definition of an object-to-object correlation: we say that the sphere with mass m_e in the model-system corresponds to the earth and the sphere with mass m_s to the sun. Now things get more involved. We have made several idealizations (that the sun and the earth are spherical, that there are no forces other than the gravitational interaction of sun and earth, etc.) and we now have to say how these should be understood. Unfortunately physics texts usually do not say much about this question, or remain altogether silent about it. So at this point we have to appeal to philosophical theories of idealization and the keys they imply. On a plausible reading of the Newtonian model, the idealizations made are taken to be ideal limits in the sense discussed in the last section. The limit is complex and involves many properties, but the leading idea is that we could—in principle—produce a sequence of systems where the forces acting on the sun and the earth become increasingly smaller and eventually converges towards zero (which would be done by taking more and more matter out of the universe). We can then also—again, in principle—produce a sequence of sun–earth systems in which the sun and the earth become ever rounder and their mass distributions ever more homogeneous. The claim then is that, first, in the limit the sequence of these systems converges towards the model-system (which is true by construction); second, the behavior of the systems in this sequence converges towards the behavior of the model-system (this is the ideal limit). Given this, we know how to translate claims about the model into claims about the target: if the actual target is less than δ away from the model-system, then the behavior of the actual target is less than ε away from the behavior of the model-system. This is (R2).

Asserting convergence between sequence and system constitutes a substantial claim that does not follow from the construction of the sequence. In fact, we cannot strictly *prove* that this is so. This illustrates the hypothetical character of keys: they are postulated as a hypothesis and not given to us as in the case of the map. However, this does not mean that any hypothesis is as good as any other. We justify the stipulation of the ideal limit key (rather than another key) in two ways. First we appeal to background knowledge: we have tested the law of gravity and Newton's equation of motion in countless situations and have good reasons to assume that it provides true descriptions in scenarios like the model-system. We derive predictions from the model-system (the trajectory of the earth) and compare them with observations. At this point the ideal limit key becomes essential. If we have in ideal limit, then we know how the behavior of the model-system relates to the behavior of the target. Assume now we can sensibly quantify such distances (*cf.* footnote 49) and, given what we know about the universe, the forces and masses are such that the actual target-system is less than δ away from the model-system, then we can compare the theoretical trajectories of the earth with the observed ones and see whether they are less than ε away from each other. If this is the case, then this confirms our hypothesis that the model-system is an ideal limit. But notice—to come back to the point made in Section "Model-Systems and Fiction"—that what the model-system represents is not data, nor is there anything in the model that is directly comparable to data. The data used to confirm the model are obtained with the aid of specific observational techniques (optical telescopes, radio telescopes, etc.) and the character of the data varies with these techniques. *Given* a particular technique (and the theories behind it), the model can be used to calculate what one would have to observe; but the result of this calculation is not in any way part of the make-up of the model.

With all this in place, we can then start translating facts about the model-system into claims about the world. For instance, calculations reveal that the model-earth moves on an ellipse, and given that the model-system is an ideal limit of the target we can infer that real earth moves on a trajectory that is almost an ellipse (or more precisely, on a trajectory that is not more than ε away from an ellipse).

This is a complete analysis of the model of the sun–earth system. Hence, we see that the fiction view of models is able to provide us with a complete account of how scientific models work, and it can do so without having to go at great length to reconstruct scientific practice in terms of a particular revisionary philosophy. First appearances notwithstanding, the fiction view of models is close to scientific practice and provides an analysis of modeling that scientists would recognize. The fiction view of models, then, is an account of scientific modeling that is both philosophically well founded and close to scientific practice—the kind of account of modeling that we have been looking for.

Conclusion

I have argued that scientific modeling shares important aspects in common with literary fiction, and that therefore theories of fiction can be brought to bear on issues in connection with modeling. I have identified six such issues and suggested that pretense theory offers satisfactory responses to them. From this discussion emerges a general picture of scientific modeling, which views scientific modeling as a complex activity involving the elements shown in Fig. 1. I have then used the analogy with maps to present a broad outline of an account of t-representation and have shown how this account can be used to analyze how a typical model in physics, the Newtonian model of the sun–earth system, represents.

Acknowledgements I would like to thank José Díez, Matthew C. Hunter, and Julian Reiss for helpful discussions and comments on earlier drafts. Large parts of this chapter have been written when I was a visiting fellow at the Sydney Center for the Foundations of Science. I would like to thank the Center for its hospitality and a travel grant that made the visit possible.

References

Balzer, W., Ulises Moulines, C. and Sneed, J. D. (1987), *An Architectonic for Science: The Structuralist Program*. Dordrecht: D. Reidel.

Bogen, J. and Woodward, J. (1988), "Saving the Phenomena", *Philosophical Review* 97: 303–352.

Boolos, G. S. and Jeffrey, R. C. (1989), *Computability and Logic*, 3rd ed. Cambridge: Cambridge University Press.

Brown, J. (1991), *The Laboratory of the Mind: Thought Experiments in the Natural Sciences*. London: Routledge.

Budd, M. (1992), "Review of 'Mimesis as Make-Believe'", *Mind* 101: 195–198.

Callender, C. and Cohen, J. (2006), "There Is No Special Problem About Scientific Representation", *Theoria* 55: 7–25.

Campbell, N. (1920), *Physics: The Elements*. Cambridge: Cambridge University Press. (Reprinted as *Foundations of Science*. New York: Dover, 1957).

Carnap, R. (1938), "Foundations of Logic and Mathematics", in O. Neurath, C. Morris and R. Carnap (eds.), *International Encyclopaedia of Unified Science*, vol. 1. Chicago: University of Chicago Press, 139–213.

Cartwright, N. (1983), *How the Laws of Physics Lie*. Oxford: Oxford University Press.

Cartwright, N. (1999), *The Dappled World: A Study of the Boundaries of Science*. Cambridge: Cambridge University Press.

Contessa, G. (2007), "Scientific Representation, Interpretation, and Surrogative Reasoning", *Philosophy of Science* 74, 1: 48–68.

Crittenden, C. (1991), *Unreality: The Metaphysics of Fictional Objects*. Ithaca and London: Cornell University Press.

Currie, G. (1990), *The Nature of Fiction*. Cambridge: Cambridge University Press.

Currie, G. (2004), "Imagination and Make-Believe", in B. Gaut and D. M. Lopes (eds.), *The Routledge Companion to Aesthetics*, London: Routledge, 335–346.

Da Costa, N. and French, S. (1990), "The Model-Theoretic Approach to the Philosophy of Science", *Philosophy of Science* 57: 248–265.

Davies, D. (2007), "Thought Experiments and Fictional Narratives", *Croatian Journal of Philosophy* 7, 19: 29–45.

Downes, S. (1992), "The Importance of Models in Theorizing: A Deflationary Semantic View", *Philosophy of Science (Proceedings)* 1: 142–153.

Duhem, P. (1906), *La Théorie Physique, son Objet et sa Structure*, 2nd ed. Paris: Hermann, 1914 (English trans. P. P. Wiener: *The Aim and Structure of Physical Theory*. Princeton, NJ: Princeton University Press, 1954).

Elgin, C. Z. (1996), *Considered Judgment*. Princeton: Princeton University Press.

Evans, G. (1982), *The Varieties of Reference*. Edited by John McDowell. Oxford: Oxford University Press.

Feynman, R., Leighton, R. B. and Sands, M. L. (eds.) (1963), *The Feynman Lectures in Physics*. Redwood City, CA and Reading, MA: Addison-Wesley, 1989.

Fine, A. (1993), "Fictionalism", *Midwest Studies in Philosophy* 18: 1–18.

French, S. (1999), "Models and Mathematics in Physics: The Role of Group Theory", in J. Butterfield and C. Pagonis (eds.), *From Physics to Philosophy*. Cambridge: Cambridge University Press, 187–207.

French, S. and Ladyman, J. (1997), "Reinflating the Semantic Approach", *International Studies in the Philosophy of Science* 13: 103–121.

Friend, S. (2007), "Fictional Characters", *Philosophy Compass* 2, 2: 141–156.

Frigg, R. (2003), *Re-presenting Scientific Representation*, PhD Thesis. London: University of London.

Frigg, R. (2006), "Scientific Representation and the Semantic View of Theories", *Theoria* 55: 49–65.

Frigg, R. (2010), "Models and Fiction", *Synthese* 172(2), 251–268.

Giere, R. N. (1988), *Explaining Science: A Cognitive Approach*. Chicago: Chicago University Press.

Giere, R. N. (2004), "How Models Are Used to Represent Reality", *Philosophy of Science* 71, 4: 742–752.

Godfrey-Smith, P. (2006), "The Strategy of Model-Based Science", *Biology and Philosophy* 21: 725–740.

Goodman, N. (1976), *Languages of Art*, 2nd ed. Indianapolis and Cambridge: Hackett.

Grüne-Yanoff, T. and Schweinzer, P. (2008), "The Roles of Stories in Applying Game Theory", *Journal of Economic Methodology* 15, 2: 131–146.

Hacking, I. (1983), *Representing and Intervening*. Cambridge: Cambridge University Press.

Hartmann, S. (1995), "Models as a Tool for Theory Construction: Some Strategies of Preliminary Physics", in W. E. Herfel, W. Krajewski, I. Niiniluoto and R. Wojcicki (eds.), *Theories and Models in Scientific Processes (Poznan Studies in the Philosophy of Science and the Humanities 44)*. Amsterdam: Rodopi, 49–67.

Hartmann, S. (1999), "Models and Stories in Hadron Physics", in M. Morgan and M. Morrison (eds.), *Models as Mediators: Perspectives on Natural and Social Science*, Cambridge: Cambridge University Press, 326–346.

Hempel, C. G. (1965), *Aspects of Scientific Explanation and Other Essays in the Philosophy of Science*. New York: Free Press.

Hesse, M. (1963), *Models and Analogies in Science*. London: Sheed and Ward.

Hitchcock, C. (ed.) (2004), *Contemporary Debates in Philosophy of Science*. Oxford: Blackwell.

Howell, R. (1979), "Fictional Objects: How They Are and How They Aren't", *Poetics* 8: 129–177.

Hughes, R. I. G. (1997), "Models and Representation", *Philosophy of Science* 64, Supplement: S325–S336.

Kivy, P. (2006), *The Performance of Reading: An Essay in the Philosophy of Literature*. Oxford: Blackwell.

Lamarque, P. (1991), "Essay Review of 'Mimesis as Make-Believe: On the Foundations of the Representational Arts' by Kendall Walton", *Journal of Aesthetics and Art Criticism* 49, 2: 161–166.

Lamarque, P. and Olsen, S. H. (1994), *Truth, Fiction, and Literature*. Oxford: Clarendon Press.

Laymon, R. (1991), "Thought Experiments by Stevin, Mach and Gouy: Thought Experiments as Ideal Limits and as Semantic Domains", in T. Horowitz and G. J. Massey (eds.), *Thought Experiments in Science and Philosophy*, Savage, MD: Rowman and Littlefield, 167–191.

Lewis, D. (1978), "Truth in Fiction", in D. Lewis (ed.), *Philosophical Papers*, vol. I. Oxford: Oxford University Press, 1983, 261–280.

McAllister, James W. (1997), "Phenomena and Patterns in Data Sets", *Erkenntnis* 47: 217–228.

McCloskey, D. N. (1990), "Storytelling in Economics", in C. Nash (ed.), *Narrative in Culture: The Uses of Storytelling in the Sciences, Philosophy, and Literature*, London: Routledge, 5–22.

Morgan, M. (2001), "Models, Stories and the Economic World", *Journal of Economic Methodology* 8, 3: 361–384.

Morgan, M. (2004), "Imagination and Imaging in Model Building", *Philosophy of Science* 71, 4: 753–766.

Morgan, M. and Morrison, M. (eds.) (1999), *Models as Mediators: Perspectives on Natural and Social Science*. Cambridge: Cambridge University Press.

Muller, F. A. (1998), *Structures for Everyone*. Amsterdam: A. Gerits & Son.

Parsons, T. (1980), *Nonexistent Objects*. New Haven: Yale University Press.

Rohrlich, F. (1989), "The Logic of Reduction: The Case of Gravitation", *Foundations of Physics* 19: 1151–1170.

Russell, B. (1905), "On Denoting", in *Logic and Knowledge*. London: Routledge, 1956, 39–56.

Russell, B. (1919), *Introduction to Mathematical Philosophy*. London and New York: Routledge, 1993.

Shapiro, S. (2000), *Thinking About Mathematics*. Oxford: Oxford University Press.

Sismondo, S. and Chrisman, N. (2001), "Deflationary Metaphysics and the Nature of Maps", *Philosophy of Science (Proceedings)* 68: 38–49.

Sklar, L. (2000), *Theory and Truth: Philosophical Critique Within Foundational Science*. Oxford: Oxford University Press.
Smith, L. (2007), *Chaos: A Very Short Introduction*. Oxford: Oxford University Press.
Smolin, L. (2007), *The Trouble with Physics: The Rise of String Theory, the Fall of a Science, and What Comes Next*. London: Allen Lane.
Sorensen, R. (1992), *Thought Experiments*. New York: Oxford University Press.
Suárez, M. (2003), "Scientific Representation: Against Similarity and Isomorphism", *International Studies in the Philosophy of Science* 17, 3: 225–244.
Suárez, M. (2004), "An Inferential Conception of Scientific Representation", *Philosophy of Science (Supplement)* 71: 767–779.
Suárez, M. and Solé, A. (2006), "On the Analogy between Cognitive Representation and Truth", *Theoria* 55: 39–48.
Sugden, R. (2000), "Credible Worlds: The Status of Theoretical Models in Economics", *Journal of Economic Methodology* 7, 1: 1–31.
Suppe, F. (ed.) (1977), *The Structure of Scientific Theories*. Chicago: University of Illinois Press.
Suppes, P. (1960), "A Comparison of the Meaning and Uses of Models in Mathematics and the Empirical Sciences", in P. Suppes (ed.), *Studies in the Methodology and Foundations of Science: Selected Papers from 1951 to 1969*, Dordrecht: Reidel, 1969, 10–23.
Teller, P. (2001), "Twilight of the Perfect Model Model", *Erkenntnis* 55: 393–415.
Thomson-Jones, M. (2007), "Missing Systems and the Face Value Practice", *Synthese* (forthcoming).
Toon, A. (2010), "Models as Make-Believe", in R. Frigg and M.C. Hunter (eds.), *Beyond Mimesis and Convention: Representation in Art and Science*, Berlin: Springer.
Vaihinger, H. (1911), *The Philosophy of 'As If': A System of the Theoretical, Practical, and Religious Fictions of Mankind* (English trans. C. K. Ogden). London: Kegan Paul, 1924.
van Fraassen, B. C. (1980), *The Scientific Image*. Oxford: Oxford University Press.
van Fraassen, B. C. (1989), *Laws and Symmetry*. Oxford: Clarendon Press.
van Fraassen, B. C. (1997), "Structure and Perspective: Philosophical Perplexity and Paradox", in M. L. Dalla Chiara (ed.), *Logic and Scientific Methods*, Dordrecht: Kluwer, 511–530.
van Fraassen, B. C. (2002), *The Empirical Stance*. New Haven and London.
Walton, K. L. (1990), *Mimesis as Make-Believe: On the Foundations of the Representational Arts*. Cambridge: Harvard University Press.
Woodward, J. (1989), "Data and Phenomena", *Synthese* 79: 393–472.
Young, H. D. and Freedman, R. (2000), *University Physics with Modern Physics*. 10th ed. San Francisco, CA and Reading, MA: Addison Wesley.

Fictional Entities, Theoretical Models and Figurative Truth

Manuel García-Carpintero

Preamble

In setting up his influential "constructive empiricist" project, Bas van Fraassen (1980, 12) characterizes realism about scientific theories by the following three claims: (i) Scientific theories should be interpreted "at face value". If the theory includes the sentence "there are quarks", it should be understood as making the same kind of claim we make when we say "there are cans of beer in the refrigerator": there is no reinterpretation. (ii) Scientific theories purport to be true. (iii) We may in principle have good reasons for believing that a scientific theory is true.

Anti-realism, on the other hand, can take two forms, according to van Fraassen. Traditional instrumentalism or empiricism is a form of reductionism, which accepts (ii) and (iii), but rejects (i), offering instead a reinterpretation of the claims made by scientific theories on which they are not about things such as quarks, but rather about, say, possible courses of perceptual experiences. By contrast, constructive empiricism accepts (i), but rejects instead (ii) and (iii). The view is a form of fictionalism. When Conan Doyle writes "Holmes lives in Baker Street", he is uttering a sentence that, taken literally, is supposed to refer to a detective, a person called "Holmes", and to ascribe a certain location in space to his lodgings. No reinterpretation is required to understand the sentence that Conan Doyle is uttering, and none would be adequate to understand *him*. He is putting forward an untrue claim, untrue for lack of reference of the singular term "Holmes". However, Conan Doyle is not purporting to assert an untrue claim of this kind, still less assuming that he could be in a position to know it. He is doing something else; the same, according to van Fraassen, applies to the proponents of scientific theories.[1]

M. García-Carpintero (✉)
LOGOS-Universitat de Barcelona, Barcelona, Spain
e-mail: m.garciacarpintero@ub.edu

[1] No matter how they themselves reconstruct their own aims in their philosophical moments; this is not a psycho-sociological claim, but a philosophical one about the nature of scientific practice (cf. van Fraassen 1994).

R. Frigg, M.C. Hunter (eds.), *Beyond Mimesis and Convention*, Boston Studies in the Philosophy of Science 262, DOI 10.1007/978-90-481-3851-7_7,

Hartry Field (1980) propounds a similar view, this time about mathematics. Nominalism is the doctrine that there are no abstract objects. W.V. O. Quine, Hilary Putnam and others used indispensability considerations to reject nominalism: our best "world theory" refers to, and quantifies over, abstract mathematical entities, so, according to Quine's well-known criterion for ontological commitment, abstract entities exist. Traditional nominalism would respond to this along the lines of traditional instrumentalism: it would try to propound reinterpretations of claims apparently involving reference to or quantification over abstract entities, on which these appearances would have vanished. In contrast to this, Field purports to retain the "standard semantics" for those claims; given his nominalistic leanings, this means that he takes them to be false. His project is to show that scientific theories can be reformulated in nominalistic terms, and that mathematics, even if convenient in practice, is not required either to draw consequences from nominalistically formulated theories.

These proposals have faced up to serious criticisms, some of which will come up later. What I propose to do here is to examine in some detail two cases for which a fictionalist treatment is, I think, less controversial: the case (to be distinguished, as it will become clear, from the Conan Doyle example just mentioned) of explicit reference to, and quantification over, fictional characters; and the case of reference to imaginary models in science and their components, frictionless planes and the rest. I will argue in the first place that an anti-realist, fictionalist reading of statements explicitly referring to fictional characters is more adequate than realist proposals, but also than other critical stances like that of Kendall Walton (1990) or Mark Sainsbury (2005), closer to the reductionist traditional antirealism about theoretical entities in science and abstract entities. In parallel, I will be contrasting the fictionalist proposal about fictional characters with a similar view about the models that many scientific theories appeal to; as will become clear, while I do not think that van Fraassen's fictionalist empiricism can be sustained for scientific claims purporting to refer to theoretical entities, a fictionalist view is defensible for apparent reference to models and their components in science. I will thus be drawing on two apparently unrelated disciplines, the philosophy of literature and the philosophy of science, aiming thus to illuminate in this way the nature of fictionalist proposals, their strength and limits.

Apparent Reference to Fictional Characters

Consider an utterance of (1) below by Vargas Llosa, as part of his longer utterance of the concrete full discourse that, with a measure of idealization, we can think constitutes the creation of his novel *Conversación en La Catedral* (CLC for short henceforth). (It is of course part of the idealization that we should rather be speaking of an utterance of the Spanish sentence "desde la puerta de *La Crónica* Santiago mira la avenida Tacna, sin amor", actually part of the story created by Vargas Llosa and published in 1969.)

(1) From the doorway of *La Crónica* Santiago looks at Tacna Avenue without love

(1) is in the declarative mood, which by default expresses in English assertion. Nonetheless, most accounts of fiction would not count such an utterance as assertoric in illocutionary force at all: the context in which it occurs overrides the default interpretation for (1)'s mood. On the account I (2007) have advanced, close to Gregory Currie's (1990) and similarly inspired by Walton's (1990) work, the utterance of (1) counts in the indicated context as a different speech act, guided by the communicative intention to lead audiences with appropriate features to imagine the propositions constituting the fiction's content. This view is in line with the main claims of the proposals by van Fraassen and Field mentioned in the preamble. Taken literally, (1) signifies an untrue proposition for lack of reference of "Santiago" (or is untrue because it does not signify any proposition), which we do not have good reasons to believe. However, this is no problem, because it has not been put forward as truth. It has been uttered with different purposes than those characterizing straightforward assertions: something such as putting us in a position to imagine an interesting and entertaining story.

Consider however a different speech act that one could make in uttering (1) with Vargas Llosa's story in mind. One who is familiar with the story could utter (1) in the context of telling someone else, or otherwise discussing, the content of the story, its plot, what goes on in it, for instance by uttering (1) after saying "the story begins telling us about the thoughts of someone called 'Santiago', a.k.a. 'Zavalita'." In such a context, the utterance does constitute a true assertion. But there is an obvious problem here: what is the contribution of those referential expressions made up by Vargas Llosa, such as "Santiago"? According to a well-known view, developed among others by David Lewis (1978), in the logical form of the relevant assertions of (1) there is an implicit operator, "CLC makes it fictional that ...", which behaves in closely similar ways to operators very much studied in contemporary semantics, like "S believes that ...". To the extent that we can invoke a semantic account of the significance of referential expressions when they occur in contexts governed by those operators on which they do not necessarily contribute their ordinary referents outside them, we avoid any problems caused by their lacking those referents.[2] Let us use "$\mathbf{F}_{\mathbf{clc}}$ (p)" as an abbreviation of "CLC makes it fictional that p"; (2) would then capture what is asserted by uttering (1) in the indicated context:

(2) $\mathbf{F}_{\mathbf{clc}}$(from the doorway of *La Crónica* Santiago looks at Tacna Avenue without love)

If we turn now, however, to a different kind of utterance we can make still with Vargas Llosa's story in mind, which (3) illustrates, we can see both that it is also an assertoric one, and that the operator strategy is of no use here:

[2]But only to the extent that we can so rely on such a neo-Fregean account of singular reference in indirect contexts. In my (2010) I argue that referentialist or neo-Russellian accounts, such as the one by Evans and Walton, cannot provide an acceptable semantics for the cases we are considering.

(3) Zavalita is one of the most memorable fictional characters created by Vargas Llosa

Peter van Inwagen (1977, 2003) has argued that an acceptable semantic account of the content of assertions like (3) requires an ontology of "creatures of fiction", fictional characters genuinely referred to by singular terms like "Zavalita", as used in it. His argument is shaped by Quine's well-known ontological views, which, as mentioned in the preamble, van Fraassen's and Field's anti-realism confronts. Van Inwagen in fact compares the Quinean considerations speaking for creatures of fiction to those speaking for mathematical entities, and theoretical entities in general, like genes or black holes. He shows how statements like (3) are inferentially related, through the positions occupied by referential expressions like "Zavalita" in (3), with existential claims, which we take to express true assertions as much as related claims involving sets, numbers, genes and black holes. Those existential claims are often very complex: In some novels, there are important characters who are not introduced by the author till more than halfway through the work. To avoid the ontological commitment apparently incurred, it is not an option, thus, to stop uttering those sentences which most of us consider appropriate with respect to apparent commitments to, say, witches or alien abductions.

Van Inwagen (2003) helpfully summarizes the main tenets of realist views, which he contrasts with Meinongian views such as the one contemporarily espoused by Terence Parsons (1980). Van Inwagen rejects the Meinongian account, on which "Zavalita" in (3) refers to something "of which it is true that there is no such thing", convincingly arguing that it is either contradictory, no matter which apparently consistent paraphrase we use of its main paradoxical claim, or requires a distinction between two kinds of quantifiers (an absolutely unrestricted one, and another restricted to those things that have being) that he claims is not forthcoming. Fictional Realism consists, according to him, of two main claims (2003, 147–148): (i) Fictional characters exist or have being. (ii) What appears to be the apparatus of predication in fictional discourse is ambiguous; sometimes it expresses actual predication, the having of properties; sometimes an entirely different relation, the three-place *ascription*, or the two-place *holding*. Thus, in uttering "Holmes is famous", we could be straightforwardly ascribing the property of being famous to the fictional character, or rather saying that in the Holmes stories he is described as being famous. These would be the common tenets of all forms of fictional realism; other than that, they can differ substantially.

Thus, on the account provided by Wolterstorff (1980), they are eternal, Platonic abstract universals constituted by all the features that the relevant fiction directly and indirectly ascribes to a pretended referent, typically (although not only so) by relying on the use of a fictional name, "Santiago"/"Zavalita" in CLC. This has several problems. It makes the activity of Conan Doyle merely one of, as it were, bringing the atemporally existing character Holmes to the attention of his readers. Similarly, on this view it is difficult to make sense of counterfactual claims about different features that the Holmes character could have had, conditional on decisions taken by Conan Doyle.

On an alternative view defended by Amie Thomasson (1999, 93–114),[3] characters are literally brought into existence by their creators, are only constituted (in addition to the act of creation by their authors) by some of the features ascribed to them in the fiction, and could even cease to exist under some circumstances. This is intuitively more plausible; it makes better sense of intuitions that the content of claims such as (3) is somehow singular, concerning specific individuals and not just general existential characterizations; that fictional characters to which the same general features are ascribed in causally unrelated fictions are different; that creatures of fiction are quasi-abstract entities, which, although they are not located along a particular line through space-time, do have a particular origin in time (the more or less definite time of the creation of the relevant "text"), and perhaps also an end; that a given character originated in a certain fiction can reappear, even if in a distorted manner, in another; and that one and the same fictional character might have somehow had different properties than the one ascribed to it in a given fiction. As van Inwagen points out (2003, 153–154), though, it is not clear that it is metaphysically possible, for it is not clear that there can be *created* abstract objects.

However, it is not enough to assume the ontology of fictional entities and posit them as the referents of expressions such as "Santiago"/"Zavalita" in (3) for realist accounts to work. There is still much more work to do, and it is unclear that it can be done without in effect invoking the apparatus of pretenses and imaginings deployed in non-realist accounts like the ones to be discussed later.[4] Thus, for instance, even if our intuitions concerning (3) might straightforwardly suggest an ontology of fictional entities, the case of "Zavalita does not exist", as Anthony Everett (2007) insists, points in the opposite direction. Going back to the two uses of (1) I mentioned before, the one by the creator of the fiction, and the one by someone uttering it in order to state the content of the fiction, we find versions of this very same difficulty. Thus, as David Braun (2005, Section 6) emphasizes with regard to Nathan Salmon's (1998) proposal, it is not clear how referential expressions in both those uses (by the fiction-creator, and by "critics" discussing its content) can refer to any entity, fictional or otherwise, if the referential intentions of their users in no way underwrite this. Similarly, as we have seen, the realist must distinguish predications in which properties are ascribed to fictional entities as such (*being famous, being a fictional entity*) from predications ascribing properties they only fictionally have (*eating inner organs*), and they should explain what in the intentions and thoughts of speakers underwrites this distinction.

A parallel problem can be put to a parallel proposal for the parallel case I would like to consider vis-à-vis that of reference to fictional characters in statements such as (3), reference to hypothetical, unreal models in science and their hypothetical constituents. Thus, consider cases such as those discussed by Adam Toon in his

[3]Related views are put forward by Currie (1990), Lamarque and Olsen (1994), Schiffer (2003) and Voltolini (2006).

[4]Friend (2007) helpfully summarizes the difficulties for realist accounts, among them the ones I am interested in, to be mentioned presently.

contribution to this volume. We want to predict the behavior of a real bob bouncing on the end of a spring. In order to do so, we provide what Nancy Cartwright (1983) calls a "prepared description" of the bouncing spring system. We use Hooke's law to formulate the equation of motion for a simple harmonic oscillator, $m\, d^2x/dt^2 = -kx$, where m is the mass of the bob, k is the "spring constant" and x is the distance that the spring has been stretched or compressed away from the equilibrium position, the position where the spring would naturally come to rest. In using this equation we make a number of assumptions, among them (4):

(4) The bob is a point mass *m* subject only to a uniform gravitational field and a linear restoring force exerted by a massless frictionless spring with spring constant *k* attached to a rigid surface

Ronald Giere (1988) has provided an account of statements such as this analogous to van Inwagen's for (3), on which expressions such as "the bob" in (4) refer to abstract objects. As Peter Godfrey-Smith (2006, 735) points out, however, this posits a similar problem to the one discussed for the abstract fictional entities account of (3): "modelers often *take* themselves to be describing imaginary biological populations, imaginary neural networks, or imaginary economies. An imaginary population is something that, if it was real, would be a flesh-and-blood population, not a mathematical object". The same applies to our example; the modeler may well take himself to be referring to an imaginary bob, which could be exactly the real bob we are studying if the idealizations we are assuming became actual fact.

The first objection is then that, even if our intuitions about claims such as (3), and the related quantificational claims that van Inwagen provides, suggest that we contemplate the ontology that the fictional realist ascribes to our discourse, the ascription of that ontology is at odds with other equally relevant facts about speakers' thoughts and intentions. A second compelling objection to both forms of realism, about fictional characters and models, derives from what I take to be the main features of the robust views on reference that Saul Kripke's (1980) influential work has made prevalent today. In a nutshell, the second objection is that the acts of reference we seem to make in cases like (3), unlike paradigm cases of referential acts (such as referring to persons and places), appear to be very easily justified as correct; it just requires a proper set of intentions, or perhaps conventions, to guarantee their success.

Relying on the prejudices defining the philosophical landscape when that work was published, Quine took for granted that it was enough to establish that use of quantificational modal logic commits one to Aristotelian essentialism, to discredit thereby serious applications of that logical theory.[5] Quine disagreed with Rudolf Carnap and other philosophers on whether there was a distinctive class of necessary truths; but he shared with them the empiricist assumption that, if it exists, it

[5]As he himself emphasized, according to Quine the commitment to Aristotelian essentialism does not lie in that a proposition stating it is a theorem of the logical theory, but depends on its use. See Burgess (1998) and García-Carpintero and Pérez Otero (1999).

coincides with those of analytic and a priori truths: necessity has a linguistic foundation, if it has any at all, which for Carnap and other empiricists meant a foundation on convention.

Kripke proposed compelling examples, and on their basis provided clear-cut distinctions and forceful arguments. He distinguished genuinely referential from descriptive denoting expressions. He argued that referential expressions like indexicals and demonstratives, proper names and natural kind terms are *de jure* rigid designators; this distinguishes them from other singular terms like definite descriptions, which might also behave *de facto* as rigid designators, but *de jure* are not so.[6] On this basis, he took away the force of the only argument that Quine had provided against essentialism, based on the claim that no object instantiates *de re* essentially or contingently any property, but only relative to different ways of referring to it. Quine argued that, even if the world's tallest mathematician is in fact the world's tallest cyclist, he is not *de re* necessarily rational or two-legged, but only *de dicto*, necessarily rational as the world's tallest mathematician, necessarily two-legged as the world's tallest cyclist. This is plausible for this case. However, in order to generalize this Quinean argument we would need to overlook the distinction between rigid and nonrigid designators. The issue is whether modal claims we make using rigid designators, as when we say that Socrates is necessarily human, or Phosphorus necessarily identical to Hesperus, are only true *de dicto*, when some appropriate description is provided, or rather, as they seem to be, *de re*, true given the natures of the entities we are talking about, independently of the particular way we choose to pick them out. Relatedly, and also importantly, Kripke distinguished epistemic from metaphysical necessity. Some truths, he argued, are a priori, but nonetheless contingent; some other truths are necessary, but nonetheless a posteriori.[7]

In this way, Kripke undermined dogmatic rejections of essentialism based more on philosophical prejudice than sound argument, vindicating a traditional anti-empiricist view. A striking manifestation of this lies in the well-known consequence of Kripke's view on reference, that there are modal illusions, propositions that are in fact necessary but appear to be contingent. Paradigm cases are instances of the schema ***if n exists, n is F***, with a rigid designator in the place of "n" and a predicate signifying a hidden essential property of its referent in the place of "F". A familiar illustration is this:

(5) If water exists, water contains hydrogen

---

[6]A *rigid designator* is an expression that designates the same entity in all possible worlds in which it designates anything at all, unlike designators such as the description "the inventor of the zip". Descriptions such as "the actual inventor of the zip" and "the even prime" are rigid designators, but, unlike proper names and indexicals, merely *de facto*, not *de jure*. Kripke does not define how he understands the latter distinction. In my view, the suggestion is that *de jure* rigid designators designate rigidly in virtue of the semantic category (*proper name*, *indexical*) to which they belong; *de facto* rigid designators are definite descriptions which, even though as such are non-rigid, designate rigidly by virtue of features of the properties signified by the NP that compose them.

[7]See Soames (Chapter 14), for an excellent presentation of these issues, on which I draw.

Of course, if one adopts a Platonistic attitude towards mathematics, one will be prepared to accept that some mathematical claims are true, and therefore necessary, without perhaps being provable unless through empirical evidence, for instance by essentially relying on the opaque calculations of computers one takes to be reliable. What is interesting in Kripke's arguments is that they do not depend on such controversial ontological assumptions as Platonism; they just rely on an intuitively well-supported view about reference, and in compelling considerations to disregard philosophical prejudices veiling them from us.

In the presence of these Kripkean views just outlined, there is another compelling objection to realism about fictional characters and theoretical models, that is, that it overlooks an important distinction. It intuitively seems that the commitment we incur when we refer to and existentially quantify over theoretical entities like genes and black holes and the one we incur when we refer to and existentially quantify over fictional characters or hypothetical bobs are rather different, in epistemologically and ontologically significant ways. Those of us sharing the realist attitudes congenial to the Kripkean views on reference will not feel that it is at all appropriate to invoke the sort of Tolerance advocated by Carnap through the famous Principle (which I will compare in the afterthought to the view I will be defending), with respect to the first commitments, involving theoretical entities like genes and black holes: there are "morals" in this case; successful reference to these entities is not just a matter of convention; it might be perfectly in order here to set up "prohibitions", in the way that further knowledge of the way the world is led us to "prohibit" reference to phlogiston. Carnapian Tolerance intuitively appears to be in order, however, with respect to the second commitments, those involving fictional characters and hypothetical bobs. It intuitively seems that, in this case, entering the appropriate conventions suffices for successful reference.

This is just an intuition, in need of theoretical articulation; let me elaborate slightly, before offering such an articulation. When we refer to, and quantify over, genes and black holes we incur a commitment to the existence of entities that we take to have a hidden essence, one that can only be discovered empirically, if at all. Typically, as props for our referential practices, we rely on reference-fixing stipulations;[8] but we do not have any a priori guarantee that they will succeed in securing reference to anything. The world has to oblige, so to say. It is in this way that, when the world does cooperate, *de re* necessary a posteriori truths such as (5) can be expressed. But none of this is the case with respect to the commitment we incur in making assertions like (3) and (4). As Stephen Schiffer (1996, 159) puts it with respect to the former sort of case, following Mark Johnston's (1988) similar proposals concerning reference to propositions in theories of meaning, while genes and black holes have hidden and substantial natures for empirical investigation to discover, "there can be nothing more to the nature of fictional entities than is determined by our hypostatizing use of fictional names. The 'science' of them may be

[8] In my (2000, 2006a) I argue that this is not just "typically" so, but conceptually necessary, and I provide on this basis a descriptivist framework for capturing the Kripkean rigidity intuitions.

done in an armchair by reflective participants in the hypostatizing practice". He characterizes this as a "*something-from-nothing* feature": A trivial transformation takes one from sentences in which no reference is made to fictional characters—sentences like (1), in both of its uses discussed above, the one by the creator of the fiction, and the one by someone uttering it in order to state the content of the fiction—to sentences containing a singular term whose referent is a fictional character—(3).

To sum up: Although, as we have seen, utterances such as (3) and (4) appear to provide a good case for fictional realism, there are also important problems with this view. In the first place, it is not clear how to provide an intuitively convincing elaboration of the view, beyond van Inwagen's two defining traits. In the second place, there are compelling intuitions at least as relevant as those afforded by (3) and (4) which are at odds with it. Finally, the success of apparent references to fictional characters seems to be suspiciously easy to achieve.

We have not yet explored, for the case of statements such as (3) and (4), the kind of anti-realist alternative to realism that van Fraassen and Field rule out, the reductionism corresponding to traditional instrumentalism and traditional nominalism: to provide non-committal paraphrases allegedly representing what is said. Walton (1990) has appealed to his influential make-believe theory of fiction to argue in favor of this alternative, and different writers, including Toon in this volume and Roman Frigg (2010, see "Fiction and Scientific Representation" this volume) have explored similar proposals for the case of models.[9] However, even if the use Walton makes of the make-believe account is illuminating, some of the paraphrases he provides are strained and ad hoc, and there is no guarantee that a paraphrase will always be forthcoming, for any claim we want to assert *prima facie* committing us to the existence of fictional characters.

Consider for instance the case of (1) when it is uttered in order to state the content of the fiction. Walton's main idea is that by making such utterances we primarily illustrate by exemplification acts made fictional by the fiction, in the present case CLC. It is not just what intuitively constitutes the content of such a fiction that is fictional, or correctly imagined when appreciating it; the fiction also makes it fictional—i.e., authorizes us to imagine—that we make correct speech acts in reaction to it, such as true assertions. By uttering (1), we are showing one of those speech acts which it is legitimate to imagine, and thereby asserting by means of this act of exemplification that it is *also* made fictional by Vargas Llosa's fiction that one who asserts in response to it that from *La Crónica*'s doorway Santiago looks at Tacna Avenue without love, asserts truly: "when a participant in a game of make-believe authorized by a given representation fictionally asserts something by uttering an ordinary statement and in doing so makes a genuine assertion, what she genuinely asserts is true if and only if it is fictional in the game that she speaks truly" (Walton

[9]Sainsbury (2005) also favors such an alternative. In Chapter 6 of his forthcoming book *Fiction and Fictionalism*, however, he adopts a more open view; the suggestion there that I find more congenial, to appeal to a relativized notion of *truth on a presupposition*, is, I take it, very close to the one I will be making, perhaps they are just notational variants.

1990, 399).[10] It is this kind of convoluted claim that we could properly assert by prefixing (1) with the "CLC makes it fictional that" operator, as in (2). Once this is in place, Walton extends the idea to account for assertions such as (3) by appealing to more or less ad hoc "unofficial games", which draw on different fictions and/or implicit ad hoc "principles of generation" (1990, 405–416).

This is an interesting suggestion, which nonetheless I do not think we should accept. Van Inwagen (2003, 137 footnote) objects that it does not seem that the typical utterer of "in some novels, there are important characters who are not introduced by the author till more than halfway through the work" is doing something different than what he does in uttering "some novels are longer than others", i.e., to make a straightforward assertion about its apparent subject-matter, as opposed to one about what it is legitimate to imagine in unofficial games given their implicit principles of generation. Similarly, Mark Richard (2000, 209–212) cannot find any good reason to think that when ordinary speakers utter (1) in the envisaged context they are performing the quite complex task of engaging in pretense in order to discuss the pretense performed, as opposed to saying, of what is said by (1), that it is "true in CLC". Even if, I am afraid, these writers would object along similar lines to the proposal I will make, I think it at least has more resources to answer them.

There is thus some motivation to look for the sort of alternative to realism that van Fraassen's and Field's proposals illustrate. In the next section I will present such an account for the case of apparent reference to fictional entities, as in (3); in the section "Scientific Models as Fictions" I will discuss the case of apparent reference to hypothetical models, as in (4). The idea I will be developing is as follows. When Romeo utters "Juliet is the sun", he is obviously not asserting the semantic content of that sentence, although we must assume that the sentence does have that semantic content, if we want to understand what he is in fact doing. As in the cases theorized in fictionalist accounts such as van Fraassen's and Fields', the sentence has its ordinary semantic content, but its utterer cannot properly be faulted on account of having made a wrong assertion, because he is not in fact asserting that semantic content. Nevertheless, Romeo is indeed asserting something, although there is no reason to assume that there is going to be a uniquely correct paraphrase of what he has in fact asserted; its determination depends on the vagaries of interpretation.

The same applies to the utterer of (3) and (4). These sentences involve *hypostasizing* or *reifying* fictional characters and fictional massless frictionless springs;

---

[10]There is a problem here posed by Walton's commitment to neo-Russellian referentialism, which I have mentioned in a previous footnote: "If there is no Gulliver and there are no Lilliputians, there are no propositions about them" (Walton 1990, 391). As Walton notes (1990, 400), the class of pretended assertions thus authorized by a given fiction should be characterized semantically, and it remains totally unclear how, under Walton's referentialist assumption, this can be done. The account should allow that a Spanish speaker who reacted to CLC by uttering a Spanish translation of (1) would thereby be making an equally true claim. Thus, Walton's account appeals to "kinds" of pretenses. But how can "Santiago" semantically contribute to characterizing any such kind of pretense, if it lacks semantic content? However, this could be solved by adopting a less radical form of referentialism, for instance one envisaging "gappy" singular propositions, as I suggest in my (2010).

I take reification to be understood so that, while the literal contents of the likes of (3) and (4) do involve purported reference to such fictional entities, this is just a figurative manner of speaking with respect to what speakers ultimately are doing. The apparently purported literal reference is doomed to fail, because (for all we need to be committed to, in order to properly account for our data) there are no such things. But the utterer cannot be faulted, because he is not engaged in asserting those contents. He is indeed asserting, but he is asserting something else, even if typically there is no uniquely correct paraphrase of the content(s) he is really asserting. In the same sense that Romeo is using metaphorically the predicate "is the sun", I will be claiming that to *hypostasize* or *reify* fictional entities as in (3) and (4) does involve a metaphorical use of the apparatus of singular reference.[11]

Genuine vs. Figurative Reference

In uttering (1) in the context of producing the discourse that constitutes CLC, Vargas Llosa, we said, was not really asserting a proposition; he was merely pretending to do so, for fiction-making purposes, i.e., to lead potential audiences to carry out some imaginings. Pretending to assert is not the only way of making fiction, against what John Searle (1975) claims; fiction can be made by arranging color patches on a canvas, or by filming people pretending to act in certain ways, and none of these requires the pretense of assertion. But in literary fiction, pretending to assert (and to ask, to request, and so on) is the usual way; and the pretended assertions usually also involve pretended references as an ancillary tool.

Speech acts like assertion do not typically occur in a vacuum, but in a cognitive background of shared knowledge, with which they dynamically interact (Stalnaker 1978). Real assertion usually involves ancillary real references, which must be understood relative to this dynamic aspect of the speech acts to which it contributes. Reference is an ancillary speech act[12], with communicative purposes such as leading the audience to attend to the referents, or having the audience use the referential expression as a label to create a "dossier" or "file" (Perry 1980) where to pile up different pieces of information about the referent. The referential expression thus serves as a sort of *anaphoric node* throughout a discourse; that is to say, it indicates co-reference throughout its different uses, and thus helps the audience to collect together the different pieces of information thus imparted

[11] If metaphor is itself a form of fiction, as Walton (1993) contends, then reference to fictional character is itself a straightforward form of fiction. However, I find Walton's assimilation of metaphor-making to fiction-making almost as much strained and *ad hoc* as his paraphrasing-away fictional characters, even if also illuminating.

[12] Speech acts such as assertions have contents, such as the asserted proposition, the proposition the belief of which the utterer expresses, or to whose knowledge he commits himself, depending on what the proper account of assertion is; reference, I take it following Searle's views on speech acts, is an auxiliary act through which "components" of those contents such as objects and properties are specified.

about the purported referent. In real reference, shared descriptive information (say, that the referent is called "Santiago", or that it is whoever uttered the relevant token of "I") is used for reference-fixing purposes, and new descriptive information obtained from unchallenged assertions adds to the relevant "file". However, on the Kripkean view I outlined before, the contribution of genuinely referential expressions to the content of the assertions and other speech acts is the object itself, with its perhaps hidden substantive nature. When we estimate the possible worlds truth conditions of those assertions, the descriptive information that is taken for granted to apply to the referent is irrelevant; it is only the object itself, with its perhaps hidden essence, which matters. This is why the contents of assertions like (5)—or, instantiating the schema with singular terms, "if Phosphorus exists, Phosphorus is-identical-with-Hesperus"—might be necessary but apparently contingent propositions.

In pretending to make an assertion with (1), Vargas Llosa also pretends to refer to someone called "Santiago".[13] But this is mere pretense; the contribution of the expression to the content of his act of fiction-making (the proposition his fiction thereby prescribes his audience to imagine) is not an object, but that of a description understood à la Russell, as a quantifier[14], collecting the information that would go into the relevant file, in an imaginary context in which the acts were not pretended but actually performed: whoever is called "Santiago", who was looking without love at an avenue called "Tacna" from the doorway of a newspaper called "La Crónica" . . .). Correspondingly, although embedded referential expressions in attitude reports might well be genuinely referential (when the reported propositional attitudes themselves involve genuine reference), those of expressions like "Santiago" in the second, assertoric use we considered before for (1)—the one whose logical form (2) captures—are merely descriptive.[15] Thus, mere pretense of reference obtains when Vargas Llosa uses "Santiago" in his own fiction-making utterance of (1); and the assertoric utterances of (1) intended to report the content of the fiction he thereby created, although not pretended at all, do not involve genuine reference to anybody called "Santiago" either.

What about the referential expression "Zavalita" in (3)? Although I share to a large extent his intuitions, I do not find Schiffer's (1996) discussion clear, for reasons

---

[13]I also think that, relative to the speech-act of fiction making, Vargas Llosa merely pretends to refer to a newspaper called "La Crónica" and to an avenue called "Tacna", even though there actually were entities answering to those descriptions in Lima at the time of the narrative and, if (1) were used literally in a relevantly corresponding context, those names would genuinely refer to them. Now, in the same way that a fiction-maker might well make genuine assertions indirectly, through his fiction-making, he can also make genuine references (in our case, to the newspaper and street)—but in my view only indirectly.

[14]I am here assuming Kripke's (1977) Russellian view that definite descriptions, when literally used, are not referential but quantificational expressions.

[15]Currie (1990, 146–162) makes a similar proposal. The main difference with the one I elaborate upon elsewhere (2007, 2010) lies in that, where Currie's account posits a fictional author who fictionally produced the token-discourse by whose production the relevant fiction was created, mine has the real author actually producing that token-text.

like those that Amie Thomasson (2001) gives. Schiffer contends that entities introduced through processes with the "something-from-nothing" feature are in some sense language-created, and also that the terms referring to them are guaranteed of reference. But, just to concentrate on the example we are discussing, none of these contentions is true of claims like (3).[16] We can imagine situations in which "Zavalita" as used there lacks reference; this would occur, for instance, if, contrary to what the utterer assumes and is in fact the case, Vargas Llosa's narrative was not fiction at all, but history. And this shows also why Schiffer's first contention is false. There is a convention, or (perhaps better put) a practice, of fiction-making; there are standard ways of indicating that one agrees to place oneself under the norms constituting this practice. It might well involve the use of language, and it typically does. But there is no interesting sense in which this is a *linguistic* practice; it is no more a linguistic practice than promising, voting or marrying are, all of them convention-governed practices that also typically involve the use of language at crucial points. The existence of this convention is a prerequisite for attempted reference to fictional characters, as in (3), to be successful; unless, by invoking the rules constituting of that practice, Vargas Llosa created CLC, the attempted reference to a fictional character would be unsuccessful. Thus, the hypostatizing use of fictional names as in (3), by itself, is insufficient to create fictional characters; and what else is needed is not in any interesting sense linguistic in character. We cannot thus make good sense of the claim that they are language-created entities.[17]

There are additional reasons to doubt that we have any entities here, created or pre-existing. "No entity without identity", the Quinean motto goes; but, as Alberto Voltolini (2006, 209) admits, "the problem with the community of uruk-hai (as well as with that of dwarves, elves, hobbits, etc.) is that the identity of these alleged characters is totally indeterminate. How many uruk-hai are there in the fictional 'world' of Tolkien?" Everett (2005) forcefully presses this point. Imagine a fiction introducing two characters, one called "pseudo-Hesperus" and another "pseudo-Phosphorus", which manifestly leaves unsettled the issue of whether or not pseudo-Hesperus is pseudo-Phosphorus. How about the fictional characters? Do we have one, or two, on account of this fiction? Similar issues arise with respect to characters from one fiction occurring in others. Is the gay Holmes of post-modernist parodies the same character as the one introduced in Conan Doyle's stories? What about Joyce's Bloom vis-à-vis Homer's Ulysses? If fictional characters exist and we do refer to them, these questions should have answers, even if we are never able to find them.

In my view, the most natural reaction to this conundrum is to reject the issue, by contending (in the Carnapian spirit outlined in the afterthought) that we stipulate fictional characters into existence, and are thereby free to answer those questions

---

[16]It is easy to see that the point also applies to other entities that Schiffer takes to be introduced in that way, like properties, events, possible worlds or propositions.

[17]Schiffer (2003) contains a new proposal, still ontologically deflationary, which is not subject to these criticisms, but it has the problems discussed in the following paragraph.

as we see fit; and the most useful theoretical proposal to account along these lines for the difference we intuitively see between reference to fictional characters and reference to genes is Yablo's (2001) suggestion to "go figure": it is only figuratively or metaphorically speaking that we refer to fictional characters. (Yablo applies his proposal to mathematical objects; here I suspend judgment on the application of the view I am advancing to this and other philosophically controversial cases, like properties, propositions or possible worlds.)

Research on metaphorical discourse is hardly in a position to provide a full-fledged account of the phenomenon, philosophically and linguistically accurate. Fortunately, we do not need that to make a plausible case for a figurativist account of reference to fictional characters.[18] It suffices that we can show that such references appear to have the main, uncontested features of paradigm metaphors that, in one way or other, the different proposals capture. In order to show that, we should use the resources of some sufficiently promising account, to the extent that they could be translated, for the cases we are interested in, onto those of other similarly plausible accounts. With that goal in view, I might as well resort to the proposal that I find most congenial.

On what I find to be the best accounts of metaphor, such as Kittay's (1987), a metaphorical piece of discourse has the following features. In outline: (i) It involves a (perhaps improper) part, the metaphorical vehicle. (ii) The vehicle has a primary literal meaning. (iii) Throughout the Gricean mechanism of conversational implicature[19], the vehicle acquires, relative to the context of the utterance of which it is part, a secondary, figurative meaning. (iv) The application of the Gricean mechanism has distinctive features, distinguishing metaphor from other figures of speech and, in general, from other conversational implicatures: the metaphorical meaning is derived so as to preclude a *prima facie* conceptual inconsistency in which the speaker would otherwise incur if he meant in the context the vehicle with its literal meaning; and (v) it is derived by keeping for the figurative interpretation of the vehicle some of the features commonly known to be associated with it, including those constituting its literal meaning, (vi) while excluding the others. Thus, in the stock example "Juliet is the sun", the metaphorical vehicle "is the sun" acquires in context a secondary meaning (say, *is something that produces pleasant sentiments*), thus evading the conceptual inconsistency of identifying an entity presupposed to be animated with another presupposed to be unanimated.

---

[18]It is slightly misleading to speak of "metaphorical reference" as I will be doing henceforth. That expression is more frequently used for ordinary reference that involves a metaphorical characterization of the referent, as when we utter "That festering sore must go", referring to a derelict house. See Bezuidenhout (2008), from where I take the example. I hope that the reader will be able to put aside the misleading associations.

[19]The mechanism brilliantly analyzed by Grice (1975), through which speakers utter sentences that, if taken with their literal meanings, would obviously flout "conversational maxims" (such as that requiring speakers not to say what they know is false, which Romeo appears to flout in saying "Juliet is the sun") hoping to convey thereby a different meaning that their audiences will be able to derive given that from the literal meaning and context.

In this way, metaphors lead us to consider a domain (that of lovers, say, in the example) in terms of concepts literally appropriate only for a different one (that of heavenly bodies, say), and thus have a cognitive function, the potential to supply knowledge; this is so even though metaphors cannot be paraphrased away with the same effect, by means of an utterance whose literal meaning exhausts the figuratively conveyed content, at least because they are open-ended (there are indefinitely many other features commonly known of the sun that could meaningfully apply to Juliet) and also because a literal utterance would lack the same potential to activate our inquisitiveness, our engaged contemplation of propositions.

Accounts of metaphor along these lines must confront well-known objections.[20] A full discussion of these objections would immerse us in contemporary debates about the semantics/pragmatics divide. Researchers with contextualist leanings would insist that metaphorical meanings belong in *what is said* and not merely in *what is implicated*, resulting (unlike paradigm Gricean conversational implicatures) from optional "primary pragmatic processes" in François Recanati's (2004) sense. Here I would just like to point out that, as I have contended elsewhere (2006b), the Gricean theorist does not need to claim, as contextualists typically assume, that literal meanings are in any way processed (at the personal or subpersonal level) at any stage in the calculation of pragmatically conveyed meanings, the metaphorical content in our case. It is enough for the literal meaning to be psychologically real if (to use Christopher Peacocke's (1989) turn of phrase) the processing mechanisms "draw upon" the information encapsulated in the literal meaning of the metaphorical utterance. The main reason to claim that metaphorical meanings are not what is literally said, on the other hand, is that we need a compositional theory to explain the productivity and systematicity of linguistic understanding; Peter Pagin and Jeff Pelletier (2007) provide a good account of how the contextualists insights can be made into a compositional meaning theory.

The expressive resources of natural languages, and therefore their potential metaphorical vehicles, do not only include words and lexemes; as linguists put it, they include not only *lexical* categories, but also *functional* categories. The difference between playing the role of an agent in a relation, and playing the role of a patient, is semantically fundamental; this difference is expressed by means of lexemes in Latin, but in English only by means of syntactic features more difficult to pinpoint. That an expression is referential is also a semantically significant expressive resource that, in English, is constituted by complex syntactic features—which I am unable to specify. No matter what they are, "Zavalita" in (3) instantiates those features, semantically indicative that it is intended to refer to an entity.

On the present view, these grammatical features indicating referentiality constitute the metaphorical vehicle in the cases we are interested in.[21] The *prima*

[20]See Romero and Soria (ms) for a helpful summary of those objections, and the responses open to its proponents.

[21]Glanzberg (2008) argues that functional categories differ from lexical ones in that they do not admit metaphorical interpretations. However, (i) Glanzberg does not provide any argument for his view, he just gives some examples of sentences which determiners do not appear to have a

facie conceptual inconsistency which gives rise to the metaphorical interpretation could be the one I have been formulating intuitively for the Quinean strategy that van Inwagen pursues, given the Kripkean assumptions about genuine reference. A metaphorical interpretation is asked for because there is no genuine reference that the speaker could be sensibly attempting in this case. In the first place (and this is perhaps the only psychologically relevant case), he cannot be attempting to genuinely refer to a person, because when we refer to a person, in the context of making another speech act, we presuppose in the first place that there is such a person, and we somehow know him, which is of course not presupposed at all in the case of the use of "Zavalita" in (3); and, even if there were, we are not presupposing, as we do in genuine cases, that our referent "is an object", i.e., has many unknown properties, in addition to those we invoke to fix reference to it, whose discovery may well later serve, as Gareth Evans (1982, 146) puts it, to establish the correctness or otherwise of the speech act to which our act of reference contributed: "a subject who has a demonstrative Idea of an object has an *unmediated* disposition to treat information from that object as germane to the truth or falsity of thoughts involving that Idea". In the second place, he cannot be attempting to genuinely refer because he is not at risk of failing to do so, as he would be if reference were not secured by the reference-fixing means deployed, but required a referent with a perhaps hidden essence.

In genuine cases of reference, the speaker knows who or what the referent is in virtue of his successfully deploying the reference-fixing features he invokes; and this knowing who or what is a genuine achievement, relying on a kind of procedure that may go wrong and does go wrong in some cases. None of this applies to any entity to which the speaker of an utterance like (3) might be attempting to refer by "Zavalita".[22] It does not make any sense to imagine that such a referent might have properties (still less, essential ones), such as being-identical-to-pseudo-Hesperus (the fictional character, in an earlier example), that no ideally cognitively well-placed human being might discover. Additionally, there might well be conflicting but equally legitimate interpretations of a given fiction (Currie 1990, 99–106), giving rise to incompatible properties for a fictional character; if so, neither of two interpreters ascribing these incompatible properties to the character would be

metaphorical interpretation; (ii) prepositions are usually regarded as functional categories, and there are whole books, such as Tyler and Evans (2003), to discuss the proper treatment of what, from the point of view I adopt here (see (iii)), are metaphorical meanings; and, last but not least, (iii) as I indicate later, the metaphorical meanings I envisage are *not* freshly baked literary metaphors, but deeply entrenched, conventionalized ones; and some remarks by Glanzberg about the case of prepositions (2008, 43 footnote 7) may suggest that his claim only concerns fresh metaphors.

[22]Or to any one to which such a speaker might attempt to refer by "La Crónica" or "Tacna Avenue", respectively; this is the ultimate ground for the view put forward in footnote 10 above. See Bonomi (2008) for elaboration.

making a mistake, which shows that, unlike discourses involving genuine reference to persons, discourses involving reference to fictional characters do not exert "cognitive command" (Wright 2002).

In summary, what—assuming a theory of metaphorical discourse such as Kittay's—triggers the metaphorical character of apparent reference to fictional characters as with "Zavalita" in (3) is the fact that it is mutually known to the speaker and his audience that there is no such entity to be referred to; or, when there is—as with "Tacna Avenue"—the fact that only its mutually known properties matter to the correctness of the relevant speech act. This assumes that, intuitively, those expressions do not refer to abstract entities; otherwise, the linguistic intuitions of theoretically unsophisticated speakers should also trace the distinction between "encoding" properties (such as being a non-existent Peruvian journalist, in our example) and exemplifying them (such as being an existing abstract fictional character). But, as I argued before, this is totally unwarranted; nothing in the linguistic behavior and attitudes of ordinary speakers warrants ascribing to them such a notion. The only psychologically reasonable candidate for a referent for "Zavalita" is an actually existing Peruvian journalist.

Apparent reference to quasi-abstract entities (such as what Currie (1990) calls a "role") in statements like (3) should hence be taken as merely figurative. What is the content that we figuratively convey by means of them? It does not of course include any such reference; the only thing that can be really memorable about Zavalita is that "he" is ascribed such-and-such properties in a particular fiction, in contrast to corresponding portraits in other fictions by the same author; i.e., ultimately, that it is fictional in CLC that Zavalita ..., that it is similarly fictional in other works by Vargas Llosa that ..., and that such and such relations of comparative impact on the audience's memories obtain among those facts. Walton's (1990, 405–419) paraphrases are thus a much better guide to the real content, except that, as is generally the case with any other metaphorical claim, we should not expect to find a literal paraphrase having exactly the same import.

What about the content of quantificational claims we can infer from them, such as "there are fictional characters created by Vargas Llosa" in the case of (3), or the convoluted ones on which van Inwagen (1977) famously based his Quinean case for the existence of fictional characters, such as "There are characters in some 19th-century novels who are presented with a greater wealth of physical detail than is any character in any 18th-century novel"? Thomas Hofweber (2005), making a proposal to which the present one is very close, usefully distinguishes an external from an internal reading of quantifiers.[23] The truth-conditions of quantificational sentences in the latter use are helpfully equated with those of substitutional

[23] The main difference lies in that he argues for polysemy, while I am arguing—following Yablo (2001)—for a figurative or metaphorical reading of apparent reference to, and quantification over, fictional characters, understood as pragmatically conveyed readings. But this apparent difference vanishes when it is acknowledged, as I will do presently, that the metaphors in question are deeply conventionalized; this is to posit a form of polysemy.

interpretations—disjunctions or conjunctions of their instances, as expressible in a previously acknowledged vocabulary.[24]

Figurative recourse to the referential apparatus is very useful. When proper names like "Zavalita" in claims like (3) are used to figuratively refer to a role, they themselves may serve as anaphoric nodes throughout a discourse, in the same way as ordinary names do, to label dossiers including the information that the speaker thereby gives. Through the logical relations existing among statements including expressions in referential positions, and quantificational statements, these figurative uses can also allow to neatly pack complex non-figurative contents by means of statements involving multiple quantifiers, like those already mentioned, on which van Inwagen (1977) focuses. But reference to those roles as in (3) is mere figurative, not genuine reference. The nature of those roles is fully determined by what a relevantly informed interpreter can derive from a fiction, on the basis of agreed procedures established by a social practice. Because of this, the two reasons given before why the speaker of (3) is not genuinely referring to a person, also establish that he is not genuinely referring to a role. The discourse does not exert cognitive command; two interpreters might define the role in terms of contradictory features, without either of them making a mistake. And it does not make sense to think that roles have features (still less essential ones) that no human being in epistemically ideal situations can discover.

Of course, if there is a metaphorical meaning here, it has to be a deeply conventionalized one; it cannot be a freshly created literary metaphor that has to be consciously derived. Starting with the pioneering work of George Lakoff, linguists have come up with different criteria to isolate primary, core meanings in the networks of related senses of highly polysemous expressions—senses in many cases derived from core meanings through essentially the procedures by means of which metaphorical meanings are derived in paradigm cases. Prepositions such as "over", with spatial meanings at their core (a "trajector" being above, or higher than, a "landmark", in this case), and "covering" senses among those derived from it (in addition of course to much more abstract senses) offer good examples[25]; so do verbs such as "crawl", whose core meanings are basic actions (*moving by muscular activity while the body is close to the ground or another surface*), and whose derived meanings include those in which it applies to traffic, and of course to servile behavior.[26] The criteria that these researchers use include[27]: (i) multiple senses can be clearly traced back (diachronically and/or psychologically, in acquisition history) to one; (ii) the set of senses permits a network-like description in which pairs of adjacent senses are related by motivated linguistic processes, such as one or another type of metaphorical mapping, that recur across the lexicon; (iii) in all such links there is a cognitive

[24]Cf. Kripke (1976) for elaboration.

[25]Cf. Tyler and Evans (2003).

[26]Cf. Fillmore and Atkins (2000).

[27]Fillmore and Atkins (2000, 100); Tyler and Evans (2003, 47).

asymmetry in that the understanding of each derivative sense is aided by knowledge of the sense from which it is derived.[28]

Yablo (2001, Section XII) makes a point in connection with his figurative account of reference to numbers that I subscribe to. The main reason in favor of the figurative account of reference to fictional characters does not come from metaphysical scruples regarding abstract entities, or to alleged special epistemic difficulties we would have if we accepted them. The main reason is that it accounts for the intuitive differences we perceive among entities to which we are otherwise equally committed, given Quinean considerations. Earlier I invoked Carnap's Principle of Tolerance to express those intuitions. Now we can see how the figurative proposal accounts for the restricted intuitive adequacy of the Principle. Given that the secondary content of a metaphorical claim is granted, to put forward the metaphor, which we are assuming satisfies the six requirements by means of which we earlier outlined the main features of that practice, is essentially to make a stipulation to which one is perfectly entitled, given the existence of the practice of speaking metaphorically.[29] For someone who accepts that Juliet does have the properties metaphorically ascribed to her by "Juliet is the sun", it would make no rational sense to reject the metaphorical claim, on the basis perhaps that in its literal meaning it is absurd. It is tolerance of this sort to which whoever invokes referential language for fictional entities, as in (3), is entitled. I believe that the obscure intuitive feeling that they are so entitled accounts for the impatience that literary critics experience when confronted with philosophical discussion as to the reality of fictional characters. (Of course, the impatience is ultimately unjustified, because philosophy is needed to transform the obscure intuitive feeling into a theoretically articulated view.)

Scientific Models as Fictions

On the account I have been assuming here, although literally taken utterances of (1) are understood to make assertions, an ancillary part of which involves reference to a person called "Santiago", a.k.a. "Zavalita"—whose correctness, on a normative account of assertion and reference, would require the speaker to know the signified singular fact, and hence to know who the person concerned is—as a matter of fact, in its context (i.e., having being produced as part of a literary fiction) the speaker is not really doing or purporting to do any such thing, but a different speech act, one (fiction-making) whose correction does not require the speaker to know such a person or such a singular fact. The speaker is rather trying to put his audience in

[28]As Nunberg (2002, footnote 15) nicely puts it, "the fact that dictionaries assign the word *crawl* a sense 'to act or behave in a servile manner' doesn't mean that people couldn't come up with this use of the word in the absence of a convention".

[29]One would also be entitled to the stipulation in a context in which the practice did not exist, but one could still count on the pragmatic rationality of one's fellow speakers.

a position to imagine a purely general, descriptive content, and the correctness or otherwise of the act he is really doing should only be judged on this basis.

On the account I have been outlining for sentences like (3), something very much like this applies. Taken literally, the speaker should be understood as making an assertion, and thereby purporting (and thus miserably failing) to know a singular fact, one about a certain non-existent entity (or rather one about an existing but non-concrete one), reference to which is understood to be an ancillary act for the understood assertion, so that he thereby represents himself as knowing which entity this is (and miserably failing here too, for obvious reasons on the non-existent entity interpretation, on the existing but non-concrete entity alternative interpretation because the knowledge he may claim to have is no achievement). But none of this is what he is really doing; as before, he is merely pretending to do this, with the real purpose of doing something else. In the present case, what he is really doing is of course not the different speech act of fiction-making, but rather one which is also typically involved (at least indirectly) in serious fiction-making: that of asserting an unspecified set of different facts, facts about the import and shape of a certain fiction.

The present proposal thus has the main features of what Mark Kalderon (2005) describes as "modern fictionalism", whose main representatives are the work of Field on numbers, and van Fraassen on theoretical entities, outlined at the beginning. In contrast to more traditional forms of fictionalism or instrumentalism, those proposals do not purport to reduce the claims made by the offending utterances to others not making reference to the problematic entities, nor suggest that those utterances do not purport to state facts. The view is rather that, although the sentences taken literally are supposed to express propositions whose success requires reference to the problematic entities, they are in fact being put forward for other goals, whose standards of correctness are different—in particular, the truth of the relevant assertions is not required, nor the success of the ancillary reference. My proposal is therefore a form of modern fictionalism about fictional entities.

The argument that I have used to defend it, however, highlights my distance from those two paradigms of modern fictionalism. I have based my arguments on the contemporary views on genuine reference of Kripke and Putnam; it is the contrast with the requirements for successful reference on those views, given the mutually known facts concerning the alleged referents of expressions like "Zavalita" in (3) that, according to my proposal, triggers the metaphorical interpretation of utterances such as (3). *Prima facie* at least, this form of argument cannot be used for the case of reference to theoretical entities in science, if Kripke and Putnam are right (as I myself think they are); for these are genuine references, in fact paradigm cases thereof. Theoretical entities such as genes and black holes play crucial explanatory roles, which van Fraassen's "constructive empiricism" does not allow us to do without. Unless we adopt an extreme form of phenomenalism (itself with its own problems, not very far away from van Fraassen's), there does not seem to be any well-motivated reason for limiting genuine reference to observable entities. The very same considerations that justify assuming that our experiences and perceptual beliefs do manage to refer to external entities beyond their intrinsic phenomenal features, on the basis that there is an inextricable causal-explanatory element in our very notion of the content of experiences and perceptual beliefs, justify the scientific realist assumption that our correct theoretical beliefs and assertions manage to successfully refer to theoretical entities. And the Kripkean considerations on which I have partly based my reasons for fictionalism about fictional entities are consistent with these externalist considerations about the contents of experiences and perceptual beliefs. With respect to mathematical entities, it is at the very least clear that the form of argument that I have invoked cannot be deployed without further ado. Numbers and sets are not less abstract than other entities we cannot similarly do without, for all Field tells us, such as expression-types and meanings.[30] Thus, I find van Fraassen's and Field's fictionalism unmotivated and wrong, unlike the limited proposal I have made here.

However, as previous authors in fact have already suggested, the present account can be usefully applied to the case of explaining by means of hypothetical (in a few cases, actual) models, illustrated by (4) above. As Frigg (2010, 251) reminds us, "The first step in tackling a scientific problem often is to come up with a suitable model. When studying the orbit of a planet we take both the planet and the sun to be spinning perfect spheres with homogenous mass distributions gravitationally interacting with each other but nothing else in the universe; when investigating the population of fish in the Adriatic Sea we assume that all fish are either predators or prey and that these two groups interact with each other according to a simple law; and when studying the exchange of goods in an economy we consider a situation in which there are only two goods, two perfectly rational agents, no restrictions on available information, no transaction costs, no money, and dealings are done in no time".

In contrast to previous writers such as Giere (1988), who (in sync with van Inwagen's proposals on fictional characters) take these hypothetical models in science to be abstract entities, and for reasons very much like those mentioned before against van Inwagen's view, Frigg (2010) and Godfrey-Smith (2006) propose to understand descriptions of hypothetical models along fictionalist lines. As Godfrey-Smith (2006, 735) puts it, in a text from which I previous quoted in part: "I take at face value the fact that modelers often take themselves to be describing imaginary biological populations, imaginary neural networks, or imaginary economies. An imaginary population is something that, if it was real, would be a flesh-and-blood population, not a mathematical object. Although these imagined entities are puzzling, I suggest that at least much of the time they might be treated as similar to something that we are all familiar with, the imagined objects of literary fiction. Here I have in mind entities like Sherlock Holmes' London, and Tolkein's Middle Earth. These are imaginary things that we can, somehow, talk about in a fairly constrained and often communal way. On the view I am developing, the model systems of science often work similarly to these familiar fictions. The model systems of science will often be described in mathematical terms (we could do the

[30] Cf. Rosen (1994), Section IV, for elaboration on these objections.

same to Middle Earth), but they are not just mathematical objects". Frigg develops this view further, proposing the analysis of the description of models in science along the lines of Walton's proposal for fiction—a view similar to the one on which I have been relying here for straightforward fictional claims, such as those made by fiction-makers with sentences like (1).

There is a crucial difference, however, between straightforward fiction-making utterances like one of (1), and the description of hypothetical models in science: although in some cases (almost always, in serious fiction), the act of producing fictions is (as Lewis (1978) expresses it) put to the service of truth, so that the fiction-maker is, at least indirectly, making claims, suggestions, etc, about human psychology, human possibilities, values, and so on, this is not, I take it, constitutive of the practice. On the other hand, the producer of a hypothetical "model system" in science, as both Frigg and Godfrey-Smith insist, typically purports thereby to be making claims—straightforward assertions, true or false—about a real "target system".[31] In this, the case of model-building in science is much closer to (3) than to (1), and, as we have seen, Walton himself accepts that in the case of (2) and (3) we have assertions, at least derivatively. Even if the utterer of (3), as I have claimed, merely pretends to refer to a Zavalita, he is in addition making straightforward assertions—about the import of a fiction with a given content, I have claimed. The same applies to the utterer of (4), who ultimately wants to make real claims about the actual bouncing bob he is studying. Because of this, I think that a fictionalist account along the figurativist lines of the proposal I have made offers better prospects for the kind of view of scientific models that Frigg and Godfrey-Smith advocate. Even if he is speaking metaphorically, Romeo is purporting to make true claims when he utters "Juliet is the Sun"; the same, I think, applies to the scientific modeler.

Frigg, as I said, provides an analysis, based on Walton's proposals, which goes beyond Godfrey-Smith's undeveloped suggestion of a fictionalist account of model-mongering in science. Of particular interest here is his discussion of what he calls "transfictional propositions", those in which fictional characters in different fictions, or fictional characters and real individuals, are compared; I take it that both our examples (3) and (4) would constitute examples of this category, but perhaps (6) and (7) are examples more to the point:

[31] In his contribution to this volume, "Models and Make-Believe", Toon makes a proposal that, precisely on account of this, I take to be only superficially similar to that of Frigg and Godfrey-Smith. He is concerned with the nature of the representation-relation which obtains between scientific models and their target systems, and contends that it is of the same kind as that obtaining, on Walton's account, between a fiction and the real entities (such as Napoleon or Russia in the early nineteenth century, in the case of *War and Peace*) which it may be said to somehow represent. Following Walton, then, he contends that model-descriptions in science prescribe imaginings about their target systems. Unlike the two-stage proposals of Frigg and Godfrey-Smith, and unlike Walton's own views about (2) and (3), which, as we have seen, admit that they are at least derivatively assertions, this proposal in my view fails to capture the essential component of truth-aptness that modeling in science involves. Fiction-making is evaluated only relative to the quality of the imaginings it prescribes; I do not think this applies at all to representation by means of scientific models.

(6) Marcus Wolf, the head of the East German secret police, was less interesting than Karla, John le Carré's fictional character based on him

(7) The period of oscillation of the bob in the model is within 10% of the period of the bob in the system

Frigg (2010, 263) acknowledges that these transfictional propositions "pose a particular problem because they—apparently—involve comparing something with a nonexistent object, which does not seem to make sense"; but he thinks that the problem is not insurmountable: "Fortunately we need not deal with the problem of transfictional statements in its full generality because the transfictional statements that are relevant in connection with model systems are of a particular kind: they compare features of the model systems with features of the target system. For this reason, transfictional statements about models should be read as prefixed with a clause stating what the relevant respects of the comparison are, and this allows us to rephrase comparative sentences as comparisons between properties rather than objects, which makes the original puzzle go away".[32]

I have been arguing here that van Fraassen's and Field's fictionalism is the best option for the anti-realist about fictional characters, in reply to the realist Quinean argument. Walton offers us a version of the traditional instrumentalist strategy, arguing that statements like (3) should not be taken at face value, but its apparent commitment to fictional entities paraphrased away. I understand that Frigg is offering us a Waltonian proposal. I have given some reasons to reject it, and pursue instead a figurativist version of the fictionalist proposal. My main concern applies unmodified to Frigg's account of (7) (and, *mutatis mutandis*, (6)): what is the justification for the claim that the transfictional statements in model-based science "compare features of the model systems with features of the target system"? I assume that many of these transfictional claims do not explicitly make such comparisons; this is implicitly acknowledged when Frigg resorts to normative terminology, saying that they "should be read as prefixed", which seems to admit that they in fact are not so prefixed. Studying a particular biological example of model-based science, Godfrey-Smith (2006, 732) says: "the currency of theoretical argument at each stage is the model. Interestingly, these are often not formal mathematical models,

---

[32]Cf. Toon (2010, 213–214) discussion of (7): "I think we may still analyze our theoretical hypotheses without commitment to any object that fits our prepared description and equation of motion. When we say 'the period of oscillation of the bob in the model is within 10% of the period of the bob in the system', we are simply comparing what our model asks us to imagine with what is true of the system. Specifically, we assert that the period of oscillation of the bob has some value T_0 and that it is fictional in our model that the bob oscillates with period T_1, where T_1 is within 10% of T_0". This paraphrase is correct, and Toon is right that it does not commit us to any object beyond the real bob. But the example raises two worries about Toon's views. The first applies equally to Frigg's proposal: how is this paraphrase generated? On my alternative proposal, the paraphrase is just one way of stating a metaphorical meaning, and, as in other cases, there probably is no systematic theory of how those meanings are generated. The second question is specific to Toon's own view, and it relates to the objection in the previous footnote. For it is clear, I think, that his paraphrase states a content to which the utterer of (7) is *assertorically* committed.

though some are. Many of the models instead proceed by describing an idealized, schematic causal mechanism, noting how it will and will not behave, and exploring plausible evolutionary paths from one situation to another". This does not suggest that the claims made in this example are in any way prefixed as Frigg says they should be. Notoriously, it is not so easy to justify semantic claims to the effect that some class of statements should be understood as containing implicit prefixes or operators.

The figurativist proposal does not commit us to such implausible assumptions. Claims such as (6) and (7) should be taken at face value; thus taken, they are untrue, for lack of reference of some of the referential expressions in them. But in uttering them, we are not committing ourselves to their truth, even less to our having good reasons for accepting the propositions they express. Paraphrases such as the ones that Frigg suggests provide a plausible indication of what we in fact purport to commit ourselves to assertorically; but their determination is subject to the pragmatic vagaries of interpretation. Thus, if the fictionalist proposal to analyze model-based science is elaborated along the figurativist lines of my own proposal for claims apparently about fictional entities, the problem for Frigg's proposal I have pointed out would be skirted, with the end result being close to the one that Frigg wants.

On most accounts of metaphors, and certainly on the one due to Kittay on which I have based my proposal, metaphorical claims are ultimately ascriptions to a target domain of some of the features associated with a source domain. In cases like ((3) and (7), the target domain is that of content-features of fictions and our emotional and cognitive engagement with them, while the source domain is that of our representational referential and quantificational dealings with ordinary objects of reference. In the case of (6), the source domain is the same, and the target domain is, typically, the real physical systems for which models posit frictionless planes. However, a proper elaboration of these suggestions concerning how to understand model-based science should be left for those more knowledgeable than I am. Instead, I will briefly conclude this section by briefly indicating how the figurativist account deals with the six desiderata Frigg (2010, 256–257, 9–10) usefully provides for accounts of models:

(1) ***Identity conditions***. *Model systems are often presented by different authors in different ways. Nevertheless, many different descriptions are meant to describe the same model system. When are the model systems specified by different descriptions identical?* The (untrue) literal contents of (3) and (6), taken at face value, can of course be expressed by different people in different utterances and context, in different languages. The literal content determines the identity conditions of these potential cases of same saying. The same applies to claims such as (7). The fact that there are no referents for the referential expressions in those utterances poses no problem.[33]

[33]Not, at least, on the assumption that Evans and Walton are mistaken in their radical referentialist assumption that no referent, no proposition expressed; see footnote 9.

(2) ***Attribution of properties.*** *Model systems have physical properties. How is this possible if model systems do not exist in space and time?* It is possible in the same way that it is possible that fictional characters, like Zavalita, have biological properties. We are only supposed to imagine the literal content of (7), according to which the (non-existent) referent of "the bob" has a period of oscillation, in the same way that in meaningfully uttering (1) Vargas Llosa is only imagining the non-existent Zavalita to have eyes.

(3) ***Comparative statements.*** *Comparing a model and its target system is essential to many aspects of modeling. We customarily say things like "real agents do not behave like the agents in the model" and "the surface of the real sun is unlike the surface of the model sun". How can we compare something that does not exist with something that does?* This is just the issue raised by transfictive statements such as (6) and (7), which we have already dealt with.

(4) ***Truth in model systems.*** *There is right and wrong in a discourse about model systems. But on what basis are claims about a model system qualified as true or false, in particular if the claims concern issues about which the description of the system remains silent?* There is right and wrong about the extent of metaphorical claims, and its implications for the serious claims people making them really want to commit themselves to, even if this is subject to the pragmatic vagaries of interpretation. The sun is something that has recently risen when Romeo has breakfast, but it is unlikely that he wants to assert that Juliet has also recently risen when he has breakfast in asserting that Juliet is the sun. That property of the source domain is irrelevant to characterizing the target domain. Even if it is a relatively indeterminate matter which properties are "transferred" from one domain to the other, there are clear positive and negative cases.[34]

(5) ***Epistemology.*** *We investigate model systems and find out about them; truths about the model system are not forever concealed from us. How do we find out about these truths and how do we justify our claims?* The previous answer dictates the one to this question: by investigating which properties the fictional bob has, and how they are relevant for the claims we really want to commit ourselves to concerning actual bobs.

(6) ***Metaphysical commitments.*** *We need to know what kind of commitments we incur when we understand model systems along the lines of fiction, and how these commitments, if any, can be justified.* The metaphysical commitments we incur are those incurred in the more or less accurate paraphrases we could provide for what we really want to commit ourselves to. For all we can tell, these do not include commitments to fictional entities (in (3) and (6) or frictionless planes (in the likes of (7)).

---

[34]If Walton (1993) is right that metaphor-making is a form of make-believe, the extent of right and wrong here is exactly the extent to which "principles of generation" are sufficiently settled in fiction: truth-in-a-model, on the present proposal, would then exactly coincide with truth-in-fiction. I have already expressed doubts about this account, though (cf. footnote 10), but of course it is not in competition with the present proposal; to adopt it I would just have to rely on this account of metaphor, instead of relying on Kittay's.

Concluding Afterthought: Carnapian Associations

Carnap famously espoused a *Principle of Tolerance: "It is not our business to set up prohibitions, but to arrive at conventions . . .. In logic there are no morals.* Everyone is at liberty to build up his own logic, i.e. his own language, as he wishes. All that is required of him is that, if he wishes to discuss it, he must state his methods clearly, and give syntactical rules instead of philosophical arguments" (*Logical Syntax*, §17). In "Empiricism, Semantics and Ontology" he expresses the advice in a different way: "Let us be cautious in making assertions and critical in examining them, but tolerant in permitting linguistic forms" (Carnap 1956, 221).

Quine's (1951) influential criticism of the deflationary attitude that the principle proposes accounts in part for the contemporary unpopularity of the Carnapian principle, whose import, following Quine, we could present in the following way. Let us focus on existential utterances of the form of "There are *X*", taken as answers to questions such as "Are there *X*?" Depending on the generality of the expression substituting for "*X*", we can distinguish (I use Quine's terms) *category* questions ("there are numbers") and *subclass* questions ("there are prime numbers about a hundred"). Now, category questions can be taken, according to Carnap, in two different ways. They can firstly be taken (in the "external" manner) as intended to make stipulations or agreement-proposals for the adoption of representational resources; with respect to them, only practical considerations (which Carnap's Principle suggests us to conduct with an open-minded, tolerant spirit) are in order. In particular, the attitude we should take with respect to a serious assertion (i.e., to study in earnest whether it satisfies relevant requirements to put us in a position to acquire knowledge from it) is in this case, Carnap claims, entirely misguided. The subclass questions are indeed, on the other hand, serious assertions, although they can only arise when the stipulations in some category questions have been adopted; and if so, the relevant category questions may also be taken (in the "internal" manner) as making serious assertions, although they would then be either trivially true or trivially false. This is why, out of context, utterances such as "there are numbers" would be taken as expressing external questions.

In the two quotations, Carnap restricts his Principle to logical or semantic issues, more in general to issues depending on matters of linguistic forms; and I have taken this into consideration in interpreting it. This is of course, as Quine (1951) sees, in harmony with his analytic/synthetic distinction, and in particular with his view that convention lies at the heart of analyticity. Correspondingly, Quine's (1953) general contention that there is no such distinction, together with his more specific criticisms of the Carnapian conventionalist version, lie at the heart of his objection. Most contemporary philosophers have been convinced by Quine's arguments that there is no such distinction, or at least that any one such that could be stated with sufficient clarity would be philosophically immaterial; and this is one of the sources of resistance to anything like the Carnapian Principle. For it supports the sentiment that there cannot be any epistemologically or ontologically relevant distinction between two forms of reference and quantification: the one in internal questions, which is serious in that the satisfaction or otherwise of its commitments depends on how the world is, independently of our thought and language; and the one in external questions, the satisfaction of whose commitments is sufficiently up to us for us to be thereby free to stipulate.

In a previous co-authored paper (García-Carpintero and Pérez Otero 2009) I argued for a limited form of Carnap's conventionalism about analyticity from Quine's criticisms. Although we agree there with what we take to be the philosophically more substantive aspects of Quine's criticism of Carnap's views on analyticity (for instance, we agree that there is no interesting sense in which we can stipulate the logical principles), we suggest that its influence in contemporary views is overdrawn.

In line with this more general previous criticism, in this paper I have in fact defended a restricted version of Carnap's Principle of Tolerance, applying to a particular kind of example, reference to and quantification over fictional entities. I have argued that a deflationary fictionalist reading of statements explicitly referring to fictional characters is more adequate than realist proposals, but also than other critical stances like that of Walton (1990) or Sainsbury (2005). To test the limits both of the vindication of conventionalism about analyticity, and its more specific application to Carnapian Tolerance, I have contrasted the fictionalist proposal about fictional characters with van Fraassen's and Field's fictionalisms about, respectively, theoretical and mathematical entities. Finally, I have suggested that the proposal could be helpfully deployed to defend a fictionalist view about the reference to hypothetical models in scientific theorizing.

I will conclude by briefly discussing a certain "Carnap's Paradox" set up by Yablo in a recent talk[35], whose resolution can be taken as a test for approaches to ontological questions sympathetic to the Carnap's suggestions summarized here. The paradox, applied to the case I have been mostly discussing, is that, while (8) entails (9), we have both (10) and (11):

(8) Zavalita is a fictional character introduced by Vargas Llosa in CLC

(9) Fictional characters exist

(10) It is clear that Zavalita is a fictional character introduced by Vargas Llosa in CLC

(11) It is controversial that fictional characters exist

My suggestion is as follows: (8) has a reading as an answer to an "internal" Carnapian question; on the present view, this is a figurative reading, on which its metaphorically conveyed content does not go beyond what different Waltonian paraphrases would capture, that Vargas Llosa wrote a novel, CLC, in which he used "Zavalita" pretending thereby to refer to a person, and so on and so forth. This is a reading on which (8) is true. It also has an "external" reading, a straightforward, literal one, in which it is untrue, for lack of reference of the subject. The same applies

[35]"Carnap's Paradox", given at the LOGOS Metametaphysics Conference, June 19–21 2008, http://www.ub.es/grc_logos/mm/inicio.htm.

to (9), with the "internal" reading being such that its metaphorically conveyed substitutional content does not go beyond a disjunction of different potential Waltonian paraphrases. It is only when the readings of the two claims are both internal or both external that the inference is acceptable (and sound, in the first case). The difference captured in (10) and (11) is explained by the fact that, uttered in normal contexts, (8) leads us to focus on the internal reading; it invites us to, figuratively speaking, assume the existence of fictional characters. (9), on the other hand, at least in the typical philosophical contexts in which it is uttered, leads us to focus on the external reading.

Acknowledgments Research for this paper has been funded by the Spanish Government's MCYT research project HUM2006-08236, and a *Distinció de Recerca de la Generalitat, Investigadors Reconeguts* 2002-8. I am very grateful to Esther Romero for very detailed comments on a previous version of this paper. The co-editor of this volume, Roman Frigg, also provided extended comments on several versions, which have helped me to clarify my views and their presentation in many ways. Thanks finally to Michael Maudsley for his careful grammatical revision.

References

Bezuidenhout, A. (2008), "Metaphorical Singular Reference", *The Baltic International Yearbook of Cognition, Logic and Communication*, at http://cognition.lu.lv/ , 3: *A Figure of Speech*.

Bonomi, A. (2008), "Fictional Contexts", in P. Bouquet, L. Serafini, and R. Thomason (eds.), *Perspectives on Context*. Stanford, CA: CSLI Publications, 213–248.

Braun, D. (2005), "Empty Names, Fictional Names, Mythical Names", *Noûs* 39, 596–631.

Burgess, J. P. (1998), "Quinus ab Omni Nævo Vindicatus", *Canadian Journal of Philosophy: Meaning and Reference*, sup. vol. 23, A. Kazmi (ed.), 25–65.

Carnap, R. (1956), "Empiricism, Semantics, and Ontology", supplement A in *Meaning and Necessity*. Chicago, IL: The University of Chicago Press, 205–221.

Cartwright, N. (1983), *How the Laws of Physics Lie*. Oxford: Clarendon Press.

Currie, G. (1990), *The Nature of Fiction*. Cambridge: Cambridge University Press.

Evans, G. (1982), *The Varieties of Reference*. Oxford: Clarendon Press.

Everett, A. (2005), "Against Fictional Realism", *Journal of Philosophy* 102: 624–649.

Everett, A. (2007), "Pretense, Existence, and Fictional Objects", *Philosophy and Phenomenological Research* 74: 56–80.

Field, H. (1980), *Science without Numbers*. Oxford: Blackwell.

Fillmore, C. and Atkins, B. T. S. (2000), "Describing Polysemy: The Case of 'Crawl'", in Y. Ravin and C. Leacock (eds.), *Polysemy: Theoretical and Computational Approaches*, Oxford: Oxford University Press, 91–110.

Friend, S. (2007), "Fictional Characters", *Philosophy Compass* 2, 2: 141–156.

Frigg, R. (2010), "Models and Fictions", *Synthese*, 251–268.

García-Carpintero, M. (2000), "A Presuppositional Account of Reference-Fixing", *Journal of Philosophy* XCVII, 3: 109–147.

García-Carpintero, M. (2006a), "Two-Dimensionalism: A Neo-Fregean Interpretation", in M. García-Carpintero and J. Macià (eds.), *Two-Dimensional Semantics*, Oxford: Oxford University Press, 181–204.

García-Carpintero, M. (2006b), "Recanati on the Semantics/Pragmatics Distinction", *Crítica* 38: 35–68.

García-Carpintero, M. (2007), "Fiction-making as an Illocutionary Act", *Journal of Aesthetics and Art Criticism* 65: 203–216.

García-Carpintero, M. (2010), "Fictional Singular Imaginings", in R. Jeshion (ed.), *New Essays on Singular Thought*, Oxford: Oxford University Press.

García-Carpintero, M. and Pérez Otero, M. (1999), "The Ontological Commitments of Logical Theories", *European Review of Philosophy* 4: 157–182.

García-Carpintero, M. and Pérez Otero, M. (2009), "The Conventional and the Analytic", *Philosophy and Phenomenological Research* 78: 239–274.

Giere, R. (1988), *Explaining Science*. Chicago, IL: University of Chicago Press.

Glanzberg, M. (2008), "Metaphor and Lexical Semantics", *The Baltic International Yearbook of Cognition, Logic and Communication*, at http://cognition.lu.lv/ , 3: *A Figure of Speech*.

Godfrey-Smith, P. (2006), "The Strategy of Model-Based Science", *Biology and Philosophy* 21: 725–740.

Grice, H. P. (1975), "Logic and Conversation", in P. Cole and J. Morgan (eds.), *Syntax and Semantics*, vol. 3, New York: Academic Press.

Hofweber, T. (2005), "A Puzzle about Ontology", *Noûs* 39: 256–283.

Johnston, M. (1988), "The End of the Theory of Meaning", *Mind and Language* 3: 28–42.

Kalderon, M. (2005), "Introduction", in M. Kalderon (ed.), *Fictionalism in Metaphysics*, Oxford: Oxford University Press.

Kittay, E. F. (1987), *Metaphor*. Oxford: Clarendon Press.

Kripke, S. (1976), "Is There a Problem about Substitutional Quantification?", in G. Evans and J. McDowell (eds.), *Truth and Meaning: Essays in Semantics*, Oxford: Oxford University Press, 325–419.

Kripke, S. (1977), "Speaker's Reference and Semantic Reference", in French et al. (eds.), *Contemporary Perspectives in the Philosophy of Language*, Minneapolis, MN: University of Minnesota Press, 255–276.

Kripke, S. (1980), *Naming and Necessity*. Cambridge: Harvard University Press.

Lamarque, P. and Olsen, S. H. (1994), *Truth, Fiction and Literature: A Philosophical Perspective*. Oxford: Clarendon Press.

Lewis, D. (1978), "Truth in Fiction", *American Philosophical Quarterly* 15: 37–46.

Nunberg, G. (2002), "The Pragmatics of Deferred Interpretation", in L. Horn and G. Ward (eds.), *Blackwell Encyclopaedia of Pragmatics*, Oxford: Basil Blackwell.

Pagin, P. and Pelletier, J. (2007), "Content, Context and Composition", in G. Peter and G. Preyer (eds.), *Context-Sensitivity and Semantic Minimalism*, Oxford: Oxford University Press, 25–62.

Parsons, T. (1980), *Nonexistent Objects*. New Haven, CT: Yale University Press.

Peacocke, C. (1989), "When Is a Grammar Psychologically Real?", in A. George (ed.), *Reflexions on Chomsky*, Oxford: Basil Blackwell.

Perry, J. (1980), "A Problem about Continued Belief", *Pacific Philosophical Quarterly* 61: 317–332.

Quine, W. V. O. (1951), "On Carnap's Views on Ontology", *Philosophical Studies* II: 65–72.

Quine, W. V. O. (1953), "Carnap and Logical Truth", in P. A. Schilpp (ed.), *The Philosophy of Rudolf Carnap*, La Salle, IL: Open Court.

Recanati, F. (2004), *Literal Meaning*. Cambridge: Cambridge University Press.

Richard, M. (2000), "Semantic Pretence", in A. Everett and T. Hofweber (eds.), *Empty Names, Fiction and the Puzzles of Non-Existence*, Stanford, CA: CSLI, 205–232.

Romero, E. and Soria, B. (ms), "Metaphor: What Is Said or What Is Implicated?", talk at the 15th Annual Meeting of the ESSP, 2007, Geneva, at http://www.ugr.es/~eromero/draft1.htm .

Rosen, G. (1994), "Gideon Rosen on Constructive Empiricism", *Philosophical Studies* 74: 143–178.

Sainsbury, M. (2005), *Reference without Referents*. Oxford: Clarendon Press.

Salmon, N. (1998), "Nonexistence", *Noûs* 32: 277–319.

Schiffer, S. (1996), "Language-Created Language-Independent Entities", *Philosophical Topics* 24: 149–167.

Schiffer, S. (2003), *The Things We Mean*. Oxford: Clarendon Press.

Searle, J. (1975), "The Logical Status of Fictional Discourse", *New Literary History* 6: 319–332.

Stalnaker, R. (1978), "Assertion", in P. Cole (ed.), *Syntax and Semantics*, vol. 9, New York: Academic Press, 315–332.

Soames, S. (2003), *Philosophical Analysis in the XXth Century, vol. 2: The Age of Meaning*. Princeton, NJ: Princeton University Press.
Thomasson, A. (1999), *Fiction and Metaphysics*. Cambridge: Cambridge University Press.
Thomasson, A. (2001), "Ontological Minimalism", *American Philosophical Quarterly* 38: 319–331.
Toon, A. (2010), "The Ontology of Theoretical Modeling: Models as Make-Believe", *Synthese*, 301–315.
Tyler, A. and Evans, V. (2003), *The Semantics of English Prepositions*. Cambridge: Cambridge University Press.
van Fraassen, B. (1980), *The Scientific Image*. Oxford: Oxford University Press.
van Fraassen, B. (1994), "Gideon Rosen on Constructive Empiricism", *Philosophical Studies* 74: 179–192.
van Inwagen, P. (1977), "Creatures of Fiction", *American Philosophical Quarterly* 14: 299–308.
van Inwagen, P. (2003), "Existence, Ontological Commitment, and Fictional Entities", in M. Loux and D. Zimmerman (eds.), *Oxford Handbook of Metaphysics*, Oxford: Oxford University Press, 131–157.
Voltolini, A. (2006), *How Ficta Follow Fiction. A Syncretistic Account of Fictional Entities*. Dordrecht: Springer.
Walton, K. (1990), *Mimesis and Make-Believe*. Cambridge: Harvard University Press.
Walton, K. (1993), "Metaphor and Prop Oriented Make-Believe", *European Journal of Philosophy* 1: 39–56.
Wolterstorff, N. (1980), *Works and Worlds of Art*. Oxford: Clarendon Press.
Wright, C. (2002), "What Could Antirealism about Ordinary Psychology Possibly Be?", *Philosophical Review* CXI: 205–233.
Yablo, S. (2001), "Go Figure: A Path through Fictionalism", in P. A. French and H. K. Wettstein (eds.), *Midwest Studies in Philosophy* XXV, Oxford: Blackwell, 72–102.

Visual Practices Across the University

James Elkins

In 2005, I was working at the University College Cork in Ireland. Visual studies, film studies, and art history were expanding, and the time seemed right for a university-wide center for the study of images. I was interested in finding out who at the university was engaged with images, so I sent an email to all the faculty in the sixty-odd departments, asking who used images in their work. The responses developed into an exhibition that represented all the faculties of the university. It only had a couple of displays of fine art: one proposed by a colleague in History of Art, and another by a scholar in the History Department. Fine art was swamped, as I had hoped it would be, by the wide range of image-making throughout the university. The result was a book, *Visual Practices Across the University.*[1] The book is largely unknown outside of Germany, because the press, Wilhelm Fink, serves the German academic book market and does not concern itself with worldwide distribution or advertizing. (The book was published in Germany because most research on non-art uses of images is in German-language publications.) In this essay, I will report on the philosophic frame of the book, and give a sample of what it contains. To date it is the one of only two books that attempt to understand the full range of image production and interpretation in all university departments, including Engineering, Law, Medicine, and even Food Science.[2]

J. Elkins (✉)
School of the Art Institute of Chicago, Chicago, IL, USA
e-mail: jameselkins@fastmail.fm

[1]See Elkins (2007a), with contributions by thirty five scholars. This book is in English, and is available on Amazon Deutschland. This essay is adapted from the Preface, Introduction, and one of the chapters of the book. The exhibition was originally intended to be published along with a conference called "Visual Literacy", in a single large book. In fact the conference will appear as two separate books. The main set of papers in the conference, with contributions by W.J.T. Mitchell, Barbara Stafford, Jonathan Crary, and others, is Elkins (2007b); a second set of papers from the conference, on the subject of the histories of individual nations and their attitudes to visuality and literacy, will be forthcoming as *Visual Cultures.*

[2]The other is Beyer and Lohoff (2006); the glossary is on pp. 467–538. Their book surveys many more technologies than mine, and groups them according to an eclectic glossary

R. Frigg, M.C. Hunter (eds.), *Beyond Mimesis and Convention*, Boston Studies in the Philosophy of Science 262, DOI 10.1007/978-90-481-3851-7_8,

The book is an attempt to think about images beyond the familiar confines of fine art, and even beyond the broadening interests of the new field of visual studies. Outside of painting, sculpture, and architecture, and outside of television, advertizing, film, and other mass media, what kinds of images do people care about? It turns out that images are being made and discussed in dozens of fields, throughout the university and well beyond the humanities. Some fields, such as biochemistry and astronomy, are image-obsessed; others think and work *through* images. The humanities—not surprisingly—are in the minority when it comes to making and using images, and—perhaps surprisingly—they are generally *less* visual, less dependent on images, than other fields.

So far visual studies has mainly taken an interest in fine art and mass media, leaving these other images—which are really the vast majority of all images produced in universities—relatively unstudied. Outside the university, scientific images crop up in magazines, on the internet, in popular-science books, and in the familiar "art meets science" exhibitions. In those contexts images are often drastically simplified, shorn of much of the significance they had for their makers. In the book, I try to pay close-grained attention to the ways people make and talk about images in some thirty fields across all the faculties of a typical contemporary university. There are examples of the study of dolphins' fins, of porcelain teeth, of Cheddar cheese. In assembling and editing the various contributions, I was less interested in what might count as art or science, or in what might be of interest from an aesthetic (or anti-aesthetic) point of view, than I was in just *listening* to the exact and often technical ways in which images are discussed.

A great deal is at stake on this apparently unpromising ground. It is widely acknowledged that ours is an increasingly visual society, and yet the fields that want to provide the theory of that visuality—visual studies, art history, philosophy, sociology—continue to take their examples from the tiny minority of images that figure as art. At the same time, there is an increasingly reflective and complicated discourse on the nature of universities, which has as one of its tropes the notion that the university is "in ruins" or is otherwise fragmented. One way to bring it together, or at least to raise the possibility that the university is a coherent place, is to consider different disciplines through their visual practices. To begin a university-wide discussion of images, it is first necessary to stop worrying about what might count as art or science, and to think instead about how kinds of image-making and image interpretation might fall into groups, and therefore be amenable to teaching and learning outside their disciplines. Above all, it is necessary to look carefully and in detail, and not flinch from technical language or even from the odd equation.

All these points are theorized in the Introduction to the book. In this essay I will restrict myself to just one subject: the quality of the existing discourse between arts and sciences.

---

of "visualization techniques" such as "Modell", "Notationssystem", "Objektklassendiagramm", "Phasendiagramm", "Piktogramm", "Prototyp", and "Radardiagramm". I find their book interesting as a resource, but I am more optimistic about organizing the material into a smaller number of conceptual units.

1

Among the things that *Visual Practices Across the University* is not, it is primarily not a contribution to the many exhibitions and books that present scientific images as art, or as possessing the aesthetic properties or even the "richness" that supposedly inhere in art. I ignored the intermittent temptation to say such-and-such an image is beautiful, and I did not present any image, no matter how luscious, as possessing any aesthetic properties that its maker or its intended audience had not already claimed for it. My interest was the particular ways of talking about images in different fields, so I avoided generalized art-science talk about "beauty", "richness", "pattern", "symmetry" and other such concepts whenever I could.

(It happens that some ways of talking about images incorporate the kinds of broad claims about art or science that I would normally want to avoid, and it happens that people call one another's images "beautiful", but reporting on other people's use of such claims is different from using them to organize the argument.)

There are a number of examples of the kind of art/science talk I tried to avoid in *Visual Practices Across the University.* The most widely publicized recent conferences on science-art themes are Felice Frankel's two "Image and Meaning Initiative" conferences, the first at MIT in June 2001, and the second at the Getty Center in Los Angeles in June 2005.[3] Frankel is a science photographer, originally trained as a landscape and garden photographer, who rephotographs scientific experiments for publication.[4] In the past her work has raised interesting questions about the relation between her artistic choices and the scientists' visual preferences, especially when her rephotographs have helped scientists discover new features of their work that they had not seen.[5] Her books *On the Surface of Things* (2008) and *Envisioning Science: The Design and Craft of the Science Image* (2004a) present accomplished, colorful photographs of various physical and chemical phenomena. Frankel's conferences and books provide a chance for art photographers to think about scientific images, and for scientists to ponder such things as the place of beauty or art in visualization. Phenomena such as iridescence on an oil surface, colors generated by opal, and patterns of crystals on a surface, are visualized in great detail and with attention to composition and symmetry. The photographs' formal properties are, however, not theorized. Frankel presents her work as scientific photography and writes only as a technical photographer. She does not articulate the artistic influences on her own work, even though that history is pertinent because it guides her choices of compositions, colors, symmetries, and textures. Frankel's books therefore lack the analysis of artistic influences that might have enabled her to account for her photographic preferences. Her compositional choices, for example, are influenced—I assume mostly indirectly, without deliberation—by Abstract

[3]See web.mit.edu/i-m/intro.htm. My review of the 2001 conference is Elkins (2001a).

[4]See web.mit.edu/felicef/

[5]In this context I am only giving the outline of the argument: an example is discussed in detail in Elkins (1999).

Expressionism, and by realist projects such as the Boyle Family's fiberglass castings. In art historical terms, her practice derives from several strands of modern painting and photography from the 1940s to the 1980s. Those precedents are not irrelevant, because they can illuminate the aesthetic decisions that appear, unexplained, simply as "beauty". And because she does not know the science except to the extent that it is explained to her, the scientific content of her images is seldom broached except in the most general terms. For the book *On the Surface of Things,* a prominent chemist provided very brief, nontechnical summaries of the relevant science—not enough to account for individual passages in Frankel's very complex and detailed images. The chemist's caption for Frankel's picture of opal, for example, describes how the colors of an opal derive from microscopic bubbles: but the photograph does not show the bubbles, and so its colors, and its very complicated planes of color and form—all of them captured in a way that would not have been possible before Symbolism and abstract painting, using modernist criteria of coherence, composition, and visual interest—are entirely uninterpreted by his commentary. The same happens when the chemist describes a pictures of a shimmering pool of oil. The description of iridescence cannot be understood by reference to Frankel's photograph, and the composition of her photograph—which is indebted, probably indirectly, to Antoní Tàpies and other abstract painters and sculptors—cannot be understood by reference to the chemical description of oil films.

As a result Frankel's projects miss the many specific connections between photographic decisions informed by the history of art, on the one hand, and by the scientists' purposes, on the other. Her photographs can only appear as mute testimony to her "eye", her unarticulated judgment of what counts as an interesting image. *On the Surface of Things* is a successful coffee-table book, because it can be read by scientists and artists; both will recognize meanings that are not spelled out, but neither will know how to make a bridge between the two domains. What is needed, I think, is an inch-by-inch analysis of her photographs, to bring out the individual artistic decisions and their histories, together with—matched line by line with—an inch-by-inch account of the scientific meaning of each form.

Frankel also writes a column called "Sightings" in *American Scientist* magazine, interviewing scientists about their images. One column is an interview with Jeff Hester of Arizona State University, who was one of the scientists who made the widely-reproduced Hubble Space Telescope image of young stars in the Eagle Nebula (1995; Fig. 1).

The interview is brief, only a few paragraphs; and because of its brevity, it is a good example of what I think of as the abbreviated, impoverished structure of much of this generalized art/science discourse. Hester tells Frankel how the image of the Eagle Nebula was combined from thirty-two images taken by four separate cameras, and how the images were stitched together, cleaned up, and given false colors. Blue, for example, stands for emissions from doubly ionized oxygen. The colors appear "representational", in Frankel's word—that is, they make it seem the photograph is a picture of mountains. Hester explains the image is more like a "map of the physical properties of the gas", but that, fortuitously, "it is also closer to what you

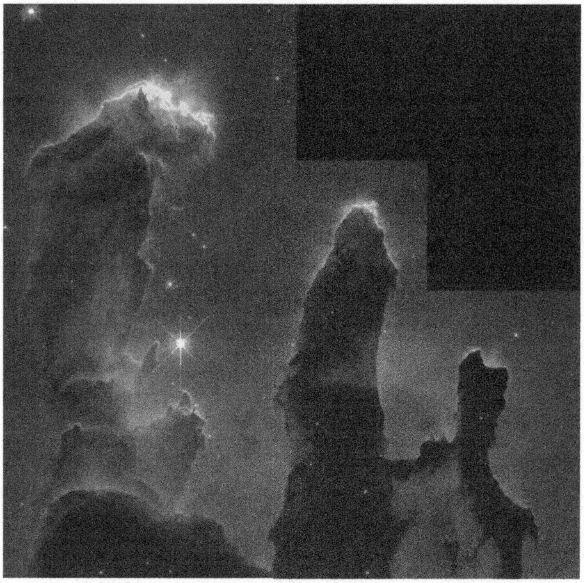

Fig. 1 Hubble space telescope image of young stars in the eagle nebula

might see through a telescope with your eye than is a picture taken with color film" (2004b, 462). Toward the end of the one-page interview, Hester says "the beauty of the image is not happenstance. When people talk about 'beauty,' they are talking about the presence of pattern in the midst of complexity".

Several things need to be asked about that claim if it is to make sense. It would be good to know why Hester felt he should mention beauty at all; I assume it was on account of the popular-science context of the interview, and the idea that beauty might serve as a bridge to a wider public. But what kind of bridge is beauty here? Instead of bringing beauty in, why not present the image as something wonderfully and unexpectedly complex—that is, after all, another alleged art-world value—by saying, as he had a moment before, that "there is one hell of a lot of information present"? But having mentioned beauty, why identify it with pattern recognition? That is not an association I think many people in art would have, unless they are following psychologists such as Rudolph Arnheim.

There are at least five assumptions at work in Hester's mention of beauty, and in Frankel's silence about it: that beauty is relevant, that the image is beautiful, that the meaning of beauty is clear, that beauty can help the image communicate to non-scientists, that beauty is an idea shared across the arts and sciences. Hester remarks that "the same patterns present in the image that make it aesthetically pleasing also make it scientifically interesting". If that were true—and to assent I would have to agree that beauty is present, and that beauty can be identified with pattern recognition—then it would have to mean something like this: If I appreciate the patterns in this image, I also appreciate the science. I think that is untrue, and it is not supported by what Hester says. He concludes that he and his collaborators "use color in the image in much the same way that an artist uses color", as an "interpretive tool". That may mean that the false colors he and his collaborators chose to represent emissions of oxygen, hydrogen, and sulfur are like the false colors artists chose, and it might also mean that artists also choose false colors that are at the same time like representational colors. Either way the parallel is too loose to do much work, and that is one of the reasons conversations like these are often so short.

An artist like Emil Nolde, who chose "false" colors as well as naturalistic ones, made his decisions for completely different reasons—and even using a different palette—than physicists who make false-color astronomical images. Scientists' choice of colors have specific histories, just as artists' choices. Some of the more garish productions of astronomical images owe their color choices to 1960s hallucinogenic art like *Yellow Submarine* or tie-dyed T-shirts. The Eagle Nebula image owes its color choices to the history of landscape painting and photography. It has a saturated, Kodachrome look that derives from nostalgic reworkings of 1950s photography, and it also owes something to the kitsch paintings popular in "starving artist" sales and exemplified for North American consumers by the painter Thomas Kinkade. (He paints tumble-down English-style thatched cottages, decorated with rainbow-colored flowers.[6]) I do not mean that any of these influences were direct, or conscious. The built-in color palettes of astronomical software, like the palettes in Photoshop, NIHImage, ImageJ, and other scientific image processing software, were often designed with certain aesthetics in mind—there are Cézanne-like palettes, and science fiction paperback-cover palettes. The salient point is that the colors are not often chosen only because they provide optimal contrast and legibility. Contemporary scientific practices are indebted to specific moments in the history of art, and it is the job of an observer in the humanities to make those connections.

In terms of forms, the Eagle Nebula image as it is presented here (it could have been cropped and oriented quite differently) belongs to the history of romantic landscape painting, from Arnold Böcklin and other German and French painters to the exaggerated mountains of the Hudson River School painters. It may even belong to the lineage of fantastical mountainscapes in Chinese painting, beginning in the Song Dynasty and continuing to the present. I do not mean any of this as a put-down: scientific images have their own lineages in the history of art, their own aesthetic histories. They are not merely or simply "beautiful"; and "pattern" has almost nothing to do with these historical lineages.

And even if artists were to agree that they use false and yet "representational" color "in much the same way", it would still be unclear what about the science has been explained aside from the fact that the colors were chosen to aid communication. Frankel's column does not explain how the image was generated, except in generalities; it does not explain the link that is proposed between art and science; and it does not explain the scientific content of the image. She asks no follow-up questions to Hester's opinions about beauty, art, and pattern.

[6] Try www.thomaskinkade.com; there are many other sites and stores.

Hester's brief comments are made in an informal context, but they follow a logic that can be found in many other places. Examples could be multiplied indefinitely. In 2005 an article in *California Monthly,* Berkeley's alumni magazine, showcased the research of Berkeley scientists (Smock 2005). In this kind of article, a "pretty picture" (the term was apparently adopted by astronomers to denote images they prepared for calendars and posters) is briefly glossed by a text identifying the scholar who produced it. A full-page photograph of a moss-covered tree, for example, is accompanied by a text describing a Berkeley scientist who recovered medicines from moss, especially "a family of chemicals called flavenoids" (Fig. 2).

Nothing more is said. In the context of an alumni magazine, all that is expected is a nice picture and a reference, and it would be assumed that anyone who wanted could follow up and find out more. But these clipped contexts are ubiquitous, so it is significant that the text explains neither the photograph (What kind of tree? What kind of moss? Was the picture used in the research?) nor the science (What are flavenoids? How are they extracted?). A reader perusing the article is treated to several dozen photographs and short paragraphs. If they are interested, they can

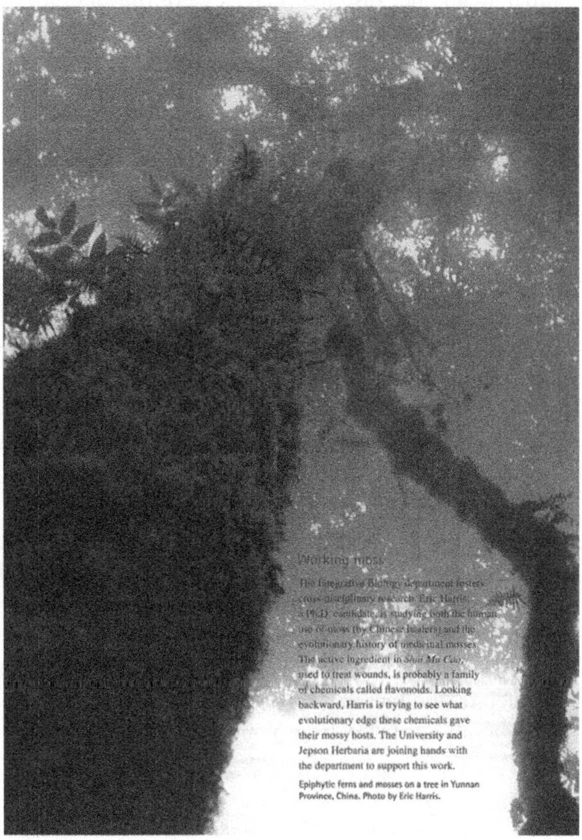

Fig. 2 Moss-covered tree: from Smock, "Picture This!" in *California Monthly* (March/April 2005), pp. 16–27. Courtesy Kerry Tremain, Editor, *California Monthly*

learn the names of the Berkeley scientists and guess at what they are doing, but the article is not really meant to teach anything. It is a wash of colorful images and new names, which suggests that lovely photographs can help laypeople understand a little science.

A few more examples will show how unquestioned this generalized art/science talk can be. In a lecture given in spring 2005 as part of the Einstein centenary, the physicist Michael Berry of Bristol University visited Ireland and gave a talk about the patterns of light that form on the bottom of swimming pools and the ceilings above swimming pools. The "caustics" and wave fronts were the object of his own scientific research, he said, and he also talked about the motion of wave packets and the physics of rainbows. He compared those phenomena to David Hockney's paintings, and to passages about reflections and light patterns in A.S. Byatt, Thomas Pynchon, and John Banville. The occasion was a "Café scientifique" sponsored in part by the British Council, and in that setting it would not have been appropriate to introduce much scientific content. Berry worked on the assumption that the audience found the images as beautiful as he did (I found them garish), and the theme throughout was that an appreciation of the beauty would provide a way to appreciate the science. The audience was appreciative because he was persuasive and animated, and because the images were full of color and light: but both the science and the art (I mean the Hockney) were done a disservice. Nothing could be gleaned about the physics of caustics from Berry's images, and his impoverished sense of artistic beauty made the parallels between artists like Hockney and the high-chroma scientific photographs unconvincing. But the event was a success—it was crowded beyond the room's legal capacity—and no questions were asked about "beauty" or scientific content.

In the art world, the same strategies of juxtaposing art and science, and implying that one seeps naturally into the other, produce work that can be taken tongue-in-cheek, as kitsch. An example at the margins of the art world is the company DNA 11, which will make framed pictures of your DNA.[7] Although their website simply identifies the images as DNA—and as "great art", and "one-of-a-kind masterpieces"—actually they are electrophoretograms, arranged in strips. They are unlabeled, making it virtually impossible to extract any scientific content from them. "The procedure we use", they write, allaying the possible objection that someone could extract information from their "art", "creates a unique fingerprint that does not provide any information about your genetic code. It is a unique, artistic representation of your genetic fingerprint". The framed prints they produce are beholden to a popularized aesthetic derived from minimalism: the color schemes they offer, and the frames that consumers can choose, all derive from second-generation minimalism in the 1990s. Their project can also be taken as just fun—which is to say as campy pseudo-science, or even kitschy sciencey minimalism. DNA 11's art credentials include the fact that it is advertised specifically as having no content: you can't learn about your DNA from your DNA art.

---

[7] www.dna11.com, accessed March 2006. I thank Curtis Bohlen for drawing my attention to this.

"Beauty" and "art" do not have much analytic purchase in any of these instances. Was Berry's use of the word that different from Ed Bell's praise of the computer graphics company Hybrid Medical Animation, when he said their animations "extend beyond the boundary of highly informative graphics: they enter the realm of high art, achieving a combination of Truth and Beauty"? Hybrid Medical Animations make Hollywood-style digital movies of proteins, antibodies, bacteriophages, and other microscopic phenomena (Fig. 3).

They use the latest textures (translucent surfaces, shining and viscous surfaces), vivid colors (magentas, lavenders) and all the bells and whistles of *Star Wars*-style action (tracking shots, zooms, fly-throughs, rapid point-of-view changes, simulated shallow focus). Their movies are like *Star Wars* or a Universal Studios theme park ride, but with molecules instead of actors. Bell is Art Director of *Scientific American;* his endorsement appears on Hybrid Medical Animation's web pages. "Beauty" would seem to mean something like "dazzling post-production-style visual effects"—different, I think, from Berry's "beautiful" which means something like "elegant curvilinear patterns not unlike Op Art", and from Hester's "beautiful" which means something like "patterns that can be universally recognized".

There is a longer history of displaying scientific images for their beauty. André Kertesz composed scientific images that way, but the most influential example was the philosopher Jean-François Lyotard's exhibition *Les Immatériaux,* which displayed bubble-chamber images as if they were analogues of gestural painters such as Tàpies or Cy Twombly (Centre de Création Industrielle 1985). Bubble chamber

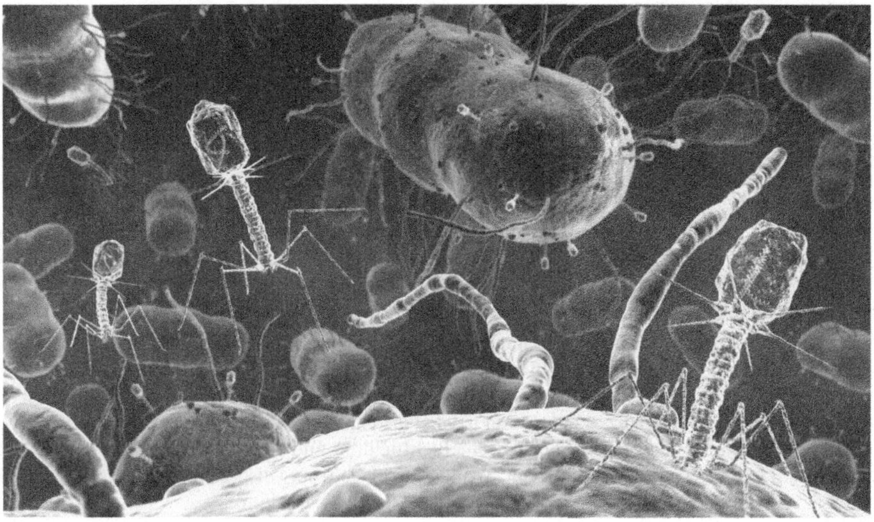

Fig. 3 Hybrid medical animations' still of microscopic phenomena (c) 2006 Hybrid medical animation. Courtesy of Geoffrey Stewart info@hybridmedicalanimation.com

images are actually intended to be *measured* and then discarded, and not appreciated for any aesthetic property. The exhibition I curated in Ireland, "Visual Practices Across the University", was intended to break with the tradition of Kertesz and Lyotard and the many people who follow in their wake. In the exhibition, each person or group of exhibitors displayed a single large image. Visitors were meant to be attracted by the large, unusual images, the way a reader of *California Monthly* might be attracted by the pictures of outer space, molecules, and mossy trees. Then when the visitors approached more closely, they found that the pictures only *appeared* to be accessible, and what little they shared with art—their compositions, their colors—was not helpful or interesting.

The opposite also happens: scientists write about artworks as if art's main interest is its scientific content. Thomas Rossing and Christopher Chiaverina's *Light Science: Physics and the Visual Arts* (1999), which finds scientific themes in pointillism, anamorphosis, and op art, is an example: it argues that a principal source of interest in the art is its illustration of basic scientific concepts.[8] Leonard Shlain's *Art and Physics: Parallel Visions in Space, Time, and Light* (1991) is a more concerted effort to find links between science and art. But Shlain is too easily satisfied by chance coincidences, metaphoric connections, and miscellaneous affinities.[9] The same could be said of other books, including John Latham's *Art After Physics* (1991) and Arturo Gilardoni's *X-Rays in Art* (1977). The common ground of these books is a dual claim: first, that art can be interesting because it demonstrates science; second, that it is not incumbent on someone writing about the science in art to account for the apparent irrelevance of the existing non-scientific interpretations of the art.[10]

A large critical and journalistic literature rose in the wake of a book by David Hockney and Charles Falco called *Secret Knowledge: Rediscovering the Lost Techniques of the Old Masters* (2001), which claims that some old masters used mirrors and other optical devices to help them make naturalistic paintings. There was an enormous conference on the theme in December 2001 at New York University, and several of the people involved continued to publish on the subject in the years

[8]For a review, see Stroke (2001); Stroke notes the asymmetry of the book, which concentrates on the influence of science on art, and notes that artists sometimes influence science. His example is Leopold Godowsky, Jr., and Leopold Mannes, who invented the Kodachrome process; but Stroke observes they both also had physics degrees.

[9]His website glosses his book by claiming that "despite what appear to be irreconcilable differences, there is one fundamental feature that solidly connects ... evolutionary art and visionary physics. [They] are both investigations into the nature of reality. Roy Lichtenstein, the pop artist of the 1960s, declared, 'Organized perception is what art is all about.' Sir Isaac Newton might have said as much for physics". It would be extremely difficult to find another artist who says that, and just as hard to define what it might mean. What art is made from "disorganized perception"? And what is "evolutionary art" anyway? Shlain, at www.artandphysics.com.

[10]The most promising project along these lines is John Onians's research at the World Art Studies Centre at the University of East Anglia, which is a patient and systematic search for things that particular branches of science—especially neurology—can say about art; see Onians (2007).

following. (My criterion of an enormous conference is that ninety seats were set aside just for journalists, and lines went halfway around Washington Square in Manhattan.) Essentially Hockney and Falco claimed that painters from Van Eyck onward had access to optical aids such as mirrors, camera lucidas, and lenses that helped them achieve the feats of naturalism that have been traditionally attributed to their innate skill. The book and conference were a sensation in the media, in part because they seemed to empower ordinary viewers—at last, so it was said, viewers do not have to listen to the increasingly arcane meditations of academics, because they can see for themselves how the paintings were made.[11]

Ellen Winner, a psychologist who gave a paper at the conference, later wrote an essay called "Art History Can Trade Insights With the Sciences", calling for a mutual respect that she felt was missing at the conference. "True", she writes, "Falco and Hockney did not speak to the meaning or beauty" of the art, but that does not imply there are no lessons to be learned by considering the science. "When art historians argue that artists did not *need* lenses because they were so talented, they seem not to realize that the argument does not rule out the use of lenses" (2004, B10). The gulf of misunderstandings I have been trying to describe is nicely contained in that sentence, because regardless of the truth of Hockney's claim, it is not true that "art historians argue that artists did not *need* lenses": they scarcely mention those things at all. The two discourses are much further apart than Winner's claim implies, and it is not likely that more than a half-dozen humanists and cognitive scientists are "going to be teaming up to study humanistic phenomena from a scientific perspective". In order for that to happen, there has first to be an agreement over the common problems, whether they are beauty or optics.

Sidney Perkowitz, another scientist who attended the conference on Hockney's book, had written a book called *Empire of Light* (1996). In the article he contributed to the conference, he says he is neither surprised nor dismayed that some artists used optical aids. "Should the use of a tool diminish the value of the art?" he asks, and he illustrates a painting by Chardin, an Op-Art abstraction, and Mondrian's *Broadway Boogie-Woogie.*[12] The question isn't wrong, but wrongheaded. To whom does it matter that Chardin or Mondrian "reflect principles of visual cognition"? That has seldom been a part of their significance, and if the idea is to find examples of visual cognition, there is no good reason to adduce art to begin with. At the conference I had a brief argument with Perkowitz. I suggested that very few contemporary artists even use science in their work—I named Vija Celmins, Dorothea Rockburne, and Mark Tansey—and he said I was wrong, that his book had many examples of "new forms of art" produced by the use of science. His essay features an artist named Dale Eldred (I had not heard of him), and his book has many more minor artists. I wonder if their marginality in the art world does not prove

[11]Notably David Stork and Charles Falco. My responses are a review of Hockney (2001), on the College Art Association review site at www.caareviews.org/hockney.html and a review of the NYU conference in Elkins (2002). The paper I delivered at the conference is Elkins (2001b); I have also rehearsed these argument in Elkins (2008a).

[12]webexhibits.org/hockneyoptics.

the point. Art that is strongly inclined to technology or science often—though not always—ends up on the margins of the art world. The large annual conferences of SIGGRAPH and ISEA are cases in point; both organizations feature digital art, and both are almost completely ignored by the mainstream art world. In some measure that is a prejudice, and a fault, of the art world: but in some measure it shows that scientific and technological themes just aren't part of the mainstreams of postmodernism.[13]

The principal humanist scholars who study the science of art, such as Martin Kemp and John Gage, have done much of what can be done on the scattered appearances of scientific content in Western art[14] (Kemp 1990; Gage 1993). The end point of such research is the fact that science has rarely constituted much of what matters in art. The complementary end point of the scientific interest in art, such as Thomas Rossing and Christopher Chiaverina's, or Leonard Shlain's, should be that scientific explanations rarely matter in humanist discourse on art. If discourse on science-art connections is rum, uninformed, unhelpfully abbreviated, unjustifiably optimistic, alienating, and generally unhelpful, then it may be time to find new ways of talking about images that are not art.

I have been arguing that public talk and journalism about art and science is a kind of faux-discourse: it has the appearance of creating meaning, but it often fails to do so because the two sites of knowledge, historical or critical and scientific or technical, are too generalized to make contact. Even the small amount of academic writing on art and science, such as Martin Kemp's, only attains its purchase by narrowing its focus to very small extracts of art history.

One way to improve this situation would be to avoid generalized tag-words like "beauty", "elegance", and "pattern", and another way would be to avoid setting up contrasts between science and art.

2

The book, *Visual Practices Across the University,* is not my first attempt to find a way of thinking that could include all sorts of images at once. The other projects are relevant here, because they form the background and justification for *Visual Practices*. The first was *The Domain of Images* (1999), which divides images first into three groups (writing, pictures, and notation), and then into a set of seven. The triad writing, pictures, and notation was intended to capture the fact that mathematical images are used and talked about differently than written language or visual images. The division into seven was partly borrowed in part from Ignace Gelb, who was Derrida's source for "grammatology". The seven included allography (calligraphy, typefaces, and the visual elements of writing), subgraphemics (writing-like

---

[13]I am not criticizing all technologically-oriented art; my main target is the perception of the mainstream art world. For a full argument see Elkins (2005).

[14]This point is elaborated in my review of Kemp (1990) in Elkins (1991) and also in Elkins (1999).

fragments of images), and emblemata (highly organized symbolic images). *The Domain of Images* is a long and complicated book, and it has the conceptual narrowness that any taxonomy imposes on itself. Its crucial limitation, as the art historian Robert Herbert pointed out, is that it has to renounce some of the history of the objects, and virtually all of their political and social contexts, in order to make sense of how they have been received. Emblemata, for example, are interpreted in distinct and definable ways—they have an inner logic, a lexicon, and protocols of reading that make them recognizable and legible—but in order to analyze the differences between emblems and other, less organized images, it is necessary to suspend an interest in the history or social contexts of individual emblems. *The Domain of Images* subordinates the purposes images serve to the ways people interpret them, and in that respect it is, in the end, a formalism.

The book *How to Use Your Eyes* (2000) took an entirely different approach. It has thirty-odd very short chapters describing such things as "How to Look at the Night Sky", "How to Look at a Twig", "How to Look at a Shoulder", "How to Look at an Engineering Drawing", and "How to Look at Sand". Each chapter gives as many names and terms as I could find about each subject: the half-dozen sources of light in the night sky aside from the moon and stars; the "leaf scars" that make it possible to identify trees in the wintertime; the names and motions of muscles in the shoulder. The book is full of pictures and unusual words. Half the chapters are objects made by people—the script Linear B, Japanese calligraphy, paintings, scarabs—and half are natural objects—moths' wings, sunset colors, twigs, grass, sand. *How to Use Your Eyes* is empirically minded, and was rightly said to depend on technical nomenclature: its methods do not work on objects that have few names or parts. As one reader said, it ends up making seeing into reading. I am not sure of the force of that claim, because it can be argued that the world only becomes visible through language, when an object has a potential name—but the book is certainly limited to visual objects that have already been extensively labeled.

Visual Practices is more technical than *How to Use Your Eyes,* and more careful about the disciplines that produce knowledge than *The Domain of Images. Visual Practices* is partly meant to be an example of what the field of visual studies might accomplish if it were to relinquish its lingering interest in art. Visual studies continues to grow very rapidly but I think it effectively remains in an academic ghetto, confined by its concerns with mass media, fine art, and politics.[15] First-year classes taught as introductions to the visual world continue to take most of their examples from Western fine art and mass media, and to a lesser extent from design, craft, and non-Western practices. When objects outside of art are considered, they are treated in a general way, as examples of production or politics. Scientific and other non-art images are adduced to enrich the cultural contexts of fine art or to explain references in individual artworks. Science is seen indistinctly, from a distance.

(This is more true in North America and the UK than in German-speaking countries and in Scandinavia. There, visual studies is frequently more attentive to

[15]The argument I am alluding to here is given in Elkins (2003).

non-art images. Examples include Gottfried Boehm's and Andreas Beyer's "Iconic Criticism" initiative in Basel, Horst Bredekamp's work at the Humboldt-Universität Berlin, and individual projects in Karlsruhe, Copenhagen, Aachen, Stockholm, Magdeburg, Leipzig, and Lund. This book fits more with German-language scholarship than with English- or French-language work, which continues to stress political, gender, and wider social meanings.)

The founding gambit of visual studies in English-speaking countries is that in a world of proliferating images, it no longer makes sense to have specialists on every conceivable kind of image, as it had once been useful for art history departments to have specialists on medieval, Renaissance, Baroque, and modern art. Visual studies posits that what matters is a more abstract, reflective concept of the production and dissemination of images, and a methodology capable of revealing the ways images are made to seem compelling, and how they reform their viewers and shape their desires. That has been a fruitful direction for several decades, and it may continue to be: but it does not address what happens in the sciences, for the simple reason that it elides the specific content of non-art images even as it pays close attention to the specific content of art and mass media. The American World War I poster with the legend "I want you!" has been analyzed in several visual studies publications, but there is still nothing in visual studies that analyzes a gene map in such a way that a student could explain what its parts signify. *Visual Studies* is intended to discover what it would sound like to pay attention to all images, art and non-art alike, with the level of detail used by their makers and their intended public. (Detailed engagement is, I think, indispensable: in the book, I made a few images myself, using scientific software and laboratory equipment. Only by operating the instruments, and learning the software, is it possible to see the limits of a humanities-based visual studies.)

The exhibition was difficult for viewers, and the book is not easy to read. Its chapters are like a collection of short stories: they have different characters and plots, but like stories by a single author, they share a number of themes, passing them back and forth, sometimes developing them, sometimes not. An editor who saw this book in manuscript said that it was too "particulate"; to her, the chapters seemed disconnected and too much concerned with the recitation of facts. This book is designed that way, instead of as a single continuous narrative, because I think that disjunctions are exactly what the field of visual studies needs in order to move forward. Texts on visual studies by W.J.T. Mitchell, Nicholas Mirzoeff, Mieke Bal, and others are limited by their strengths, as it were: they offer continuous theorizations in non-technical prose, but in doing so they exclude ideas that cannot be accommodated by humanities-style narration. What is at issue here, from the standpoint of visual studies, is the sense of appropriate theorization. The thirty practices in my book embody a number of themes, but the individual visual practices are not subsumed by those themes. Discontinuous, "inappropriately" factual, surprisingly technical, "particulate", apparently under-theorized visual encounters are exactly what I think will produce a genuine advance in theorizing the visual, an advance that will propel visual studies out of the humanities and into the wider practices of the university.

One more project needs to be added to this sequence. From 1998 to 2008 I wrote a book called *Six Stories from the End of Representation: Painting, Photography, Astrophysics, Microscopy, Particle Physics, Quantum Physics 1985–2000* (2008b). It considers six fields, two in the arts and four in the sciences, and studies them in six separate chapters. I make no connections at all between the six fields, and I do not present any over-arching theme. The idea is to let each discipline speak in its own words, in full technical detail, and not to popularize anything. *Six Stories From the End of Representation* is a kind of reductio ad absurdum of *Visual Practices Across the University* : it goes at great length into just six fields, instead of sampling thirty fields, and it declines all opportunities to make connections. *Six Stories* is intended to display the weaknesses of popularizing and abbreviating, and to pay whatever cost may be entailed in terms of readability, while *Visual Practices Across the University* contains an analysis—which I am omitting here—of the common themes of image-making that bind the university, improbably, into a coherent whole.

Those are the projects that led up to *Visual Practices Across the University,* which takes a more radical and thoroughgoing stand on these issues. I hope I have said enough to indicate why the book cannot be condensed or summarized. Instead I will close with a sample chapter.[16] I choose a chapter on the visualization of viruses, but like the other twenty-nine chapters, it stands on its own as an image-making and image-interpreting practice that is every bit as rich, difficult, and rewarding as discourse on paintings or sculptures. I will end with a brief conclusion.

3

The biologist Stephen Harrison wrote an essay called "What Does a Virus Look Like?" (1991). In it he considered over ten different kinds of images of viruses, made with different instruments. They are not all compatible—they cannot be assembled into one perfect picture. Harrison concluded that viruses don't "look like" anything except the sum total of those images.

William Wimsatt, a philosopher of science, has called this problem the "thicket of illustration": no one strategy will do, he notes, when it comes to picturing things as complex as DNA. Here we consider five different ways of producing images of viruses.

The Plaque Assay

Phages are obligate parasites of bacterial cells (Fig. 4). They have no intrinsic metabolism and are totally inert in the absence of their bacterial hosts. They attach

[16]This is chapter 29 in Elkins (2007a), titled "Visualising Viruses"; it was co-written by Stephen McGrath, University College Cork.

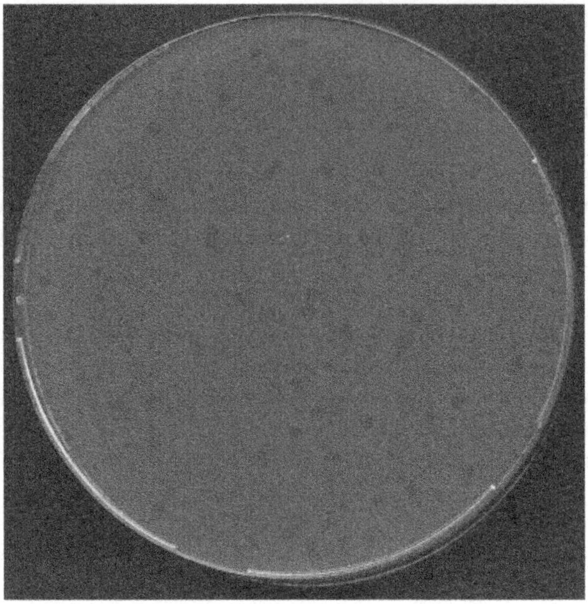

Fig. 4 An agar plate with bacterial cells and phages. Acknowledgements to Dr. Stephen McGrath, Microbiology Department, University College Cork

to the bacterial cells in a tail-first orientation, triggering the release of the DNA from the phage head, where it has been held under immense pressure.

The *plaque assay* is a method used in the laboratory to visualize the bacteriophage life cycle. An agar plate is seeded with a "lawn" of bacteria that has been mixed with some phages. The clear spots on the plate show where a phage has infected a bacterial cell and the progeny phages have killed the cells around it causing a clear zone or "plaque".

At this stage, no special optical equipment is necessary to locate the phages.

Transmission Electron Microscopy

The main structural features of phages can be seen in the large TEM image (Fig. 5). This is the lactococcal bacteriophage Tuc2009. Toward the top is the head, containing the DNA; then the tail; and at the bottom the structure that recognizes the host cells and contains the adsorption apparatus.

TEMs work on the analogy of light microscopes, but they shine a beam of electrons through the specimen. Whatever part is transmitted is projected onto a phosphor screen for the user to see. This is a typical, full-resolution TEM image; the original is 1280 $\times$ 1024 pixels in 16-bit grayscale—these images do not need to have ultrahigh resolution.

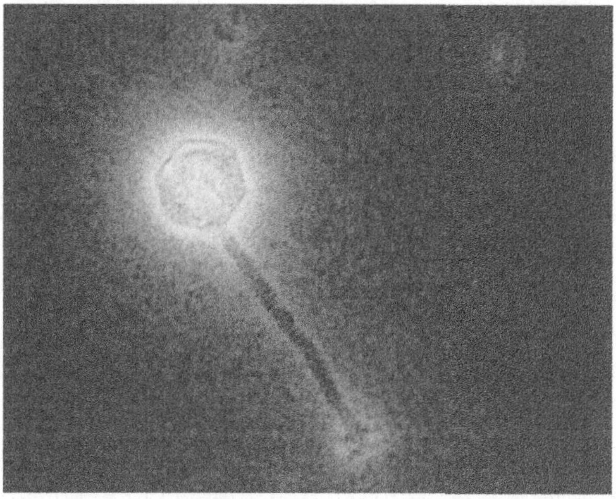

Fig. 5 Bacteriophage as visualized by transmission electron microscopy. Acknowledgements to Dr. Stephen McGrath, Microbiology Department, University College Cork

Gene Mapping

The first step in gene mapping is sequencing (Figs. 6, 7, and 8). The familiar base pairs of DNA—the rungs in its ladder—are sequenced. The graph that results is called a chromatogram (Fig. 6). The names of the base pairs can be read off the graph; the heights of the peaks show the confidence level of the analysis.

Figure 7 illustrates the genome of the bacteriophage Tuc2009. Its complete genome sequence has been determined and the individual genes contained within identified using a set of criteria based on the recognition of patterns and signatures in the DNA sequence. Each of the arrows represents an individual gene. The arrows are arranged in three rows, just to make them more visible. At the top of the image is a map of the parts of the phage that are formed by the different genes.

The colored arrows indicate genes coding for proteins to which physiological functions have been assigned. Red indicates that a function has been assigned on the basis of experimental work, whereas green denotes that a function has been assigned on the basis of the similarity of that protein to experimentally verified proteins encoded by other phages. Computer analysis allows us to predict which proteins will form part of the bacteriophage structure, but the actual visualization of these proteins is the only definitive proof.

The gene sequence in the Tuc2009 can then be compared with genes in other bacteriophages (Fig. 8). The genes occur in slightly different places, but they can sometimes be correlated, making it possible to determine some of their functions.

Electrophoresis

The electrophoresis technique is used to separate and visualise individual proteins in a biological sample (Fig. 9).

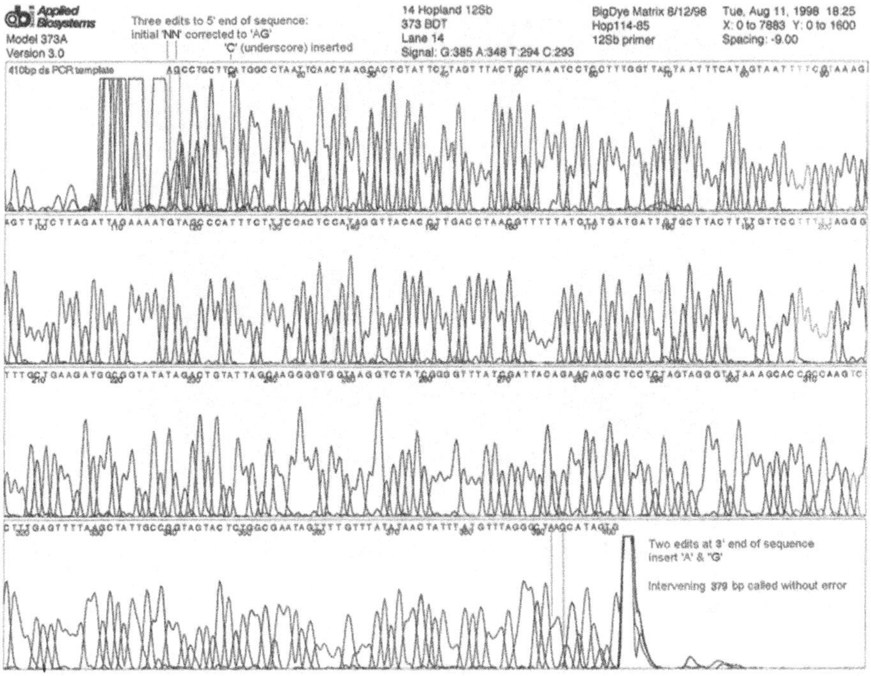

Fig. 6 Chromatogram of DNA sequence. Acknowledgements to Dr. Stephen McGrath, Microbiology Department, University College Cork

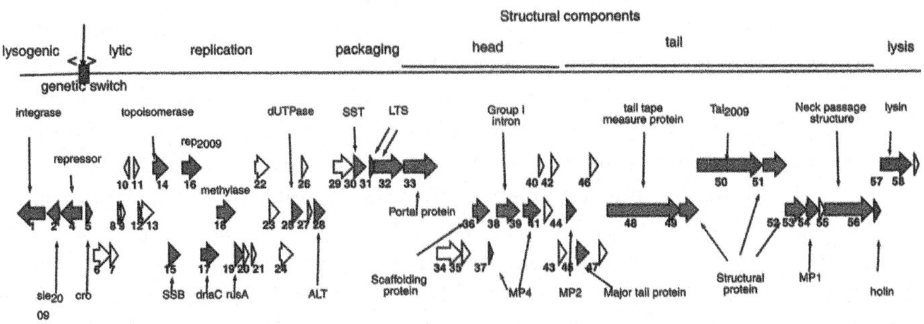

Fig. 7 Genome of the bacteriophage Tuc2009. Acknowledgements to Dr. Stephen McGrath, Microbiology Department, University College Cork

The protein bands in lane 1 represent a standard mixture of proteins of known size to which test proteins are compared. Each of the bands in lane 2 represent individual proteins that constitute the bacteriophage. Single bands representing individual proteins may then be cut from the gel and further analysed in order to determine the sequence of amino acids that they contain.

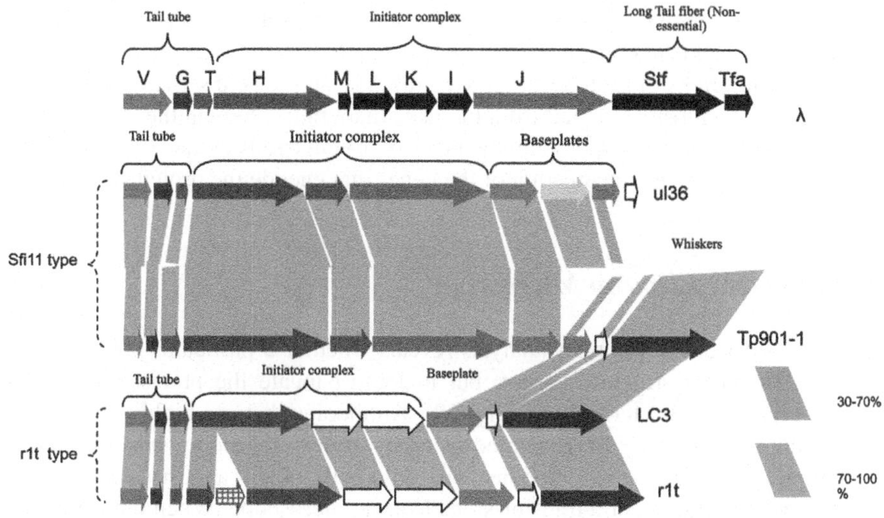

Fig. 8 Comparative genetic sequences of bacteriophages. Acknowledgements to Dr. Stephen McGrath, Microbiology Department, University College Cork

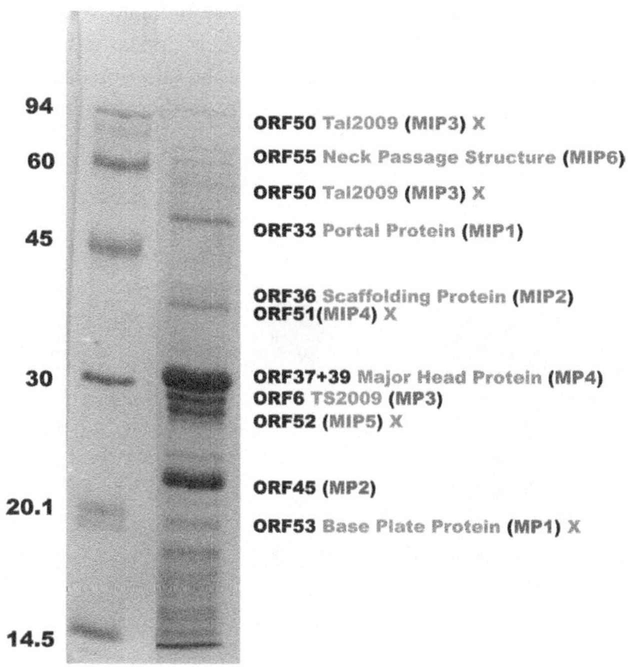

Fig. 9 Visualization of proteins by electrophoresis. Acknowledgements to Dr. Stephen McGrath, Microbiology Department, University College Cork

This type of analysis is dependant on the successful separation of the individual protein constituents into discrete homogenous bands as well as the presence of sufficient concentrations of proteins in these bands. The amino acid sequences may then be compared to those predicted from the gene map, thus allowing the identification of the structural proteins. Compare the labeled protein bands in lane 2 to the arrows in the gene map to see the location of the genes that encode the proteins.

Immunogold Electron Microscopy

Data from the electrophoresis analysis reveals whether a particular protein forms part of the phage structure or not, but it doesn't locate the precise location of the protein on the bacteriophage (Fig. 10). Antibodies that are highly specific for individual proteins may be generated using a variety of genetic and biochemical techniques. Labeling these antibodies with gold makes them appear as dense black spots when viewed under a transmission electron microscope. When the antibodies are mixed with the bacteriophage they specifically recognise and "tag" their cognate protein on the bacteriophage structure, thus marking the precise location of the protein.

The first panel is a TEM of the Tuc2009 bacteriophage without the addition of gold-labeled antibodies. Gold-labelled antibodies specifically recognizing individual proteins are added in the other pictures and are indicated on the panels. Their encoding genes are also included—the same numbers appear in Fig. 7.

The process of generating these antibodies can be laborious and expensive, and the success of the tagging of the specific protein on the phage is dependant on a

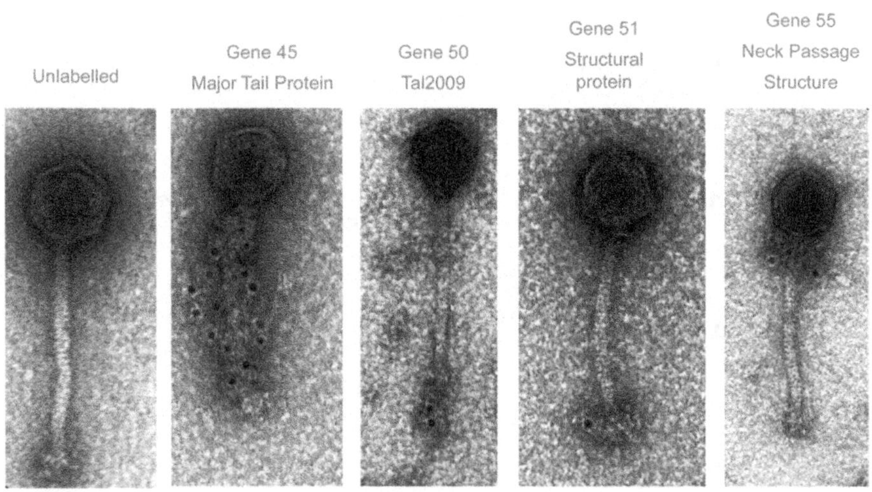

Fig. 10 Bacteriophage Tuc2009 "Tagged" and "Untagged". Acknowledgements to Dr. Stephen McGrath, Microbiology Department, University College Cork

number of critical factors such as the quality of the antibody and the accessibility of the protein on the phage structure to the antibody.

Other Kinds of Pictures

In addition to these kinds of images, virologists also make extremely detailed images of all the atoms in parts of the bacteriophages (Fig. 11). At the other end of the scale of detail, virologists find it useful to make schematic pictures of the different parts of the virus, to model how they might be put together (Fig. 12). Ideally, each part corresponds to a known gene (Fig. 13).

Conclusions

These are just eight of the ten or more methods of visualizing viruses. Clearly, no single representational method is sufficient. The opposite of the "thicket" of representation is the assumption, common in fine art, that a single image—say, the *Mona*

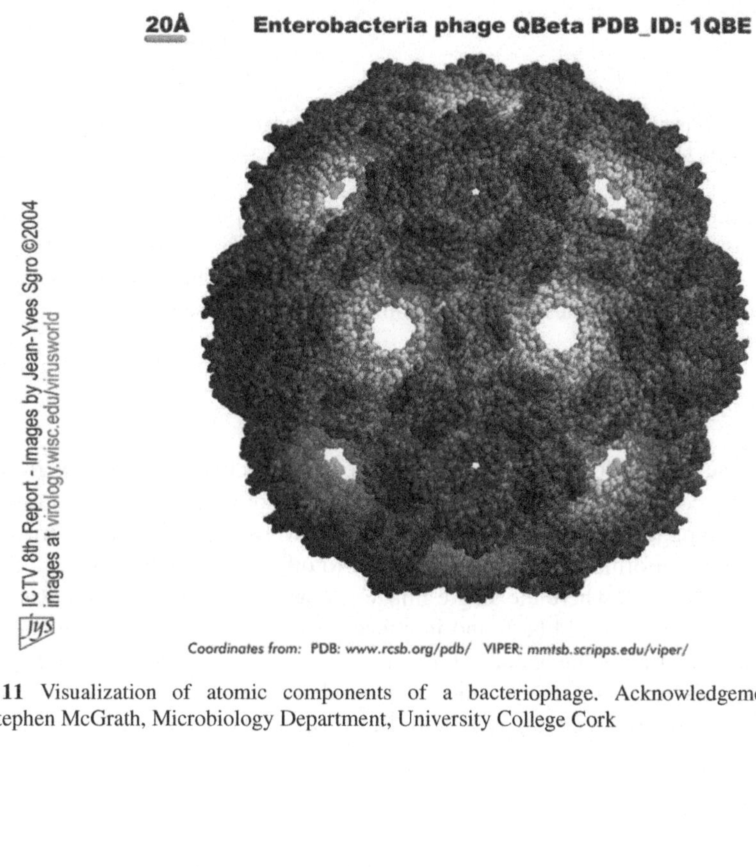

Fig. 11 Visualization of atomic components of a bacteriophage. Acknowledgements to Dr. Stephen McGrath, Microbiology Department, University College Cork

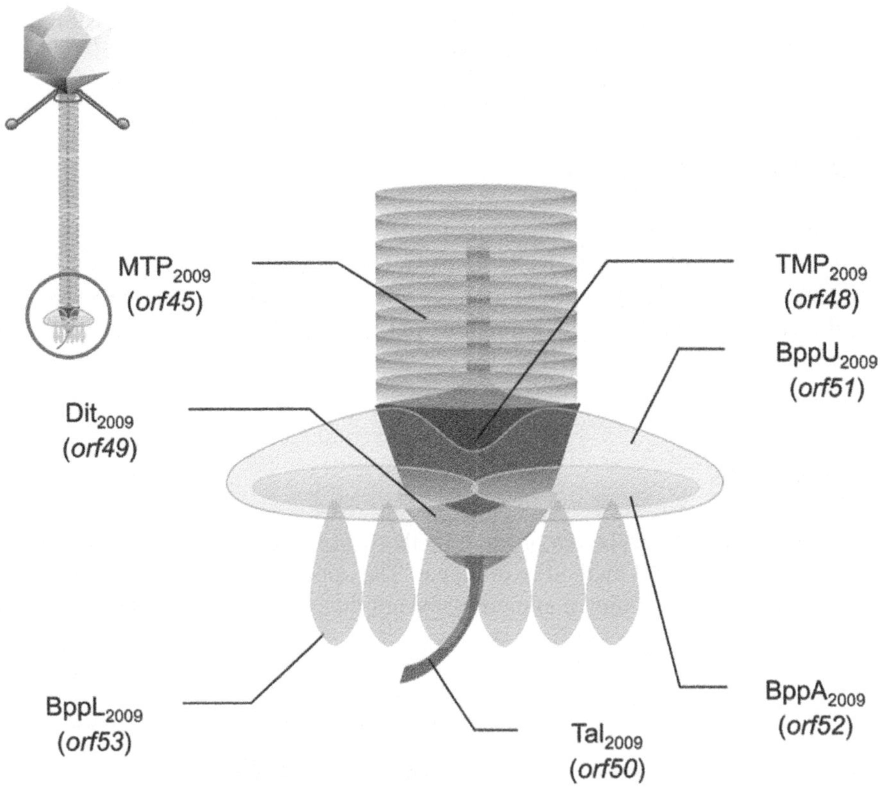

Fig. 12 Schematic model of virus structure. Acknowledgements to Dr. Stephen McGrath, Microbiology Department, University College Cork

Lisa —is not only sufficient but definitional for its subject. No further representations can even be imagined, except pastiches. In this case, however, the object does not exist except as a series of partly incommensurate representations.

*

This is the entirety of the chapter on viruses. Some chapters in *Visual Practices* have more connections to other chapters, but I did not force the links. In this case, the fascinating idea that some fields see the visual world as a "thicket" of structurally incompatible information could be extended to other fields, and contrasted against the case in fine art, where the single image is considered sufficient and even ideal. (Counter-examples could be found in conceptual art such as Art & Language, but they would be rare in the history of art.) People interested in the study of diagrams, graphs, and charts, and their relation to naturalistic representations, might find the study of viruses an especially rich field. But I would like to stress an abstract point:

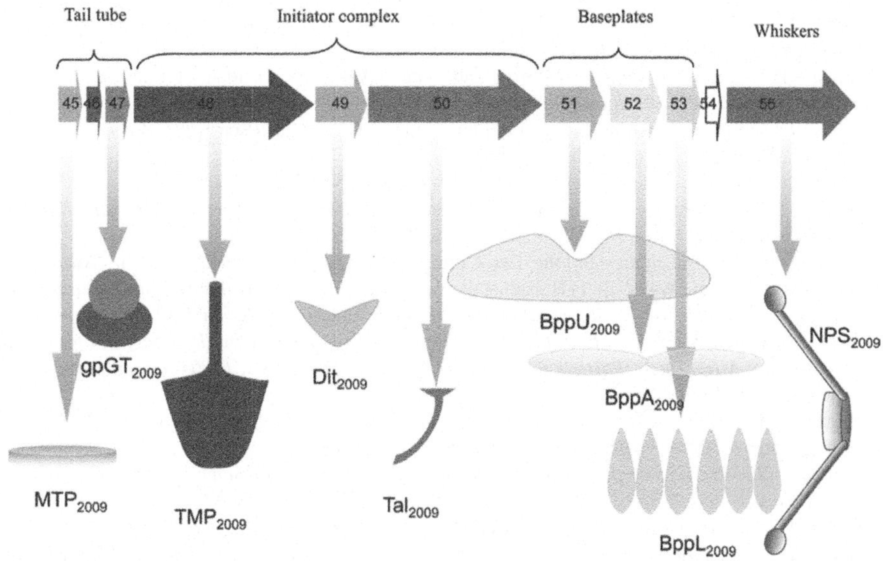

Fig. 13 Visualizing virus of known genes. Acknowledgements to Dr. Stephen McGrath, Microbiology Department, University College Cork

what matters here is the *exact* language of viral representation. A chromatrogram is different from an electrophoresis gel, and both are different from the Powerpoint animations scientists use to present their results. These are specific image technologies, and when they are subsumed under general philosophic categories such as resemblance, or general aesthetic categories such as beauty, or general formal categories such as pattern, or even general notational categories such as diagrams, their specificity—their language—is lost. The way forward through the impasse of generalized talk about art and science, is to bite the bullet and study technical and scientific imagery as it presents itself, in its own languages. Only then will it be possible to see how rich the field of images is, and only then will it become apparent that philosophy and art history do not own the interpretive tools to understand all of visuality.

References

Beyer, A. and Lohoff, M. (eds.) (2006), *Bild und Erkenntnis: Formen und Funktionen des Bildes in Wissenschaft und Technik*. Munich: Deutscher Kunstverlag.

Centre de Création Industrielle (1985), *Les Immatériaux*: *épreuves d'écriture*. Paris: Centre Georges Pompidou.

Elkins, J. (1991), "Review of Martin Kemp, *The Science of Art* ", *Zeitschrift für Kunstgeschichte* 54, 4: 597–601.

Elkins, J. (1999), *The Domain of Images*. Ithaca, NY: Cornell University Press.

Elkins, J. (2000), *How to Use Your Eyes*. New York: Routledge.

Elkins, J. (2001a), "Who Owns Images: Science or Art?", *Circa* 97: 36–37, online at recirca.com/backissues/c97/elkins.shtml
Elkins, J. (2001b), "Optics, Skill, and the Fear of Death", online at webexhibits.org/hockneyoptics/post/elkins.html
Elkins, J. (2002), "Plucking at a Popular Tune", *Circa* 99 (Spring): 38–39; online at recirca.com/backissues/c99/elkins.shtml
Elkins, J. (2003), *Visual Studies: A Skeptical Introduction*. New York: Routledge.
Elkins, J. (2005), "Preface", in E. Kac, *Telepresence and Bio Art: Networking Humans, Rabbits, and Robots*, Ann Arbor, MI: University of Michigan Press.
Elkins, J. (2008a), "Aesthetics and the Two Cultures: Why Art and Science Should be Allowed to Go Their Separate Ways", in F. Halsall et al. (eds.), *Rediscovering Aesthetics: Transdisciplinary Voices from Art History, Philosophy, and Art Practice*. Stanford, CA: Stanford University Press.
Elkins, J. (2008b), *Six Stories from the End of Representation: Images in Painting, Photography, Astronomy, Microscopy, Particle Physics, and Quantum Mechanics, 1980–2000*. Stanford, CA: Stanford University Press.
Elkins, J. (ed.) (2007a), *Visual Practices Across the University*. Munich: Wilhelm Fink Verlag.
Elkins, J. (ed.) (2007b), *Visual Literacy*. New York: Routledge.
Frankel, F. (2004a), *Envisioning Science: The Design and Craft of the Science Image*. Cambridge: MIT Press.
Frankel, F. (2004b), "Seeing Stars", *American Scientist* 92, 5 (September–October): 462.
Frankel, F. (2008), *On the Surface of Things: Images of the Extraordinary in Science*. Cambridge: Harvard University Press.
Gage, J. (1993), *Colour and Culture: Practice and Meaning from Antiquity to Abstraction*. London: Thames & Hudson.
Gilardoni, A. (1977), *X-Rays in Art: Physics, Technique, Applications*. Mandello Lario: Gilardoni.
Harrison, S. C. (1991), "What Do Viruses Look Like?", *The Harvey Lecture Series*, vol. 85: 127–152; online at crystal.harvard.edu/lib-sch/HarrisonS~91-HLect-85-127.pdf
Hockney, D. (2001), *Secret Knowledge: Discovering the Lost Techniques of the Old Masters*. New York: Viking.
Kemp, M. (1990), *The Science of Art: Optical Themes in Western Art from Brunelleschi to Seurat*. New Haven, CT: Yale University Press.
Latham, C. (1991), *Art After Physics*. Oxford: Museum of Modern Art.
Onians, J. (2007), *Neuroarthistory: From Aristotle and Pliny to Baxandall and Zeki*. New Haven, CT: Yale University Press.
Perkowitz, S. (1996), *Empire of Light: A History of Discovery in Science and Art*. Washington DC: John Henry Press.
Rossing, T. and Christopher C. (1999), *Light Science: Physics and the Visual Arts*. New York: Springer.
Shlain, L. (1991), *Art & Physics: Parallel Visions in Space, Time, and Light*. New York: Morrow.
Smock, W. (2005), "Picture This!", *California Monthly* 116, 1 (March/April): 16–27.
Stroke, H. (2001), "Light Science: Physics and the Visual Arts", *Physics Today* 54, 5 (May 2001): 60.
Winner, E. (2004), "Art History Can Trade Insights With the Sciences", *The Chronicle Review (Chronicle of Higher Education*) 50, 43 (July 2): B10.

Experiment, Theory, Representation: Robert Hooke's Material Models

Matthew C. Hunter

Robert Hooke's *Micrographia* of 1665 is an epochal work in the history of scientific representation. With microscopes and other optical devices, Hooke drew and then oversaw the engraving of *Micrographia*'s plates, images that amount to little less than revelations from beneath the range of human vision (Fig. 1). In bristling detail, molds flower into putrid bloom, crystals protrude like warts from mineral skins and, for the first time in history, cells are brought to the eyes of a general viewership. So historical scholarship has shown us, Hooke was especially well equipped to make these wondrous images. A product of Oxford's lively scientific community of the 1650s and a protégé of the chemist Robert Boyle, he possessed intimate knowledge of the "new sciences" of the seventeenth century and a particular gift as an experimentalist. Indeed, from 1662 until nearly the end of his life, Hooke held the post of "Curator of Experiments" to England's premier scientific institution, the then newly-formed Royal Society of London. But, Hooke also had an additional advantage. Following some remarkable, juvenile feats of drawing, he had previously been apprenticed to Peter Lely, leading portrait painter of later seventeenth century England. Combining scientific training with tutelage in the art of portraiture—that most detail-attentive of pictorial genres (at least as practiced in seventeenth century England)—Hooke would seem to have commanded the ideal skills for rendering the sights made perceptible through microscopes. Not surprisingly, Hooke's *Micrographia* has served as an important point of reference in recent studies of the interactions of art and science.

Yet, as the plates and pages of *Micrographia* attest, Hooke's investigations of nature also made use of representations that were neither pictures nor clearly picture-like. Directly below his elegant rendering of crystals in *Micrographia*'s seventh plate, Hooke presents the viewer with a sequence of eleven incremental combinations of circular forms. So he explains, these diagrams denote not anything seen by a microscope, but patterns of crystalline vegetation he had generated by making groups of spherical "bullets" vibrate together. "If put on an inclining plain,

M.C. Hunter (✉)
California Institute of Technology, Pasadena, CA, USA
e-mail: mchunter@caltech.edu

R. Frigg, M.C. Hunter (eds.), *Beyond Mimesis and Convention*, Boston Studies in the Philosophy of Science 262, DOI 10.1007/978-90-481-3851-7_9,

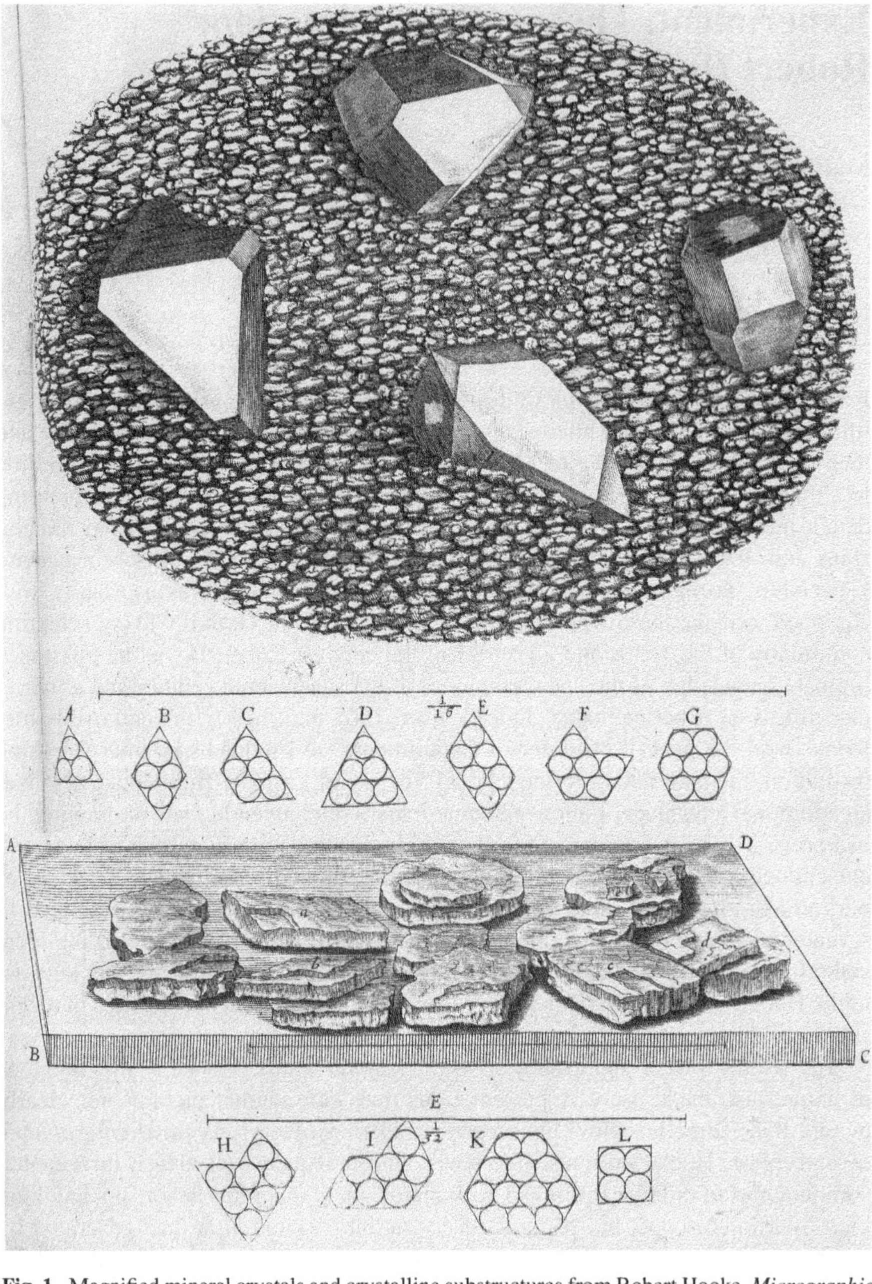

Fig. 1 Magnified mineral crystals and crystalline substructures from Robert Hooke, *Micrographia* (London: Jo. Martyn and Ja. Allestry, 1665), Scheme VII. This item is reproduced by permission of *The Huntington Library, San Marino, California*

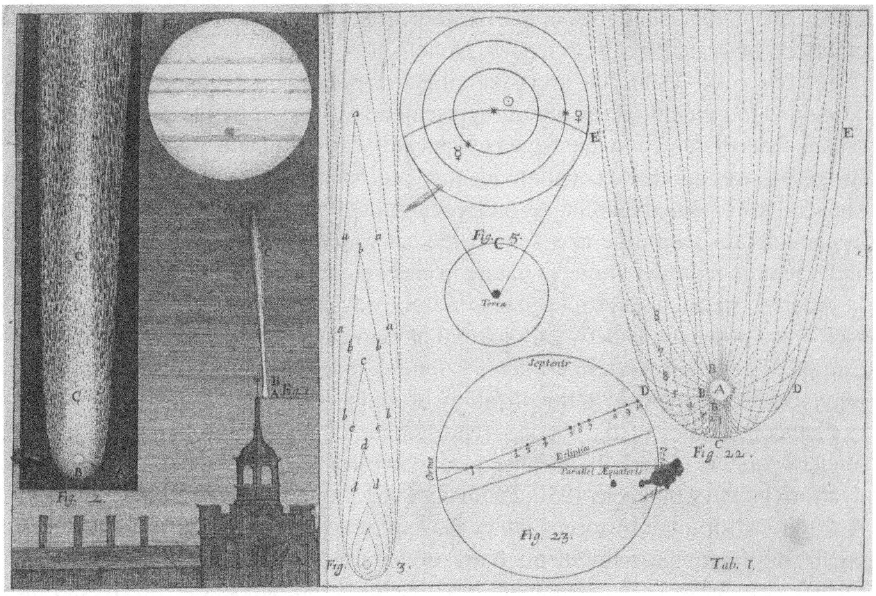

Fig. 2 Visible aspect and anatomy of comets from Robert Hooke, *Lectures and Collections* (London: Printed for J. Martyn, 1678), Table 1. This item is reproduced by permission of *The Huntington Library, San Marino, California*

so that they may run together", Hooke claimed, these bullets would "naturally run into a triangular order, composing all the variety of figures that can be imagin'd to be made out of aequilateral triangles" (1665, 85). A little over a decade later, Hooke published a treatise on a comet that had appeared over Northern Europe in the spring of 1677 (Fig. 2). Variously representing the comet's flight in accompanying prints, Hooke again detailed an action by which the meteoric object could be known. The reader is to suspend a wax ball covered with iron filings into a long beaker that has been filled with a solution of diluted sulfuric acid. Thus, Hooke proclaims, "you may plainly observe a perfect representation of the Head, Halo, and Beard [or tail] of the Comet" (1678a, 31).

What are these actions that Hooke proposes with agitated bullets and balls of wax in acid? How do these procedures, which are ubiquitous in Hooke's enterprise but rarely analyzed in historical studies, relate to his graphic representations? And what might art historians or philosophers of science learn from them?

By focusing upon these two particular cases from Robert Hooke's oeuvre, this essay aims to pursue a broader problem. That is, I suggest how researchers in the humanities and social sciences might learn from recent work in analytic philosophy of science to reconsider practices of representation shared between art and science. The time is particularly ripe for such rethinking. As philosophers in the analytic tradition have begun to look to the arts to understand the complexities of representations used in science, so art historians have increasingly sought to examine images made beyond the boundaries of the Western artistic tradition, especially those visual practices generated by the sciences. Nonetheless, for studies of the early modern period (ca. 1400–1800), the art of painting and modes of depiction proper to it have continued to guide thinking about representation. Spurred by the remarkably naturalistic feats of depiction that began to appear in early fifteenth century Florence and Bruges, researchers have sought to identify profound shifts in the orders of scientific knowledge embodied therein. As one recent scholar has asked: "Why did naturalism in painting arise with the new science? What was the relationship between artistic and scientific representations of nature in early modern Europe?" (Smith 2004, viii).

Address to such questions has certainly advanced our understanding of the vital cross-pollinations between pictorial art and empirical science in early modern Europe. Significant work in art history, for example, has demonstrated how—from Leon Battista Alberti's "rationalization of sight" to Vesalius' anatomies and on to Galileo's studies of the moon—representational techniques generated by early modern painters materially advanced techniques of scientific illustration and investigation (Ivins 1938, Pächt 1950, Panofsky 1962, Edgerton 1984, Bredekamp 2000). A reciprocal strain of historical study has explored how optical sciences and instruments informed the naturalistic turns of painting in Renaissance and Baroque Europe (Lindberg 1976, Steadman 2002, Kemp 1990, Hockney 2001). And more recent, interdisciplinary literature aligned with "science studies" has emphasized how the mimetic naturalism exemplified in early modern painting might be seen as a general model for the aspirations of the emergent natural sciences. "The picture in general, and painting in particular", so one such study has claimed, "... emerges as a dominant paradigm for the whole system of modes of representation constitutive of early modern philosophy, religion and science as well as literary or aesthetic culture" (Braider 2004, 46).

Robert Hooke has figured significantly in the formulation of these positions. In a hugely influential work from 1983, art historian Svetlana Alpers (1983) cast Hooke as a leading exemplar of the "descriptive impulse"—the penchant for the detail-attentive, naturalistic "picturing" of appearances—that she identified in the still-life and genre paintings of Dutch art, and ascribed generally to the science and culture of seventeenth century Northern Europe. Posed by Alpers as a heuristic corrective for viewing the Northern European pictorial tradition outside of the hegemonic standards of Italian art, descriptive picturing has itself become a norm. Especially in studies of the early modern period, talk of copying or picturing nature has become paradigmatic for discussions of representation in scientific contexts (Shapin and Schaffer 1985, Ogilvie 2006).[1] In turn, this apparent sympathy of aims between artistic and scientific representation has given a new encouragement to studies of art in Hooke's native Britain. Increasingly, researchers have looked to the Royal Society, and to Hooke specifically, for sources of an empirical bent that can be traced into the rising tradition of eighteenth century British painters such as William Hogarth and John Constable (Bermingham 2000, Gibson-Wood 2000). If artistic

[1] Interesting variations upon this direction are Freedberg (2002) and Daston and Galison (2007).

training informed the gaze of scientists like Hooke, this story suggests, so their empirical ethos should be seen as underpinning the pictorial achievements of the Enlightenment.

The objective of this essay is neither to take issue with these readings nor to re-stage old debates over the adequacy of Alpers' notion of "description" for understanding early modern painting (de Jongh 1984, Marin 1986). Indeed, if my point of departure is to ask whether this approach offers compelling terms for understanding the material models of Robert Hooke, my argument aims to complement the broader rethinking of art in later Baroque Britain. I do so by showing how recent philosophy of science enables us to apprehend the representational sophistication and sheer imaginative virtuosity of Hooke, Christopher Wren and their colleagues in the early Royal Society with new clarity and vigor.

Certainly, there is enough in Hooke's work to encourage the reading already available to humanities-based scholarship. Beyond his apprenticeship to the Netherlandish painter Lely, Hooke was a keen advocate of accurate representation whose scientific writings deploy various concepts from the "mimeticist tradition" (Halliwell 2002). Nonetheless, recent historical research has identified two important reasons for reconsidering this dominance of the pictorial. The first reason is a matter of focus. As historians of the built environment have shown, Royal Society Fellows like Hooke and Wren simply produced a huge body of visual work that was not pictorial. Central agents in the rebuilding of London after the fire of 1666, their collective endeavors like engineering the dome of St. Paul's Cathedral or designing telescopic observatories and mental hospitals are fascinating intersections of artistic and scientific endeavor; but they have little obvious relationship with the terms of pictorial representation (Cooper 2003, Stevenson 2005, Jardine 2003a, b). The second reason for reconsidering the interpretive appeal to painting is a matter of relative value. Historians of art have long lamented that painting was significantly underdeveloped in seventeenth century Britain, especially in comparison with Continental models (Waterhouse 1953, Pears 1988). While numerous, competing painting schools flooded the sophisticated art markets of the seventeenth century Netherlands and painters received royal patronage of their academy in Louis XIV's France, indigenous pictorial traditions in Britain prior to the eighteenth century are, by contrast, notoriously fragmentary.[2] Hardly an unalloyed good let alone a paradigm of knowledge, the art of painting was, moreover, a practice from which many English scientists sought to distance themselves and the representations they did employ. If his scientific colleague William Cole dismissed painting and sculpture as "things uselesse" pursued only for the "lusts of pride and ostentatious vanity" (ca. 1692, f 159), Hooke himself treated pictures with caution.[3] "The Pictures of Things which only served for Ornament or Pleasure", he warned, are "... rather noxious than useful, and serves to divert and disturbs the Mind" (Hooke 1705, 64). And while advocated by some of the Royal Society's gentlemen-amateurs,

[2]For a revision of this argument, see Gibson-Wood (2002).

[3]On these points more broadly, see my (2010).

evidence suggests that the kind of naturalistic pictures most valued by recent interpreters is precisely that which Hooke and Wren performed early in their careers and subsequently delegated to their assistants (Hunter 2007).

If a focus upon painting thus feels like an increasingly arbitrary imposition upon the visual activities and values of Hooke and his circles, the conceptual situation becomes even worse when their expressly scientific representations are examined in detail. Ostensibly, this would be the business of the history of science. But, despite the fact that they were performed and studied at the very center of scientific communities like the Royal Society, historians of science have had very little to say about the representational structure of events like Hooke's bullet manipulations and his effervescent wax comet. Instead, studies have tended to focus upon various socio-political objectives accomplished through such performances or via images related to them (Shapin and Schaffer 1985, Fyfe and Law 1988, Golinski 1989, Lynch and Woolgar 1990). Without disputing the interest of such work, the complementary proposal of this essay is simple. Before we reduce these largely-uncharted seas of visualization to the terms of mimetic naturalism—and before we art historians construct elaborate pre-histories of Enlightenment art upon them—it behooves students of science and art alike to first analyze those representational operations in which Hooke's community invested so much epistemological and financial capital. To do so, researchers in the humanities and social sciences can learn much from emerging work in analytic philosophy of science.

To this project, Hooke's aforementioned performances present at least three significant, interpretive obstacles; I will call these concerns methodological, categorical and quasi-existential. To an art historian, the major methodological problem is obvious: when considering procedures such these, frequently no object survives around which to organize analysis. At the very least, we would want to know if the spherical bullets or the glass beaker Hooke claimed to use for his actions possessed (or, as we will see, could have possessed) some unusual properties that made them uniquely capable of representing his targets of investigation. Surely it is true that, as the remit of art history has expanded in recent decades, the graphic resources, theoretical writings, and other kinds documentation upon which I will draw in this analysis have eroded the privileged evidentiary position once commanded by the art-object. But, given the discipline's residual methodological orientation toward objects (Koerner 1999), the approach I employ here has been to attempt to supply, as it were, replacement objects, using a modest version of the strategies of replication developed in the history of science.[4] And here, the methodological conservativeness of art history may actually become a virtue as it forces us to focus upon exactly how Hooke's models were supposed to have worked and what roles physical objects could have (or could never have) played in them.

More substantial is the second, categorical concern. To some readers, the interpretation ventured here might be read as committing a category error by treating as representations what should really be understood as *experiments*, the central means

[4]For a recent application of this approach with a useful bibliography, see Heering (2008).

of intervening into reality advanced by Hooke and his colleagues. Because representation and experiment are not only distinguished from but often opposed to one another in philosophical accounts, address to this categorical concern must be central to this and other studies of experimentalist representation. It is with this worry that I will begin. The third concern, though, is almost an existential one. That is, why should art historians care about the strange performances of brilliant but eccentric characters like Robert Hooke? What does this tell us about art? I will engage these quasi-existential charges directly only in the conclusion; but my analysis follows from the conviction that how we answer these questions powerfully reveals what we want explorations of the art/science conversation to do. Building from work by scholars like James Elkins and Peter Galison, my contention is that humanities-based studies of visual materials only become more interesting and intellectually rigorous as we increase our engagement with science. (Elkins 1999, 2007, 2008, see "Visual Practices Across the University" this volume, and Galison 1997) Therefore, if we want to understand how representation in art and science might speak to one another—indeed, if there are more than passing coincidences between naturalism in art and empiricism in science—then we need to scrutinize the representational procedures that were central to emerging science with as close attention as has been paid to practices of representation in art.

This, then, is not just a call for interdisciplinary dialogue for its own sake. For, if these methodological, categorical and quasi-existential worries can be allayed, what becomes available to interpretation is an excitingly open, but absolutely central, field of inquiry wherein representation may be approached anew. Released from the powerful gravitational pull of painting, art historians might begin to reckon more successfully with the visual achievement of Hooke, Wren and their colleagues who remain highly problematic to available accounts. Beyond learning from the flexibility, stylization and deep inaccuracies of scientific representations as they appear in recent philosophy of science, moreover, I show how historians can productively draw from this literature to reconsider what kinds of cognitive work representations were being asked to perform in early scientific contexts; why diverse styles of representation could have been useful; and what modes of knowledge they might be said to embody. Reciprocally, historically-based contributions such as this one may bring to philosophical consideration how play between graphic imagery, performance with material models, and theory deserves to be integrated into more generalized accounts of representation as practiced in the arts, sciences, and beyond.

Gross Similitudes

I want to begin by returning to the categorical concern sketched above. That is, are the procedures Robert Hooke described with bullets or his operations with wax balls in acid really representations at all? The question deserves to be posed because an important tradition within philosophy of science has seen Hooke as exemplary of a significant shift within the sciences, one defined by the differentiation of experiments from representation. Thomas Kuhn has counted Hooke among those who inaugurated this qualitative shift in the enterprise of experimentation in the seventeenth century. From antiquity through the Renaissance, Kuhn argues, everyday observation and the exercise of reason had been sufficient grounds for competence in major fields of physical science. Experiment in this pre-modern context was properly thought experiment, which aimed at demonstration of known principles or exposition of their particulars. By contrast, Kuhn claims, when "men like Gilbert, Boyle, and Hooke, performed experiments, they seldom aimed to demonstrate what was already known or to determine a detail required for the extension of existing theory. Rather they wished to see how nature would behave under previously unobserved, often previously nonexistent, circumstances" (1977, 43). In this new, "Baconian" definition of the seventeenth century, experiment was radically productive of data and, by that measure, not *re*-presentational at all. Ian Hacking has influentially endorsed a similar view of Hooke the experimenter. In Hacking's memorable words, Hooke was "a crusty old character who picked fights with people—partly because of his own lower status as an experimenter" (1983, 151). Because of the field's bias towards theories and representations, Hacking claims, philosophers of science give scant attention to experimentalists like Hooke who were committed to manipulating reality. By these views, Hooke is not only to be strongly identified with experiment, but he figures among those crucial, historical agents who brought into being practices of experiment that could be meaningfully differentiated *from* representation for the first time.

If his work abounds with examples, Hooke's theoretical writings shed only limited light on these boundaries of experiment. In a famous paper from the early 1660s, for example, Hooke defines the "Reason of making Experiments" as the very general aim of "Discovery of the Method of nature in its Progress and Operations" (Hooke 1726, 26). What available literature there is on Hooke's experimentalism also encourages softening philosophers' categorical distinction between representation and experiment. Social historians of science have emphasized how the experiments performed at the Royal Society's meetings in the later seventeenth century were rarely the bald confrontations with nature as envisioned by Kuhn. Experiments would be tried extensively in private laboratories before their demonstration to the scientific fellowship. So Steven Shapin has contended, "the weekly meetings of the Royal Society required not trials [of experiments] but shows and discourses" (1999, 497). In this reading, a public experiment was always a kind of representation insofar as it was a demonstrative replication of results previously obtained elsewhere. But, in turning specifically to analysis of Hooke's trials, I want to consider if and how a project like the bullet manipulation can be seen to participate in experiment's celebrated intervention into nature at all.

In *Micrographia*, Hooke introduces the bullet manipulation in the context of his microscopic observations of flint, casserite, alum and other mineral crystals. [See Fig. 1] Why, Hooke asks, do minerals like these betray remarkable formal consistencies? By way of explanation, Hooke appeals to a significant component of his physical thought, the theory of congruity. As Mary Hesse (1966a) has noted, Hooke understood diverse physical phenomena disclosed by his experiments to be products of particulate matter in vibrating motion. In turn, his theory of congruity stipulated that bodies of the same (or proportional) mass or vibrating frequency would attract one another; "incongruous" bodies, which have different masses and non-proportional frequencies, would repulse. In his later writings, Hooke could formulate this theory in economical terms as "nothing else but an agreement or disagreement of Bodys as to their Magnitudes and motions" (1678b, 7). But, in early works like *Micrographia*, congruity and incongruity are often suggested through a catalogue of vibrating phenomena. The cohesion of congruous bodies, for example, is explained in the following terms:

> I suppose the pulse of heat to agitate the small parcels of matter, and those that are of a like bigness, and figure, and matter, will hold, or dance together, and those which are of a differing kind will be thrust or shov'd out from between them; for particles that are all similar, will, like so many equal musical strings equally stretcht, vibrate together in a kind of Harmony or unison (1665, 15).

Although they generally agree on the importance of Hooke's theory of congruity to his broader mechanical philosophy, historians of science have been divided on its implications. John Henry (1989) and Penelope Gouk (1999) have read Hooke's materialism as a continuation of Renaissance natural magic, while Mark Ehrlich (1992) and Michael Hunter (2003) see his matter theory as characteristic of the rationalizing tendencies in seventeenth century science which would form the basis of classical mechanics. Most interestingly, Ofer Gal (2002) has argued that because Hooke's theory of congruity was a key component in his thinking on attraction at a distance, it might be seen as having material consequence for the theory of universal gravitation elaborated by his sometime-interlocutor and later great enemy, Isaac Newton.

However its influences and intricacies may be parsed out, the key point here is that Hooke's theory of congruity closely shadows his bullet operation. Because of the force of congruity, Hooke explains, homogenous matter in its most fluid, agitated form would be "driven . . . and forc't into as little a space as it can possibly be confined in" (1665, 17). When highly agitated, this congruous matter would form into spheroids, which he calls "globules". Hooke's contention is that crystal patterns in minerals can be explained by appeal to "three or four several positions or postures of Globular particles, and those the most plain, obvious, and necessary conjunctions of such figur'd particles that are possible" (1665, 85). Support for this claim is then offered by the bullet trial itself. So Hooke explains in full:

> I have ad oculum demonstrated with a company of bullets, and some few other very simple bodies . . . that there was not any regular Figure, which I have hitherto met withal, of any of those bodies that I have above named, that I could not with the composition of bullets or globules, and one or two other bodies, imitate, even almost by shaking them together. And thus for instance we may find that the Globular bullets will of themselves, if put on an inclining plain, so that they may run together, naturally run into a triangular order, composing all the variety of figures that can be imagin'd to be made out of aequilateral triangles (1665, 85).

At the most basic level, then, bullets vibrated on an inclined plane are claimed to yield the kinds of formal configurations observable in mineral crystals.

So, is this an experiment? An informative way into this question is simply to press upon how Hooke's procedure was supposed to have worked. Even the most fundamental aspects of this action are problematic. Hooke contends that bullets (a term, according to the OED, derived from the diminutive of the French *boule*, thus a small round ball) in his manipulation would move into the geometrical forms he diagrams "even almost by shaking them together" (1665, 85). Yet, such behavior runs counter to the major works of seventeenth century physics, which Hooke knew well. In *Two New Sciences* (1974, 87–88), Galileo had outlined how balls moving on an inclined plane (the trial situation Hooke stipulates) would attain identical velocities if the resistance of air and friction are eliminated. In the terms formulated by Newton some twenty years after *Micrographia*, the bullets would be expected remain in rectilinear motion until acted upon by other forces, reacting equally and oppositely to their encounters with other bullets (Newton 1989, 14–24). Hooke's bullets behave otherwise. They do not scatter or project off the edges of the trial surface, but gather into regular groups. [See Fig. 1]

For his part, Hooke is extremely vague about the exact nature of the trial, explaining nothing of the friction, agitation and angle of the plane nor the masses, diameters, or possible velocities of his bullets. Perhaps it is possible that the patterns of attraction between bullets that Hooke describes could have been achieved had his spheroids possessed some degree of magnetism, a property on which Hooke experimented and clearly saw as related to his notion of congruity (1665, 31). Yet, no such property is ever stipulated for the bullets in the trial and Hooke even suggests that the specified results can be achieved with non-magnetic objects. Although the frailties of Hooke's experimental contrivances have become well known to recent historians (Shapin and Schaffer 1985), the only success I have had at replicating the stipulated behavior with non-magnetized "bullets" has come from introducing the spheroids into a bowl and not on the inclined plane Hooke describes (Fig. 3).

Baffling as it is, the physical difficulties, if not impossibility, of Hooke's bullet operation helps to clarify its objectives. Rather than seeing it exclusively as an experimental intervention that produces new data from a natural target, the trial might be better conceived as a mechanism through which a theoretical precept (namely, the theory of congruity) can be visualized to understand a phenomenon (here, the formal regularity of mineral-crystal formation). In this capacity, Hooke's trial has a clear representational aspect. Parsed in crude terms, the bullets represent theoretical entities called globules, while the agitation of the inclined plane simulates the vibrating motion of congruity; I want to return momentarily to the procedure's semantic dimensions and particularly to what might be called its "enigma of representation". According to the representation's logic, incremental addition of bullets is claimed to reveal the possible field of formal permutations available to crystals. By using "25, or 27, or 36, or 42, &c." bullets, Hooke insists, the scientist can "find out all the variety of regular shapes, into which the smooth surfaces of [a mineral like] Alum are form'd" (1665, 86). Thus, if we disregard its practical mechanics for a moment, the bullet manipulation might be read as both a visualization of the rudimentary component particles and forces yielding crystalline structures and a means for generating rules of combination with which to predict the target's possible patterns at higher

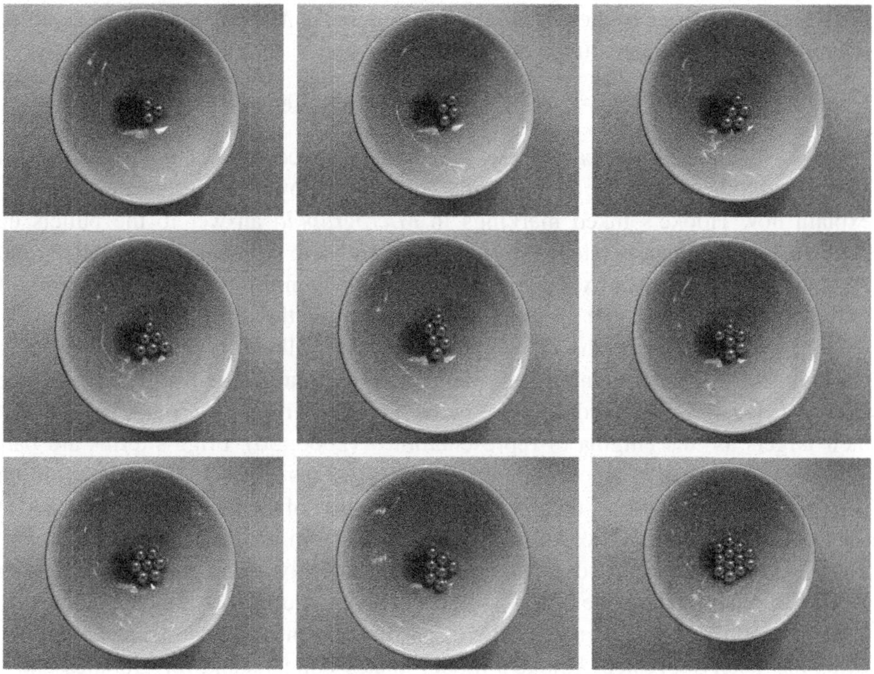

Fig. 3 Author's reconstruction of Hooke's bullet manipulation. This replication was produced by incrementally introducing stainless steel ball-bearings (diameter: 1 cm) into the curved surface of a shallow bowl (roughly parabolic curvature, diameter: 14 cm, depth: 6 cm)

levels of formal complexity. The bullet trial is a representational process that produces data from an artificial situation as a means to understand a natural target. By this reading, Hooke's bullets can be understood as a *model* of crystallization (Frigg and Hartmann 2006).

Arguably, art historians are better equipped to study the data produced by this representation than to interpret Hooke's crystallization model itself. For, in this case, the data are graphic images. Hooke has transcribed the model's informational yield in *Micrographia*'s figures A–L where the resulting bullet-patterns are rendered as sequences of spare, circular forms circumscribed within geometrical solids. (Fig. 1) No doubt an interesting art-historical account might be written by narrating how Hooke's denotation of the spherical bullets as abstract geometrical entities fits in histories of crystallographic representation, the larger development of diagrammatic notation, or the anti-naturalistic tendencies of later seventeenth century scientific illustration.[5] Yet, what is crucial to underscore are the two stages of representation

[5]On these topics, see respectively Elkins (1999, 13–30); Wilson (2002); and Freedberg (2002).

disclosed by attention to these diagrammatic figures in *Micrographia*. This doubled reference might be schematized in the following way:

Figures –––(depict)–––> Bullets ––––(represent)––> "Globules"

As is signaled by the parentheses, no particular accounts of reference are yet subscribed to here. But, the fundamental point is this: by whatever means we might explain how Hooke's inked markings in *Micrographia* answer to the bullets he claimed to have manipulated, the vexing relationship between those bullets and "globules" still demands explanation as well. It is upon this second, neglected half of the schematic figure that I will focus in the analysis that follows.

Let us return, then, to the "how" of Hooke's crystallization model. In outlining directions for expanded thinking on scientific representation, philosopher of science Roman Frigg has described what he calls "the problem of how models represent their targets as 'the enigma of representation'" (2006, 50). Frigg's terms are particularly appropriate to Hooke's perplexing crystallization model. Indeed, it is both perplexing and mysterious; there is no documentation of the model's performance at the Royal Society and all we know about it comes from the pages of *Micrographia*. There, Hooke had claimed that crystals are naturally formed by the vibrating motion of matter as it gathers into particles called globules. Governed by congruity and incongruity, globular matter then consolidates into regular crystal patterns. If, as has been suggested, Hooke's model makes bullets stand for globules and a vibrating inclined plane represent the conditions of congruity, by virtue of what is this a representation?

Following the dominant interpretive approach, we might account for these enigmatic properties by appealing to criteria of depiction as borrowed from the model's representation in *Micrographia*'s plate. Depicting and representing, as indicated in the scheme above, would thus be the same. And following Hooke's earlier appeal to the bullets' "imitative" capacity, we might read the whole enterprise through the central vein of *mimesis* in which early modern European learned cultures understood human arts. In this tradition inherited from Classical antiquity, such artifice followed from a universal human compulsion to mimic. Where Aristotle had identified the sources of these *techne mimetike* in the pleasures of making and decoding imitations, artisans across pre-modern Europe put these pleasures to work as copying of schemata made by master craftsmen became the literal core of apprenticeship and the prolegomenon to study of the privileged subject of art, the human body (Aristotle 1987, Gombrich 1960, Muller et al. 1984). But, among intellectuals eager to secure the elevated status of painting and sculpture, the mimesis proper to what would come to be called the "fine arts" was understood to be based in imitation of ideas generated in the mind of the artist (Panofsky 1968, Belting 1996). As a work of genius, this artistic imitation was to originate in but transcend observed, imperfect nature by reconciling it with idealized conceptions. "Noble painters and sculptors", so claimed Hooke's contemporary Giovan Pietro Bellori, "... form in their minds an example of higher beauty, and by contemplating that, they emend nature without fault of color or of line" (2005, 57). Rich and various as its permutations are, imitation in this ennobling tradition of early modern artistic academies was centrally concerned with idealization (Lee 1940).

Academic idealization was, of course, not the only option to which a figure like Hooke could turn; part of what motivated claims like Bellori's was the perceived influence of apparently non-idealizing modes of imitative depiction. Notorious in artistic circles were painters like Michelangelo Merisi da Caravaggio whose putative commitment to the imitation of nature in extremis threatened the supposed dignity of art (Marin 1995). As recent scholarship has emphasized, these naturalistic currents can be seen in instructive dialogue with the cultures of science emerging across sixteenth and seventeenth century Europe (Crombie 1994, Smith 2004). While numerous examples might be mustered from Hooke's activities to corroborate his interest in such naturalistic imitations—from picture-making with the camera obscura to casting carp from life—his own writings are most succinct. Nothing, Hooke would observe in a planned introduction to a universal atlas, is "more conducive to the assistance of the memory understanding and memory then a plaine simple cleer and uncompounded Representation of the Object to the Sense" (ca. 1680, f 2). It is this non-idealizing "descriptive" mode of depiction that has served to characterize Hooke's representational activity and the central visual concerns of the Royal Society more broadly.

The problem with this account is that it is simply difficult to see what light it sheds on representations like Hooke's bullets. Descriptive picturing and naturalistic copying are supposed to be founded upon the production of telling resemblances between an observed target and the representation. But, there was no perceptible target that Hooke's bullets could possibly resemble. Micro-level globules were wholly invisible, theoretical entities whose properties could only be known by rational inference. Worse, far from being some deft resemblance that Hooke had newly caught with his keen, microscopic eye, the rendering of quasi-atomic particles as spheroids was a central convention—even cliché—of physical thought reaching well back to classical antiquity (Lüthy 2000, Meinel 2004). Patently un-seeable, Hooke's globules could "resemble" bullets only when this convention of atomist thought was in place. And this was a matter of faith, not of observation. Therefore, if we insist upon finding a period, pictorial analogue for Hooke's crystallization model (and this is an option I exercise only rhetorically), we might look less to the still-life paintings perfected in the early modern Netherlands and instead think with contemporaneous Spanish renderings of religious visions (Stoichita 1995). Like those painters in seventeenth century Spain who drew upon a rich vocabulary of pictorial conventions to represent marvelous visions accessible to saints' eyes alone, so (this tortured reading might propose) Hooke utilized a stock atomist convention whereby elementary particles were spherical so as to visualize the imperceptible sub-structures of crystalline matter.

By clarifying this theoretical ontology of globules, Hooke's crystallization model brings us to the limits of interpretative utility for the available terms of pictorial depiction; there was simply no visible entity it could copy. Instead, this analysis indicates that a very funny thing had happened. Hooke certainly invokes the terminology resemblance or imitation to make his theoretical entities comprehensible; globules are bullet-like. But, to be "like" globules, these bullets—the key components of Hooke's crystallization model—had then to become unlike any actual, physical bullets available to familiar apprehension. How are we to understand the enhanced bullets that seem to populate Hooke's crystallization model? And what exactly is the nature of their "likeness" to the theorized globules?

Hooke, as we have seen, understood globules to be nearly spherical particles of matter governed by forces of congruity and incongruity that form into regular, geometrical configurations through vibrating motion. So we have also noted, physical bullets agitated on an inclining plane could not actually have generated the geometrical patterns that Hooke had claimed and depicted in *Micrographia*. Therefore, it is instructive to think of the bullets envisioned in this crystallization model not as actual, physical objects, but as continuous with the frictionless planes, spherical planets and other central stylizations deployed in scientific modeling. Indeed, such a view of models and their components as imagined physical entities has recently been advanced by in philosophy of science. Like literary fiction, so Roman Frigg argues, scientific models instantiate varieties of serious make-believe, fictionalizing their components to yield truths about the representational worlds they generate and enabling comparisons between those fictions and reality. Models, in this analysis, are "hypothetical entities that, as a matter of fact, do not exist spatio-temporally but ... would be physical things if they were real" (Frigg 2009, 3). Read in this way, the bullets of Hooke's model might be seen as fictionalized or imagined so that they share relevant properties with the theoretical globules. Hooke's model asks us to imagine, in other words, that if the bullets were real, they would behave like globules. And because globules are theorized to form into regular, geometrical configurations when agitated, the vibration of these fictionalized bullets would yield the geometrical patterns we see depicted in *Micrographia*.

Framed in this way, the relationship of "likeness" noted between Hooke's imagination-enhanced bullets and his globules can be apprehended more precisely. A useful clue in this direction is supplied by historian Penelope Gouk (1999, 218) who has described Hooke's musical, mechanical and other trials at the Royal Society as:

> ... attempts to prove, or at least render plausible, his theory of vibrating matter through experimental demonstration. It was on the basis of such simple and verifiable experiments that Hooke claimed analogous principles were operating beyond the range of ordinary sense perception.

If we bracket her "simplicity" and "experimental verifiability", Gouk's attention to principles of analogy is surely useful for understanding Hooke's enterprise.[6] That is, the imagination-enhanced bullets are not the same as globules; but they can be

[6] Even if we have no specific endorsement of this line from Hooke for the crystallization model, this style of thinking certainly finds support in his contemporaneous writing. Earlier in *Micrographia*, Hooke had noted: "It seldom happens that any two natures have so many properties coincident or the same ... and to be different in the rest" (1665, 14). Therefore, he continues, "I think it neither impossible, irrational, nay nor difficult to be able to predict what is likely to happen in other particulars also ... if the circumstances that so often very much conduce to the variation of the effects be duly weigh'd and consider'd" (1665, 14). Appealing to classical induction, in other words, patterns observable in the bullets and numerous other vibrating phenomena the encourage inference about the properties of those imperceptible physical structures undergirding them all.

seen as analogically related to them. As Mary Hesse has argued (Hesse 1966b), analogical models like this proceed by identifying properties shared between systems and eliminating their differences or negative analogies. Exploration of the better known system is then used to make predictions about the more obscure one. Therefore, we could say, Hooke's understood his enhanced bullets and his globules to share the following positive analogies: both were nearly spherical in shape; in vibrating motion; governed by forces of congruity and incongruity; and capable of forming into regular geometrical patterns. Properties they did not share might include their differences in size and frequency of vibration, or the shininess, salty taste or other accidental properties of the bullets in their possible improved state. What the model claims to offer, then, is a mechanism based on trials with the better-known system (the enhanced bullets) through which to predict patterns generated by the obscure, theorized particles called globules at increasing levels of complexity. The data yielded by this model is what we see depicted in the figures from *Micrographia*.

In this light, our schematic account of the model might thus be updated in the following way:

Figures – – –(depict)– – –> Imagined Bullets – – – represent by analogy – – –> "Globules"

Through imaginatively stylizing its putative physical company of vibrated bullets, Hooke's analogical model creates a mechanism with which to study the behavior of theorized entities. What we see represented in *Micrographia* are data yielded by this model.

This is intended to be a charitable reading of how Hooke's model was supposed to work. More fundamentally, it is a reading pursued as a means of rethinking both the opposition of representation versus experiment and the grip of pictorial mimesis in which Hooke's visual activities have been repeatedly plotted. Even under such limited analysis as this, however, the constraints of Hooke's model appear strikingly and tellingly acute. Rather than being too closely related to experiment as had been worried at the outset, Hooke's crystallization model ends up appearing overly distanced from it. With the bullets fictionalized into analogy with theoretical globules and no longer answering to the physical behavior of actual bullets in the trial situation Hooke had stipulated, it is hard to know how much information could possibly have been yielded by work with the model—or even what such work might have looked like. Did manipulation of actual bullets maintain any relevance to the project? Or, had the model entirely become a kind of thought experiment?[7] Indeed, it is instructive to remember here how Kuhn himself observed that seventeenth century experimentalists like Hooke were actually closest in spirit to the older traditions of theory-illustrating experiment precisely in

[7] Further pursuit of these points could productively engage with the stimulating reading of thought experiments and fictions proposed by David Davis (see "Learning through Fictional Narratives in Art and Science" in this volume).

those trials claiming to "reveal the shape, arrangement, and notion of corpuscles" (1977, 43).

In turning from the bullets to the arguably more successful model that Hooke devised to represent a comet with a wax ball and sulfuric acid, it is nonetheless worth stressing the representational complexity involved in the production of a seemingly humble material model like this, which Hooke himself had called a "gross Similitude". Hooke's bullets are convincingly explicable neither as the imitation of nature nor as the illustration of theory. Instead, projects like this aimed at the construction of a species of serious make-believe that could yield meaningful insight into obscure or imperceptible entities through work with a stylized representational proxy. Conventional stipulations, imaginative enhancements, analogy, and possibly deep fiction—all contributed to Hooke's seemingly innocuous study of crystals. Thus, however we wish to understand the varieties of representation instantiated by its images, the crucial point is that the pictures found in *Micrographia* are data from Hooke's modeling enterprise, not the privileged interpretive key to it. If anything, pictures were but one facet of the experimentalist's representational approach as it moved between theory, performance and material practice.

In Some Things Analogous to the One, and Somewhat to the Other, Though not Exactly the Same with Either

In the last weeks of April 1677, a comet became perceptible in the skies above northern Europe. From his observation turret in London's Gresham College, Robert Hooke studied the comet from April 21 until it disappeared a week later (see Hooke 1935, 286–287). Even without the assistance of the six and fifteen foot reflecting telescopes that Hooke used in his private observatory, the comet's teardrop tip and broom-like tail must have cut an impressive figure above the nocturnal cityscape of later seventeenth century London (Fig. 2). So the illustrative plate prepared by engraver Francis Lamb from Hooke's own drawings suggests, the Curator was fascinated by comets and committed significant energy to their study. But while he cast a jaundiced eye upon the millenarian prognostications that they elicited amongst the early modern European public, Hooke also had doubts about the calculations of comets' orbits and parallax motions as produced by his scientific contemporaries. Instead, Hooke took a typically pragmatic course in his own studies. Recognizing the limits of available instruments to provide accurate information about comets' speed, distance from the Earth, and possible orbits, he concentrated on what could be learned about comets from observation. Based upon his studies of the 1677 object, *Cometa* of 1678 set out an impressive account of how comets come to exhibit their characteristic features: an antisolar tail, luminosity and erratic motion. Briefly elucidating the theory he set out in 1678, I want to turn to the sequence of models Hooke contrived to reconcile this theory with his observations.

In *Cometa*, Hooke postulates that a comet begins as a semi-solid, spheroid body and gradually decomposes due to its significant internal instabilities. Utilizing the style of reasoning we have seen him deploying in his earlier crystallization model, Hooke found evidence for comets' instability through analogy with the behavior of the Earth. Although it seems to be "generally very dense, compact, and very closely and solidly united", Hooke's pioneering lectures on the Earth's volcanic eruptions and magnetic variations had shown that the planet "may be notwithstanding more loose, and ununited, and moveable from certain causes" (1678b, 11, Drake 1996). Comets, he proposes, are similar, albeit in a more extreme form: "It seems very probable to me, that the body of Comets may be of the same nature and constitution with that of the internal parts of the Earth, that these parts may by the help of the Aether, be so agitated and blended together, as to make them work upon, and dissolve each other" (1678b, 11–12). Susceptible to the reagent aether because of this internal agitation, the comet's disintegration accelerates, causing it to lose mass and gravitational force. And because he understood gravitation through the aforementioned dynamics of congruity and incongruity, Hooke was provided with an explanation of the formation of the comet's tail:

> The parts thus dissolved are elevated to a greater distance from the center of the Star or Nucleus, or the superficies of it, whose gravitating or attractive principle is much destroyed, ... but having given those parts leave thus far to ramble, the gravitating principle of another body more potent acts upon it, and makes those parts seem to recede from the center thereof, though really they are but as it were, left behind the body of the Star, which is more powerfully attracted that the minuter streaming parts (1678b, 12).

As the head of the comet inclines towards the gravitating body of the sun with which it is congruous, so the more incongruous particles of the tail trail behind. In this way, Hooke's theory of internal agitation compounded by reaction to aether could explain the comet's characteristic, observable trait of the anti-solar tail, which had been depicted so elegantly in *Cometa*'s plates.

Hooke's theory could also offer an account of comets' peculiar celestial motion. Once destabilized, he argues, the comet's magnetic relations become disturbed, no longer holding it in "that circular way" of a stable orbit (1678b, 13). Instead, the comet "flies away from its former center by the Tangent line to the last place, where it was before this confusion was caused in the body of it" (1678b, 13). Projecting tangentially outward from its former orbital trajectory, the comet enters into the gravitational fields of other bodies in its new path. Such attractions only intensify its disintegration, thereby lengthening its tail to upwards of seventy telescopic degrees (1678b, 13). Combined with the reaction to aether and compounded by the attraction of neighboring celestial bodies, comets' internal agitation informs Hooke's account of their enigmatic orbital behavior.[8] What *Cometa* effectively offers, then, is a theoretical template for explaining the observed form and unusual trajectories of comets, while elucidating their genesis from the deterioration of stable celestial bodies.

In turning from this theory to the material models Hooke would use to reconcile it with observation, I want to draw more explicitly upon studies of modeling from

---

[8]For Hooke's broader understanding of the internal motion of planetary bodies, see Hooke's *Lectures and Discourse of Earthquakes* in Hooke (1705, 149–190).

recent analytic philosophy of science, which remain largely unknown in art history and visual studies. Since the early 1960s, the study of models has occupied center stage in the philosophy of science, and both their relation to theory and to their respective targets have been the subject of heated debate. One crucial argument of this literature has been that models do not simply illustrate or instantiate abstract theories. Instead, they frequently depart in important ways both from the theories they ostensibly embody and the worldly targets they are used to explore. This view has received its most advanced statement within the so-called Models as Mediators project (Morgan and Morrison 1999). Multifarious in form and often intractable in function, models might thus be said to possess "lives of their own". Because of their partial independence or "autonomy", this literature argues that we see models as standing between—thus, mediating—theory and experimental engagement with nature. Although it is not above critique[9], this "models-as-mediators" approach is particularly useful for elucidating how Robert Hooke worked with his material representations of comets in the late 1670s.

Once he had set out his theory of their physical form, Hooke offered the reader of *Cometa* a way to "make a perfect representation of the body, and beard [i.e. tail] of the Comet" (1678b, 31). As he directs:

> Take a very clear long Cylindrical Glass, which may hold about a quart of water; fill it three quarters full with water, and put into it a quarter of a pound of Oyl of Vitriol [sulfuric acid], and in the midst of this suspend by a small silver wire, a small wax-ball, rould in filings of iron or steel, and you may plainly observe a perfect representation of the Head, Halo, and Beard of the Comet (1678b, 31).

Although I have not been able to replicate this action even to the modest degree of Hooke's crystallization model, the chemistry it requires is relatively simple. The iron in the filings covering the head of the "comet" reacts with the sulfuric acid to create hydrogen gas. These hydrogen bubbles rapidly rise to the surface of the acid solution, which has been diluted with water presumably to control the rate of reaction.[10] In a general sense, Hooke's account might be read to suggest that the reaction of the acid and the ferrous particles in the wax ball yields a visual effect resembling his target system; the bubbling ball looked like a comet. Yet, Hooke's model repays consideration in different sense—one wherein observation and manipulation of this strange, effervescent cocktail leads to knowing about extraterrestrial bodies. For, this model departed in important ways not only from Hooke's observations of meteoric bodies in April 1677, but from his theory of comets more broadly. How and exactly what this mediating model represented thus needs to be examined carefully.

To elaborate these points, I want to make use of the "DDI" (denotation, demonstration, and interpretation) analysis put forward by philosopher of science R.I.G.

[9]For a critique, see Giere (1999).

[10]I thank David Tirrell and Tony Jia for discussing this action with me.

Hughes (1997). Although but one of several approaches to the study of models available within recent philosophy of science, Hughes' account is particularly useful here insofar as it specifically avoids appeal to mimesis. Instead, integrating Nelson Goodman's claim that resemblance is neither a necessary nor sufficient condition for representation, Hughes' analysis can help us to peel back the veneer of plausibility that attends to Hooke's model and to schematize its structure. First, following Hughes' approach, we need to isolate what the model denotes. The wax ball in the model denotes the solid core of the comet, which Hooke had theorized "to be made of solid matter, not fluid; that the body of it especially, is considerably dense, but that the haziness or Coma about it is much more rarified, and the tail thereof is most of all" (1678b, 9). Secondly, the dramatic reaction of the comet to surrounding aether is denoted in the model by the evolution of hydrogen gas from the iron and sulfuric acid. As with comets, Hooke observed, "the menstruum falling on, or dissolving the iron, there is a continual eruption of small bubbles, and dissolv'd particles from the sides of this body" (1678b, 31–32). Finally, the force of solar gravitation that produces the comet's characteristic tail is denoted in the model by the gravitation of the earth upon the glass tube and its contents. "Being of a much lighter consistence than the anbient liquor", Hooke explains, bubbles in the glass tube denote the particles that "are by the greater gravity of that, continually protruded upwards" to simulate the tail of the comet (1678b, 32).

In the second stage of schematic analysis that Hughes calls "demonstration", we set out how the representational terms of the model can lead to new understanding of the target. Hooke explains this dynamic in the following way: "If we suppose the Aether to be somewhat analogous to a menstruum, and that there is a gravitation towards the center of the Sun, if the Nucleus or head of the Comet be supposed such a dissoluble substance, the phaenomena of the shape of the Comet may, I think, be rationally explained" (1678b, 32). Having appointed denotational values to humble materials and forces, Hooke's model provides a scenario in which the consequent effects may be observed. Visualizing the comet as a field of ferrous particles reacting with sulfuric acid, the material model creates an opportunity to observe the simulated forces of gravitation and aether-resistance upon elusive meteoric bodies, which could never be examined "first hand".

What makes this model especially interesting are the ways in which Hooke sought to gain cognitive purchase on comets through reconciling study of this materialization with observational data. Although our only surviving evidence of Hooke's actual work with his model comes from the following remarks, he makes clear that observation and manipulation of the bubbling wax ball could enable the scientist to "interpret" (in Hughes' terminology) the relations between model and target phenomena. So Hooke claims:

> By this Hypothesis [i.e. the model] the phaenomena of the Comet may be solved; for hence 'tis easie to deduce the reason why the Beard grows broader and broader, and fainter and fainter towards the top: why there is a Halo about the body; for this will appear clearly in the experiment: why the Beard becomes a little deflected from the body of the Sun; for if the dissolving Ball be by the wire mov'd either this way or that way, the arising steam or

> bubbles will bend the contrary: ... by this supposition also 'twill be easie to explicate why the beard is sometime bended, and not straight, and why it is sometimes brighter upon the one side than upon another? why the bottom of it is more round, and the other sides more undefin'd; and divers of the like phaenomena (1678b, 32).

By Hooke's analysis, observation and intervention into the behavior of the material model—including moving the wax ball "this way or that"—calls attention to phenomena observable in comets themselves. The bent stream of bubbles caused by manipulation of the model allows the investigator to hypothesize the presence of similar effects in the target system and to draw inferences about their causes. In this way, the model possesses what Hughes calls an "internal dynamic" that enables the user to draw "hypothetical conclusions about the world over and above the data we started with" (1997, S331).

How exactly did this chemical cocktail thereby represent Hooke's comet? Ingenious as this material model was, it stood in uncomfortable relation both to crucial aspects of Hooke's theory of comets and to what he had actually observed in April 1677. As we have seen, Hooke made much of the ability of his bubbling wax ball to model the reaction between aether and the meteoric body that created the comet's tail. Yet, by privileging factors that could be admirably visualized in the model such as dissolution in a reagent and its response to the force of gravity, Hooke had to compromise a crucial piece of his comet theory. After all, he had claimed that what made comets exhibit behavior so notably different from other satellites similarly exposed to the corrosive effects of aether was their extreme internal agitation.[11] In concert with the action of the aether, it was this internal activity that Hooke theorized as causing the destabilization of the proto-comet's gravitational and magnetic properties, while completely altering its orbital trajectory. In his material model, however, the decomposition of the comet was simulated as an exclusively and literally superficial process. The reader had been told how the solid wax core should be "rould in filings of iron or steel" (1678b, 31). It would be fascinating to know if and how Hooke might have attempted to engineer a model closer to his theory that could simultaneously deteriorate from discrete, yet complementary, internal and external causes. Nonetheless, the evidence we have suggests that the materiality of Hooke's made-model not only simplified but significantly departed from this crucial component of his comet theory.

More problematic for Hooke was the fact that the wax ball also failed to match a key feature of observed comets: the model could not generate light.[12] Here too the philosophical literature on mediating models is instructive. What this literature has emphasized is that because models can represent their targets only partially, scientists frequently compensate by generating numerous different models of any given system under examination. The various different models of the nucleus used in physics are exemplary. As Margaret Morrison and Mary S. Morgan observe: "Each

[11] Hooke did not know that the earth too possesses an antisolar ion tail; see Yeomans (1990, 352).

[12] In 1682, Hooke described a revised version of this material model that could produce light; see Hooke (1705, 167).

individual model fails to incorporate significant features of the nucleus, for example, the liquid drop [model] ignores quantum statistics and treats the nucleus classically. While others ignore different quantum mechanical properties, they nevertheless are able to map onto technologies in a way that makes them successful, independent sources of knowledge" (1999, 23–24). Hooke's response to the limits of his wax-ball comet model is telling in this way. Conceding its inability to explain the important, observed feature of luminescence, Hooke concludes *Cometa* by canvassing a wide field of other possible models for comets' generation of light. "Decaying fish, rotten wood, glow-worms, &c." are all offered as possible analogues before Hooke introduces a new set of models (1678b, 46). A comet's luminous head, he postulates, is like a torch or a battery of cannons whose "blazing Granadoes or Fire-balls" follow the parabolic motion of projectiles as established by seventeenth century physics—and so we see visualized in a compelling diagram also provided in *Cometa* (1678b, 46, 48) (Fig. 2).

Although these postulations are given little further treatment, Hooke's tactical or pragmatic approach to representation becomes increasingly clear over the course of *Cometa*. None of his various comet models can promise to fully reconcile theory and observation. But each can denote discrete, appointed features and thereby offer to bring aspects of cometary phenomena into demonstration and interpretation. Stating a veritable motto of this approach to representation, Hooke concludes of his models that comets are "in some things analogous to the one, and somewhat to the other, though not exactly the same with either" (1678b, 47). By way of conclusion, I want to suggest how historians of art might productively learn from this representational pragmatism, particularly as we study visual practices generated at the boundaries of early modern art and science. Beyond the important insight it offers to the historical context of Robert Hooke and his colleagues, though, this analysis also allows us to reconsider the integral problems shared by students of the visual and philosophers of science on a larger scale.

It Behove Them, Who Professe the Knowledge of Nature or Reason, Rightly to Apprehend the Severall Waies Whereby They may be Expressed

Trained as a painter and gifted as an experimenter, English philosopher Robert Hooke has risen to prominence in recent historical studies that have celebrated the connections between visual art and the "new sciences" of seventeenth century Europe. The lavish plates of Hooke's *Micrographia* have been repeatedly cited as evidence of this union. Made from observations with optical instruments, they suggest both the keen-eyed attentiveness to optical detail seen in seventeenth century painting and the guiding imprint of a novel conception of experiment—the production of new facts about nature through what Francis Bacon called the "vexations of art". By contrast, as has been the case more broadly (Hopwood and de Chadarevian

2004), Hooke's material models have received markedly less attention.[13] Reasons for this neglect are perhaps not difficult to find. Unlike the stunning illustrative plates of *Micrographia*, *Cometa* or Hooke's numerous other publications, no direct physical evidence is known to survive from his material models. In this way, they challenge both the time-honored methods of art-historical analysis and the favor for material culture exhibited in recent history of science (Galison 1997, Daston 2004). To make matters worse, no physical evidence may *ever* have existed of these models. As we have seen, it is difficult to know if and how Hooke's crystallization model—a representation wherein bullets with imagined properties were used to generate behavior of theoretical entities—ever actually required physical objects. Treading such uncomfortable ground between categories of experiment and theory, Hooke's models were strange, intermediary enterprises that could answer exactly to neither category and that departed in important ways from both.

With these doubts in mind, we might return to the quasi-existential question sketched at the outset. Why exactly should art historians or other students of the visual bother with these baffling activities which only seem to complicate the attractive, available view of Hooke and his colleagues as able copyists of natural facts? As is implicit in the foregoing argument, what I see as at stake in engaging with the evidence of material models are matters essential to the historical understanding of early scientific visuality and to the conceptual vitality of the art/science conversation. I will treat the historical argument first. We know that early scientific bodies like the Royal Society of London were organized around and gave particular privilege to experimental trials. However, as is revealed in the work of Robert Hooke, the Royal Society's central experimental performer and theorist, trials that initially appear to be clear-cut cases of experimentation may actually be better understood as varieties of representation. If glimpsed only fragmentarily through the modest sampling presented here, these models were various in form and diverse in function; they deployed varieties of representational strategy and were allotted different degrees of cognitive value. Now, such interest in employing a broad range of representations and commanding an expanded field of visual activity are importantly commensurate with the evidence of recent historical studies, which are altering our apprehension of visuality in the early Royal Society. If recent studies have shown how Hooke and Wren were polymathic masters of drawing, architecture, surveying and numerous other visual practices, their scientific colleagues in the Royal Society's ambit were no less inclined to experimenting with representation; they contrived ingenious of modes of encryption, pictographic writing, and automated notation along with forays into optical projection and anamorphic wizardry.[14]

The crucial, historical point to be apprehend here is that those in the early scientific community identified such polymorphous visual fluency as a *virtue*. Not long before he served as a mentor to Robert Hooke at Oxford, catalyst of seventeenth century English science John Wilkins published a text on cryptography. Therein,

---

[13]A rare exception here is Iliffe (1995, esp. 293–299).

[14]For extended discussion, images and further bibliography, see Hunter (2007).

Wilkins claimed: "As it will concerne a man that deals in trafficke, to understand the severall kinds of money, and that it may be framed of other materialls besides silver and gold, so likewise do's it behove them, who professe the knowledge of nature or reason, rightly to apprehend the severall waies whereby they may be expressed" (1641, 11). If Wilkins' dictum is keenly pertinent for understanding Hooke's approach to modeling as exposited here, it is more broadly instructive for what has emerged as an important direction in recent studies of early modern art and science. As we have seen with Hooke's models of the comet, being able to harness a range of representations culled from the imaginative interpretation of physical processes was critically advantageous to the experimental philosopher. But, this broad-ranging knowledge of physical materials and their imaginative, representational potential was simultaneously crucial to the architectural and other visual activities that Hooke, Wren and others practiced in later Baroque London. Thus, drawing tools from philosophy of science, we may better analyze the diverse representational techniques actually deployed and valued by early experimental philosophers. More fundamentally, we can simultaneously apprehend how diverse forms and functions of visual practice were essential to the science and art engineered by figures like Hooke and Wren. Rather than just reinforcing the familiar linkage of naturalism in painting and empiricism in science, this interpretation would advance by analyzing the performances and procedures at the very center of their scientific community's attention.

This leads to the second, conceptual point. For, what recent work in philosophy asks us to recognize in scientific representations are degrees of complexity, sophistication and, above all, degrees of *distance* from natural targets that are almost entirely absent from humanities-based accounts. In his contribution to this volume, for example, Anjan Chakravartty treats the contention that "descriptions of entities and processes afforded by scientific representations are generally false, strictly speaking", as so uncontroversial a claim that it necessitates no further argument. Cutting directly against the grain of much received wisdom in humanities-based art/science studies, such philosophical work ask us to see scientific models as stylized artifacts invested with cognitive value and modified by varieties of imaginative intervention. Introduced into serious games of make-believe, these models can mediate between observables and theory, generating meaningful insight into real-world systems even as they are highly indifferent to particular facts about their targets. To art historians and humanists more generally, questions of how ostensibly fictional objects can be invested with imaginative values and take on "lives of their own" are not marginal matters. As only the seminal volumes of David Freedberg (1989), Hans Belting (1994) and W.J.T. Mitchell (2005) need indicate, such questions are absolutely central to the Western artistic tradition.

In thinking with this research in philosophy of science, then, historians of art might reconsider both the conception of scientific representation now dominant in the humanities and the archive from which that conception has been drawn. As noted at the outset, pictures and illustrations have long served humanists as the crucial evidence of representation in science. These pictures have also come to be seen not only as the key archive of scientific representation but also the acme of its aspirations. So

Robert Hooke's work examined here suggests, though, pictorial artifacts constitute only a fragmentary component of the highly imaginative, stylized ways in which objects were being manipulated, fictionalized and performed as representations to advance scientific understanding. Examining exactly what these "clever objects" are and how they embody, direct or inform imaginative thought are questions we might begin to ask. But, these are questions we can also share. For if we can learn from the methods and ethos of recent philosophy of science, so art historians can bring to discussion the discipline's rich tradition of thinking about the properties of the aesthetic object and the various powers over the imagination latent to it. Our conversation need not be to explain "Art" by virtue of "Science" (or vice versa), but to theorize the representational practices that run between them and beyond them.

Acknowledgments Thanks to Moti Feingold, Tarja Knuuttila and, especially, to Roman Frigg for comments on previous drafts of this essay.

References

Alpers, S. (1983), *The Art of Describing: Dutch Art in the Seventeenth Century*. Chicago: University of Chicago Press.

Aristotle (1987), *The Poetics of Aristotle*, trans. S. Halliwell. London: Duckworth.

Baird, D. (2004), *Thing Knowledge: A Philosophy of Scientific Instruments*. London: University of California Press.

Bellori, G. P. (2005), *The Lives of the Modern Painters, Sculptors and Architects*, trans. A.S. Wohl. Cambridge: Cambridge University Press.

Belting, H. (1994), *Likeness and Presence: A History of the Image before the Era of Art*, trans. E. Jephcott. Chicago: University of Chicago Press.

Bennett, J., et al. (2003), *London's Leonardo: The Life and Work of Robert Hooke*. Oxford: Oxford University Press.

Bermingham, A. (2000), *Learning to Draw: Studies in the Cultural History of a Polite and Useful Art*. New Haven: Yale University Press.

Braider, C. (2004), *Baroque Self-Invention and Historical Truth: Hercules at the Crossroads*. Aldershot: Ashgate.

Bredekamp, H. (2000), "Gazing Hands and Blind Spots: Galileo as Draftsman", *Science in Context* 13, 3–4: 423–462.

Chapman, A. (1996), "England's Leonardo: Robert Hooke (1635–1703) and the Art of Experiment in Restoration London", *Proceedings of the Royal Institution of Great Britain* 67: 239–275.

Cole, W. (ca. 1692), MS 1078. Wellcome Library, London.

Cooper, M. (2003), *'A More Beautiful City': Robert Hooke and the Rebuilding of London after the Great Fire*. Sutton: Thrupp-Stroud.

Crombie, A. C. (1994), *Styles of Scientific Thinking in the European Tradition*. vol. II. London: Duckworth.

Daston, L. (ed.) (2004), *Things That Talk: Object Lessons from Art and Science*. New York: Zone.

Daston, L. and Galison, P. (2007), *Objectivity*. New York: Zone.

Drake, E. T. (1996), *Restless Genius: Robert Hooke and his Earthly Thoughts*. New York: Oxford University Press.

Edgerton, S. Y. Jr. (1984), "Galileo, Florentine 'Disegno,' and the 'Strange Spottednesse' of the Moon'", *Art Journal* 44, 3: 225–232.

Ehrlich, M. E. (1992), "Mechanism and Activity in the Scientific Revolution: The Case of Robert Hooke", *Annals of Science* 52: 127–151.

Elkins, J. (1999), *The Domain of Images*. Ithaca: Cornell University Press.

Elkins, J. (2007), *Visual Practices Across the University*. Munich: Wilhelm Fink Verlag.

Elkins, J. (2008), *Six Stories from the End of Representation: Images in Painting, Photography, Astronomy, Microscopy, Particle Physics, and Quantum Mechanics, 1980–2000*. Stanford: Stanford University Press.

Freedberg, D. (1989), *The Power of Images: Studies in the History and Theory of Response*. Chicago: University of Chicago Press.

Freedberg, D. (2002), *The Eye of the Lynx: Galileo, His Friends and the Beginnings of Modern Natural History*. Chicago: University of Chicago Press.

Frigg, R. (2006), "Scientific Representation and the Semantic View of Theories", *Theoria* 55: 49–65.

Frigg, R. (2009), "Models and Fictions", *Synthese*; preprint available at http://www.lse.ac.uk/collections/CPNSS/projects/ContingencyDissentInScience/DP/DPFriggOnline0508.pdf

Frigg, R. and Hartmann, S. (2006), "Models in Science", in E. N. Zalta (ed.), *Stanford Encyclopedia of Philosophy*, http://plato.stanford.edu/archives/spr2006/entries/models-science/

Fyfe, G. and Law, J. (eds.) (1988), *Picturing Power: Visual Depiction and Social Relations*. London: Routledge.

Gal, O. (2002), *Meanest Foundations and Nobler Superstructures: Hooke, Newton, and the "Compounding of the Celestiall Motions of the Planetts"*. London: Kluwer.

Galilei, G. (1974), *Two New Sciences*, trans. S. Drake. Madison: University of Wisconsin Press.

Galison, P. (1997), *Image and Logic: A Material Culture of Microphysics*. Chicago: University of Chicago Press.

Gibson-Wood, C. (2000), *Jonathan Richardson: Art Theorist of the English Enlightenment*. New Haven: Yale University Press.

Gibson-Wood, C. (2002), "Picture Consumption in London at the End of the Seventeenth Century", *Art Bulletin* 84, 3: 491–500.

Giere, R. (1999), "Using Models to Represent Reality", in L. Magnani et al. (eds.), *Model-Based Reasoning in Scientific Discovery*, London: Kluwer, 41–57.

Golinski, J. (1989), "A Noble Spectacle: Phosphorous and the Public Cultures of Science in the Early Royal Society", *Isis* 80, 1: 11–39.

Gombrich, E. (1961), *Art and Illusion: A Study in the Psychology of Pictorial Representation*. London: Phaidon.

Goodman, N. (1968), *Languages of Art: An Approach to a Theory of Symbols*. New York: Bobbs-Merrill.

Gouk, P. (1999), *Music, Science and Natural Magic in Seventeenth-Century England*. New Haven: Yale University Press.

Hacking, I. (1983), *Representing and Intervening: Introductory Topics in the Philosophy of Natural Science*. Cambridge: Cambridge University Press.

Halliwell, S. (2002), *The Aesthetics of Mimesis: Ancient Texts and Modern Problems*. Oxford: Princeton University Press.

Henry, J. (1989), "Robert Hooke, the Incongruous Mechanist", in M. Hunter and S. Schaffer (eds.), *Robert Hooke: New Studies*, Woodbridge: Boydell, 149–180.

Hesse, M. (1966a), "Hooke's Vibration Theory and the Isochrony of Springs", *Isis* 57, 4: 433–441.

Hesse, M. (1966b), *Models and Analogies in Science*. Notre Dame: University of Notre Dame Press.

Hockney, D. (2001), *Secret Knowledge: Rediscovering the Lost Techniques of the Old Masters*. New York : Viking Studio.

Hooke, R. (1665), *Micrographia: or, Some Physiological Descriptions of Minute Bodies made by Magnifying Glass*. London: John Martyn and James Allestry.

Hooke, R. (1678a), *Lectures de Potentia Restitutiva, or of Spring Explaining the Powers of Springing Bodies. To which are Added some Collections*. London: J. Martyn.

Hooke, R. (1678b), *Lectures and Collections*. London: J. Martyn.

Hooke, R. (ca 1680), MS Sloane 1039. London: British Library.
Hooke, R. (1705), *The Posthumous Works of Dr. Robert Hooke*, in R. Waller (ed.), London: S. Smith and B. Walford.
Hooke, R. (1726), *Philosophical Experiments and Observations of the Late Eminent Dr. Robert Hooke*, in W. Derham (ed.), London: W. & J. Innys.
Hooke, R. (1935), *The Diary of Robert Hooke, 1672–1680*, in H.W. Robinson and W. Adams (eds.), London: Taylor & Francis.
Hughes, R. I. G. (1997), "Models and Representation", *Philosophy of Science*, Vol. 64, Supplement. Proceedings of the 1996 Biennial Meetings of the Philosophy of Science Association. Part II: Symposia Papers: S325–S336.
Hunter, M. (2007), *Robert Hooke Fecit: Making and Knowing in Restoration London*. PhD Dissertation: University of Chicago.
Hunter, M. (2010), "The Theory of the Impression According to Robert Hooke", in M. Hunter (ed.), *Printed Images in Early Modern Britain: Essays in Interpretation*, Aldershot: Ashgate, 164–193.
Hunter, M. (1981), *Science and Society in Restoration England*. London: Cambridge University Press.
Hunter, M. (2003), "Hooke the Natural Philosopher", in J. Bennett et al. (eds.), *London's Leonardo: The Life and Work of Robert Hooke*, Oxford: Oxford University Press, 105–162.
Iliffe, R. (1995), "Material Doubts: Hooke, Artisan Culture and the Exchange of Information in 1670s London", *British Journal for the History of Science* 28 (March 1995): 285–318.
Ivins, W. M. Jr. (1938), *On the Rationalization of Sight; with an Examination of Three Renaissance Texts on Perspective*. New York: Metropolitan Museum of Art.
Jardine, L. (2003a), *The Curious Life of Robert Hooke: The Man Who Measured London*. London: Harper Perennial.
Jardine, L. (2003b), *On a Grander Scale: The Outstanding Career of Sir Christopher Wren*. London: HarperCollins, 2003.
Jongh, E. de (1984), "The Art of Describing: Dutch Art in the Seventeenth Century", *Simiolus* XIV/1: 51–59.
Kemp, M. (1990), *The Science of Art: Optical Themes in Western Art from Brunelleschi to Seurat*. New Haven: Yale University Press.
Koerner, J. (1999), "Factura", *Res* 39 (Autumn): 5–19.
Kuhn, T. (1977), *The Essential Tension: Selected Studies in Scientific Tradition and Change*. Chicago: University of Chicago Press.
Lee, R. (1940), "*Ut Pictura Poesis*: The Humanistic Theory of Painting", *Art Bulletin* 23: 197–269.
Lindberg, D. (1976), *Theories of Vision from Al-Kindi to Kepler*. Chicago: University of Chicago Press.
Lüthy, C. (2000), "The Invention of Atomist Iconography", *Max-Planck-Institut für Wissenschaftsgeschichte*, Preprint 141. Berlin.
Lynch, M. and Woolgar, S. (eds.) (1990), *Representation in Scientific Practice*, London: MIT Press.
Marin, L. (1986), "In Praise of Appearance", *October* 37: 98–112.
Marin, L. (1995), *To Destroy Painting*, trans. M. Hjort. Chicago: University of Chicago Press.
Meinel, C. (2004), "Molecules and Croquet Balls", in S. de Chadarevian and N. Hopwood (eds.), *Models: The Third Dimension of Science*, Stanford: Stanford University Press, 242–275.
Mitchell, W. J. T. (2005), *What Do Pictures Want? The Lives and Loves of Images*. Chicago: University of Chicago Press, 2005.
Morgan, M. S. and Morrison, M. (eds.) (1999), *Models as Mediators: Perspectives in Natural and Social Science*. Cambridge: Cambridge University Press.
Muller, J., et al. (eds.) (1984), *Children of Mercury: The Education of Artists in the Sixteenth and Seventeenth Century*. Providence: Brown University Press.
Newton, I. (1989), *Mathematical Principles of Natural Philosophy*, trans. A. Motte. Chicago: University of Chicago Press.
Ogilvie, B. (2006), *The Science of Describing: Natural History in Renaissance Europe*. London: University of Chicago Press.

Pächt, O. (1950), "Early Italian Nature Studies and the Early Calendar Landscape", *Journal of the Warburg and Courtauld Institutes* XIII, 1–2: 13–47.
Panofsky, E. (1962), "Artist, Scientist, Genius: Notes on the 'Renaissance-Dämmerung'", in W. K. Ferguson et al. (eds.), *The Renaissance: Six Essays*, New York: Harper & Row, 121–182.
Panofsky, E. (1968), *Idea: A Concept in Art Theory*, trans. J.S. Peake. New York: Harper & Row.
Pears, I. (1988), *The Discovery of Painting: The Growth of Interest in the Arts in England 1680–1768*. New Haven: Yale University Press.
Shapin, S. (1999), "The House of Experiment in Seventeenth-Century England", in M. Biagioli (ed.), *The Science Studies Reader*, New York: Routledge, 479–504.
Shapin, S. and Schaffer, S. (1985), *Leviathan and the Air Pump: Hobbes, Boyle and the Experimental Life*. Princeton: Princeton University Press.
Smith, P. H. (2004), *The Body of the Artisan: Art and Experience in the Scientific Revolution*. Chicago: University of Chicago Press.
Stafford, B. M. (1991), *Body Criticism: Imaging the Unseen in Enlightenment Art and Medicine*. London: MIT Press.
Steadman, P. (2002), *Vermeer's Camera: Uncovering the Truth behind the Masterpieces*. New York: Oxford University Press.
Stevenson, C. (2005), "Robert Hooke: Monuments and Memory", *Art History* 28, 1: 43–73.
Stoichita, V. (1995), *Visionary Experience in the Golden Age of Spanish Art*, trans. A.-M. Glasheen. London: Reaktion.
Waterhouse, E. (1953), *Painting in Britain, 1530–1790*. New Haven: Yale University Press.
Wilkins, J. (1641), *Mercury; Or, The Secret and Swift Messenger*. London: I. Norton.
Wilson, S. (2002), *Information Arts: Intersections of Art, Science and Technology*. Cambridge: MIT Press.
Yeomans, D. K. (1990), *Comets: A Chronological History of Observation, Science, Myth and Folklore*. Chichester: Wiley Science Editions.

Lost in Space: Consciousness and Experiment in the Work of Irwin and Turrell

Dawna Schuld

[A]part from the experiences of subjects there is nothing, nothing, nothing, bare nothingness.
Alfred North Whitehead, 1929

On several occasions during the years 1968–1971 artists Robert Irwin and James Turrell, an experimental psychologist named Ed Wortz, and a number of UCLA student volunteers spent hours depriving themselves of light, sound and human contact. They were engaged in a series of experiments involving an anechoic chamber used for psycho-physical experimentation by the Garrett Corporation, a contractor to the National Aeronautics and Space Agency (NASA). The interior of the chamber was soundproofed, suspended to minimize the effects of the earth's rotation and utterly darkened. Self-projected sounds like speech were deadened. Sitting in these reduced surroundings was exhausting; rather than depriving the subject of the senses of sight and hearing, the lack of focal markers proved to heighten them, causing the subject to strain his eyes and ears, searching for something upon which to focus his attention. Most startling were the effects upon leaving the chamber when the body re-adjusted to the overwhelming array of stimuli in daily life and the world became intensely bright, loud and noticeable.

Thus, in the experimental psychology laboratories of Southern California Irwin and Turrell would explore the possibilities of ambiguity, with profound implications for their subsequent artistic practices. What came to be known as "light and space" art arose from a focus on the contingencies of the art experience in contrast to a media-centric approach advocated by modernist critics such as Clement Greenberg and Michael Fried. This essay addresses the ways in which the parameters of critical analysis—in the fields of both art and psychology—were tested and/or altered by the introduction of Irwin and Turrell's experiments and their development of a situational art. I use the terms "situational art" and "situational form" deliberately so as foreground the contingent nature of the work, where site, temporality, viewer

D. Schuld (✉)
Indiana University Bloomington, Bloomington, IN, USA
e-mail: dlschuld@indiana.edu

R. Frigg, M.C. Hunter (eds.), *Beyond Mimesis and Convention*, Boston Studies in the Philosophy of Science 262, DOI 10.1007/978-90-481-3851-7_10,

experience, and the created "object" cohere as art. These artists approached a work of art as an event of engagement, rather than as any particular object. Their work—in the laboratory and in the studio—provides us with the means to recognize that conscious awareness binds together site, viewer and art into what John Dewey (1934, 48) calls *an* experience: "In short, art, in its form, unites the very same relation of doing and undergoing, outgoing and incoming energy, that makes an experience to be an experience. ... The artist embodies in himself the attitude of the perceiver while he works". Following a brief prologue in which I contextualize the terminology through which I am reading the relationship of these artists to contemporary trends in neuropsychology, my argument is presented in two parts. The first part is an historical account of the circumstances which led to the conflation of artistic and scientific experiment in the anechoic chamber. The second part examines how the situational art of Irwin and Turrell exposes the explanatory gap created by the parallel discourses of modernist criticism and behaviorist psychology both of which exclude the material role of conscious thought in aesthetic experience.

In discussing his work, Robert Irwin (1977) uses the phrase "posture of inquiry" as a means of describing an individual's open-ended and open–minded questioning stance, a situated mindset that begets creative thinking regardless of and prior to disciplinary distinctions such as "artistic" or "scientific". This is basically a phenomenological approach of "bracketing" experience so as to consider it on its own terms.[1] It is important to note, however, that Irwin began reading phenomenologists like Husserl, Merleau-Ponty, and Schütz only *after* he had been experimenting with situational art—or "art in response"—for many years. In this regard, the language of phenomenology can be seen as a useful means of articulating an already well-developed and *practiced* posture of inquiry. Such a posture is kind of pragmatic naiveté wherein the inquiring artist or scientist remains open to investigating unforeseen, idiosyncratic, and/or deeply subjective data. I make this point here so as to distinguish the more recent and pragmatic hybrid of phenomenology practiced by Irwin from the philosophy that came beforehand. Likewise neurophenomenology, described in more detail shortly, resembles Irwin's practical application as much if not more than it does the phenomenological philosophy to which it owes its roots: it is phenomenology naturalized. In creative questioning—or curiosity—Irwin saw a convergence between the work of artists and that of scientists. His scientific counterpart, and longtime collaborator in questioning, Ed Wortz agreed. For the younger Turrell, whose artistic and psychological interests developed in tandem, the conflation of scientific and artistic inquiry was self-evident (Adcock, xix). For all three there were immediately apparent correspondences between the psychologist's concerns with the disorientations of space travel and the artists' interests in perception as medium. In both cases, investigators enacted a phenomenological

[1]"Bracketing" (Einklammerung) is a term first posited in Edmund Husserl (1991, original 1913). The term is succinctly defined by Husserl scholar David Woodruff Smith (2007, 429) as "the method or technique of turning our attention from the objects of our consciousness to our consciousness of those objects", an awareness of being aware, so to speak.

shift in methodology: from an emphasis on observed reality to felt reality, or subjective feeling. Each set of questions necessitated renewed consideration of first-person experience—whether that of the astronaut or of the art viewer—in one's experimental methodology. And for each, experiments in sensory deprivation were means of essentializing the perceptual processes in question.

Sensory deprivation, it should be understood, is not exactly what the phrase suggests. The kind of tests Irwin, Turrell, and Wortz wished to conduct involved the extreme reduction of sensory stimuli. The senses remained intact; there was simply very little to which one could physically attend. This is the difference between being able to see nothing and *looking at* nothing, or hearing nothing and *listening to* nothing. Subjects were not deprived of their senses; they were instead asked to attend to a lack. Indeed, it can just as easily be claimed that Irwin was interested in sensory *enhancement*, for this was effectively the end result of limited exposure to the "deprivation" chambers. This is a significant point as it underscores the relativity of perceptual experience, a key aspect in both Irwin's and Turrell's subsequent art practices and in emerging theories of consciousness that rely on similarly subjective (though more tightly controlled) experimentation.

Irwin and Turrell's experiments with Wortz took place within developments in psychological experiment that were displacing the previously pre-eminent methods of radical behaviorism as led by J.B. Watson and B.F. Skinner. The behaviorists prioritized the scientific description of observed behavior, and rejected introspection as unreliable data. But in the 1960s the field of psychology was undergoing a shift similar to that in the art world, where methods once considered objective—or disinterested—were emerging as contextually conditioned. This parallels a development in physics in the first decades of the twentieth century. Einstein's theory of relativity and the process of measurement as understood in mature quantum mechanics allocate a central role to the observer in that they see the results of observations as essentially determined not only by how the world is, but also by the observer's situation and actions. This paradigm shift in physics had a de-stabilizing effect throughout the sciences (Kuhn 1962). In light of these findings, to what degree was objectivity or certainty attainable? The experiments at Garrett were small indicators of a broader overall subjectivization of experiment—or phenomenological turn—taking place in neuroscience and psychology (or cognitive sciences). This turn was marked by willingness on the part of scientists to revisit the philosophy of mind, especially as put forth by William James in the previous century. Writes Patricia Churchland (1986, 250): "With William James ... the revered presumption that science, and knowledge generally, required foundational certainties began to seem questionable. If, agreeing with Kant, our sensory experience is interpreted experience, then the 'certainties' of sensory evidence are only as good as the infused interpretation".

Within the interdisciplinary field of "neurophilosophy" (Varela 1996) arose a plethora of new cognitive sciences, including contemporary "neurophenomenology", as practiced by Francisco Varela, Bernard Baars and like-minded colleagues. For neurophenomenologists, "the experience of being a body, and not just having a body ..., forms part of the primary existential conditions of our becoming in the world" (Flores-González 2008, 188). This emphasis on the feeling of what happens was characteristic of the cognitive revolution as it took place in California; and it was particularly applicable to artistic practice, in that it could accommodate the artists' intuitive approach. As Stephen Pinker points out, "East Pole/West Pole" divisions arose in the world of cognitive science, with the MIT-centered "eastern" axis of Jerry Fodor and neo-Chomskyites arguing for the essential nature of concepts, while "those at the West Pole suspect they begin as small innate biases in attention and then coagulate out of statistical patterns in the sensory input" (2002, 35). Both "poles" make up what is collectively referred to as the "cognitive revolution", a wholesale backlash against behaviorist methods encompassing concerns in neurology, anthropology, philosophy, and linguistics (Miller 2003).The phenomenological "branch" of cognitive studies constitutes only one element in the "West Pole" faction. It also has self-evidently strong links to continental philosophy (especially phenomenologists Husserl and Merleau-Ponty), although American pragmatism (notably in the work of William James) has its place in neurophenomenological theory as well (Thompson et al., 2005).

The parallels between light and space art and neurophenomenology are compelling, as each practice emphasizes the primacy of lived experience. A modified—or pragmatic—phenomenology provides a means for understanding the cognitive materiality of aesthetic experience, while light and space art demonstrates what neurophenomenology asserts: i.e. you have to be there.[2] Cognitive scientists obviously must apply more rigorous constraints to their testing methods than did Irwin and Turrell; however, the scientists and artists share an open-minded posture of inquiry at the outset of each experimental endeavor. This state of "pure research" is what both Irwin and Wortz assert that they were sharing when they embarked upon their Collaboration (Weschler 1982, 131–133).

At issue for Irwin and Turrell were the idealizing standards of modernist formalism in the New York-centered art world and the objectivizing principles of behaviorist psychology. While the modernist critic appeals to a cognitively held *a priori* meaning, the behaviorist restricts his studies to observable reality. Both approaches necessitate a Cartesian distinction between meaning and experience, mind and body. In so doing they also facilitate key distinctions between life and art, psychology and philosophy. The essential continuities between these categories—while often understood as given—were sacrificed for the sake of disciplinary autonomy.

By facilitating situations of perpetual disorientation and re-orientation, Irwin and Turrell pre-empted the postures of disinterestedness required both by behaviorism and by the idealizing modernist exhibition space, now known as the "white

[2]This is my primary reason for not including images with this text. While I acknowledge that my own descriptions are limited, they are less likely to be mistaken for "the thing itself" than are photographs, which at best offer only severely limited versions of the work in question, and prioritize visual perception at the expense of other sensory percepts present in the immediate experience. For years, Robert Irwin held the same position and refused to allow his work to be photographed, though he has since relented.

cube" (O'Doherty 1976). The black box and white cube are effective metaphors for exclusion—of a materially productive consciousness—and for disciplinary exclusivity. When participants attempted to negotiate the eerily dense nothingness of the anechoic chamber, however, behavior and thought coalesced into an experiential continuum. What follows is a consideration of the ways in which a literal black box—the anechoic chamber—exposed the integral role of the figurative black box—the conscious mind—in composing a coherent reality and a conditional art.

Entering the Black Box: Irwin, Turrell and the Anechoic Chamber

In late 1967, Maurice Tuchman, then senior curator for the Los Angeles County Museum of Art, set out to marry contemporary art and the technology in the Los Angeles area by inviting corporations to become both financial and technical resources for a number of artists whose work already tended toward or somehow engaged industrial materials and/or production (Tuchman 1971, 9). One of the first artists Tuchman approached was Robert Irwin. Irwin was on the verge of doing away with the art object entirely in favor of investigating its circumstances: the subjective body states of the viewer and the supposedly neutral gallery space. After several years as a successful painter, Irwin was nevertheless bothered by what he saw as the arbitrary limitations of the frame: "I no longer felt comfortable with that sense of confinement. It no longer made sense to me".[3] Rather, by the early 1960s he was developing an interest in making the experiential transition between the work of art and its context as fluid as possible. At the time of Tuchman's initial contact, Irwin was working on a series of disc-paintings out of machined aluminum—and later acrylic—approximately 60 inches (152.4 cm) in diameter and knife-edged, which he painted in extremely subtle shadings of white and gray. The discs were Irwin's first foray into industrial materials and therefore also his first collaborative efforts (Weschler 1982, 98–109, Gilbert-Rolfe 1993). The work was then mounted several inches away from the gallery wall and cross-lit so that the discs seemingly dematerialized in aureoles of light and shadow. The resulting effect was one where the shadows had as much material presence as the painting, if not more. In his biography of Irwin, Lawrence Weschler (1982, 111) writes: "He began to wonder how it might be possible to make an art of the incidental, the peripheral, the transitory—an art of things not looked at (indeed, invisible when looked at directly) yet still somehow perceived". In blurring the boundaries between object and subject, Irwin challenged the rationalizing separations of perception (physical) and conception (non-physical).

That Irwin believed a perceptible, yet non salient art object was even a possibility tells us that he thought of perception as more than a simple one-way conduit for information. By rejecting the object as the singular locus for aesthetic inquiry, Irwin

[3]Robert Irwin, quoted in Weschler (1982, 99).

began to see a role for his own work alongside a scientific community investigating the relationship of perception to cognition. Tuchman's *Art and Technology* project was the beginning of Irwin's enduring relationship with Ed Wortz, an experimental psychologist whose work at the time was predominantly concerned with the constraints and experiential idiosyncrasies of astronautics (e.g. see Robertson and Wortz 1969). Along with fellow artist James Turrell, who was subsequently invited to join in on the collaboration, they began to design a series of experiments demonstrating that even when the senses are given nothing to work with, the mind insists upon creating a relationship between the body and its environment. That is, to the mind's eye nothing is just as substantive as something. Under such conditions—where we perceive meaning in the event of our encounter—it becomes preposterous to situate meaning in objects alone. Wrote Irwin (1985, 28): "*Circumstance* ... encompasses all of the conditions, qualities and consequences making up the real context of your being *in* the world. There is embedded in any set of circumstances and your being in them the dynamic of a past and future, what was, how it came to be, what it is, and what it may come to be".

In Southern California, economic and geographical factors created the perfect laboratory for linking psychology to technology to the development of a conditional aesthetic such as Irwin's or Turrell's. For a Southern Californian, there is a disorienting incongruence between the immediacy of nature and ever-proximate technology in the defense industries, aerospace, and film (a curious hybrid of art, technology, and commerce). The art world was not exempt from these influences. The Los Angeles art scene—small, fluid, and burgeoning—allowed for constant experimentation, definition, and redefinition; there the New York-centric modernist critical culture had only tangential (and sometimes ironically perverse) effect. In this regard, I echo Cécile Whiting's (2006) claim regarding the Los Angeles art community in the 1960s: that it sought to invent an identity for itself drawing on aspects of life distinctive to Los Angeles, developing an art cross-pollinated with technological or commercial elements proliferating in the area.

In these circumstances where artists found themselves rather ambivalently positioned vis-à-vis the established art world, artists such as Irwin and Turrell were nonetheless adhering to an alternative art tradition of sorts, best represented in the work of composer John Cage. Cage himself had experienced the anechoic chamber environment at Harvard University in 1951, with important implications for subsequent incorporation of silences in his compositions (Cage 1961). Turrell in particular was well-versed in the philosophy of Cage, who was a fellow Pomona College alumni (1928–1930, Turrell attended from 1961–1965), and went to hear him speak when the composer gave a distinguished alumnus talk, likely in 1962 (Emmerik 2003–2007).

The formidable presence of the California Institute of Technology (Caltech) in Pasadena (with names on its faculty that at different times included Einstein, Oppenheimer and Feynman) ensured that the physical sciences played an important role in educational and community developments. In its wake were drawn businesses and organizations that included the Jet Propulsion Laboratory (JPL),

Lockheed Air Corporation, the Rand Institute and Garrett Corporation, all of which contributed to the Los Angeles County Museum of Art's ambitious and controversial *Art and Technology* program and provided key materials for light and space artists.[4] As Michael Compton pointed out in a 1970 catalogue discussing the work of light and space artists Larry Bell, Robert Irwin and Doug Wheeler: "The aerospace industry . . . is not only orientated to rapid obsolescence but therefore also to technological extemporisation and to free access to outside experts, techniques and information. The preoccupations with precision, environmental and sensory control are naturally shared [by these artists] with this industry" (1970, 13). Against this background, it should also be noted that Turrell's father at one time trained as an aeronautic engineer (he subsequently worked almost exclusively as an educator), and that this has influenced the artist's ongoing interest in scientific instruments, methods, and measures, not least of all in his role as a pilot. As Craig Adcock (1990, 1) describes it: "He regards time spent in the air as time spent in the studio".

In Southern California, where "physics and meta-physics continued to rub shoulders in a variety of weird circumstances" (Davis 1992, 58) the new developments and challenges that arose with the space race and atomic physics stretched the parameters of what had been considered reality to its breaking point. Uneasily situating itself between physics and metaphysics was the developing field of cognitive science. The cognitive scientists were perhaps even more rigorously experimentalist than the behaviorists but at the same time, drawing from the quantum model, they transformed psychological methodology from one of stimulus/response to one of integrated processes or networks. If reality were a matter of integrated processes, it was difficult to maintain a position, as behaviorism would have it, that thought is extrinsic from behavior—or epiphenomenal. The position that conscious states are epiphenomenal is in no small part due to a behaviorist reaction to what it considered the misleading and unscientific methods of psychoanalysis: "The good Freudian attributes observable behavior to a drama played in nonphysical space by an immanent triumvirate scarcely to be distinguished from the spirits and demons of early animism" (Skinner 1964, 482). In this regard, the "cognitive revolution" is more properly understood as a counter-revolution, and—as has been pointed out—the cognitive scientists' empirical stance is much closer to behaviorism than it is to psychoanalysis (Miller 2003). For one contingent of psychologists, predominantly on the East Coast, this re-configuration supported the burgeoning development of computer science and attendant artificial intelligence (A.I.) models of cognition, thereby effectively removing the sticky wicket of lived consciousness from psychological study.

In Southern California a number of cognitive psychologists chose to return to the psycho-physical roots of their discipline, which emphasizes the *relationship* of the world "out there" to its correlates—or percepts—in the brain (i.e. how does mental

---

[4]Davis (1992, 54–62); and Tuchman (1971). For more extensive histories on the development and presence of the scientific community and specifically the aerospace industry in Southern California see: Newell (1980) and Koppes (1982), whose work relies upon an in-depth knowledge of the inner workings of the Jet Propulsion Laboratory as gleaned from its (de-classified) records.

imagery arise from engagement with the physical world?). In the work of cognitivist radicals William James' theory of volition gained notable new currency, wherein the direction of attention begets willed action (Neisser 1967, Mandler 1975). James was modest in his claims for willed action, stipulating that experience begets conscious thought and action, rather than the other way around. Nevertheless, he laid the groundwork for an understanding of consciousness as effective rather than mere affect. "We learn all our possibilities by the way of experience. When a particular movement, having once occurred in a random, reflex, or involuntary way, has left an image of itself in the memory, then the movement can be desired again, proposed as an end, and deliberately willed" (James 1890, 1099). In this context, the study of consciousness, rather than being beyond the psychologist's purview, was seen as "respectable, useful and probably necessary" (Mandler 1975). This new group of cognitive psychologists included Ulric Neisser and D.W. Hamlyn, both of whom are listed in a bibliography Turrell compiled for the *Art and Technology* project. Turrell first learned of their work when he earned his B.A. at Pomona College in perceptual psychology in 1965 (Adcock 1990). His work as an artist therefore developed alongside an interest in the language and practice of psychological experiment, a situation that differs from Irwin's, whose scientific and philosophical investigations emerged in the wake of developments in his artistic practice.

Tuchman's *Art and Technology* project afforded the perfect opportunity for Irwin and Turrell to study the nature of attention in the form of a series of "sensory deprivation" experiments. At first Irwin was matched with Lockheed Aircraft and later with Turrell introduced at the Garrett Corporation where the artists were interested in psychological experiments being performed at these facilities. Lockheed's Rye Canyon research facility proved promising for investigating sense and orientation. There, staff used anechoic and other "sensory deprivation" chambers to test human reactions to sensory stimuli in controlled environments. Because an anechoic chamber is so heavily insulated—for both sound and light—any sensory input would (at least theoretically) have to be introduced and perhaps more importantly, could be controlled. For Irwin in particular, who had been expending a great deal of time and effort attempting to reduce contextual distraction in his recent experiments with the disc-paintings, these chambers represented a clean slate in which to investigate experience (Tuchman 1971, 127).

Irwin's "wish list" to Lockheed included "investigations necessary to determine what perceptual awarenesses [sic] are necessary for basic orientation and stabi lity . . . human prowess . . . [and] basic necessities for maintaining sanity" (Tuchman 1971, 127). The question of orientation clearly was foremost in Irwin's mind. Orientation is the phenomenon that allows us to establish our position relative to the circumstances in which we find—or become—ourselves. It is central to any idea of the self in space. When we find ourselves in familiar circumstances orientation is maintained beneath our conscious notice. A subtle interrelation of sensory data and neural adjustment allows us the luxurious illusion of constancy as we go about our days. What is most notable about orientation is its fluidity; it must remain unstable in order to seem consistent. Without this paradoxical structure we could ostensibly lose our balance every time we turn our heads. Neurologist Alain Berthoz

(2000, 91) delineates how the brain constructs this remarkable stabilizing framework through a complex system of checks and balances between sensory input and neural adjustments:

> Perception, is an interpretation; its coherence is a construction whose rules depend on endogenous factors and on the actions that we plan. The difficulty in building a theory of coherence is that there is most likely not one single coherent theory for all of perception. . . . This range of possibilities is probably a key to the way illusions are manufactured.

Furthermore, as Berthoz (2000, 29) asserts, the maintenance of sensory coherence relies on input from the brain (still unconscious) that "[modulates] sensory information at its source, to adapt it to the requirements of movement . . .".

This adaptive mechanism of the neural networks is key to understanding the outcome of some of Irwin and Turrell's tests. Distraction—that which catches us unaware while attending to something else—became the focus of Irwin and Turrell's experiments. Unlike the normal "silence" in our lives, which might nevertheless include the hum of machinery or chirping of birds, the silence of the anechoic chamber even blocks out the sounds you make yourself in an odd way. "[I]t was suspended so that even the rotation of the earth was not reflected in it, or any sounds being bounced through the earth", said Irwin (Weschler 1982, 128), ". . . Nothing went into that space. And no light at all". Without reverberation, "outside noises" that we may make such as snapping our fingers become overly internalized. There is no *there* in which the snapping can occur. You are well aware that you are snapping your fingers but the sound of that snapping has no resonance. "When I clicked my tongue", stated one subject, "it had a dull, faraway sound" (Tuchman, 136).

These descriptions indicate that the experience within the chamber was somehow at once incoherent and yet distinctive for that very reason. The question of what constitutes meaningful engagement therefore becomes paramount: how does the experience become *an* experience? This is also a key issue in contemporary consciousness theory; without discernable stimuli, how does coherence in experience come about? Bernard Baars (1997) has proposed that our brains engage in *contrastive phenomenology*, wherein fields of possibilities on a conscious/unconscious continuum (such as "normal versus subliminal perception" and "novel versus routine and automatic processes") enable us to differentiate perceptual entities and establish orientation. So Baars (1997, 166) argues:

> Consciousness appears to be the major adaptive faculty of the brain. Our personal experience of the world is the subjective aspect of that adaptive activity. Philosophical arguments *against* the adaptive function of awareness rely on a little verbal magic, in which we pretend to suck out all the real features of consciousness—usually the ones that happen to be externally observable today—and ask, is anything left after we take away everything, except the last residuum of subjectivity?

In other words, consciousness is not about fixing qualities to perceived objects or categorizing objects according to a rationalized schema; but rather, it is a continuous cycle of adaptation of the percept to an illusionary constancy that keeps us oriented to our surroundings. Situational art forestalls that illusion by intervening with uncertainties. The resultant deferral of perceptual certainty—what *are* we looking at, through, in?—allows us the luxury of observing the physiological shift that otherwise seamlessly enables adaptation in more quotidian circumstances.

In Irwin and Turrell's experiments it was therefore essential that the subject begin with at least a momentary disorientation, as becomes evident when we consider their plans for building an anechoic environment for the museum exhibition. To establish a base-line disorientation they proposed several interventions both in and out of the anechoic chamber: the chamber was "obscured by either a blind wall or curve"; the chair in which the subject sat would "slowly flatten" and rise on hydraulics so that he was ultimately lying flat on his back in the middle of the room; "sub-threshold light flashes" would be introduced to induce a sensation of hallucination. This project was never realized, but stemmed from the artists' findings with Wortz at Garrett.[5]

Doing nothing was extremely disconcerting to subjects new to the project and they would report feeling uncomfortable after very brief periods (fewer than 10 minutes) while the artists would happily spend hours in the chamber. In the early 1960s Irwin had already been experimenting with a form of self-imposed sensory deprivation by locking himself in his studio for days at a time. There he spent long hours contemplating the perceptual properties of his "line" paintings (large canvases of saturated color interrupted by one, two, or three horizontal lines of another tone). The relentless boredom helped him reduce his art to its essential matter, which ultimately turned out to be his own conscious response. In his biography of Irwin, Lawrence Weschler (1982, 77) describes this eventuality: "Back at home, you may remember what it felt like to stand before the painting, the texture of the meditative state it put you in, but the canvas itself, its image in your mind, will be evanescent". Throughout this development the work became ever more ethereal and conditional. By these meticulously reductive means, Irwin was slowly but persistently breaking down the divide between subject and object.

Considering Irwin's long experience attending to very little, it is possible to presume that he came to the experiments already adapted somewhat to the situation. Irwin and Turrell's extraordinary involvement with the anechoic chamber (according to Irwin, six- to eight-hour stints compared to several minutes for most subjects) was possible because as artists they had already developed attentional faculties that saw more in less, from paying disproportionate attention to what would otherwise be filtered out by a constantly self-regulating perceptual system. In his apology for consciousness as a viable field of study, George Mandler (1975, 30) addresses the exceptional sensory capacities of someone with heightened attentional capacities, usually in meditation:

[5]Some have interpreted this proposal as especially disturbing and manipulative, seeing the viewers as playing the part of unsuspecting test subjects (Perchuk 2006). Though the experience would likely have been discomfiting, I can't agree entirely. Considering that the option to participate would have been solely the viewer's, and that no evidence indicates that the results would be "classified" or that one viewer would be prevented from discussing the procedure with the next, participants could hardly accuse the artists—who if anything were looking for ways to *communicate* the experience—of coercive tactics.

> ... the relationship between the object or event and ourselves is changed continuously by our mutual relations with the rest of the world. The new information, in a way, is always acquired in new contexts. ... This restriction of possible relations presumably provides not only the illusion but possibly also the reality of depth of perception which the special experience provides. *In contrast, artists and scientists, for example, apparently achieve the same depth of perception of special objects or events without the meditative experience*". [my emphasis]

In other words, as physicists had been asserting for half a century, there may be a whole lot more to nothing than first meets the eye, if one can find a way to reduce the distractions of things sufficiently to attend to it. Asserts astronomer Sten Odenwald (2002, 6):

> Space enters our perceptible world only in an oblique way. Because of this, we have to look carefully into our daily experiences to remind ourselves that there is something to wonder about. You need look only as far as the page of this book you are now reading to experience one of the most ancient and puzzling mysteries of the Void. You see the page and its letters; you do not, however, see the space that separates the page from your eyes.

In normal situations we fail to recognize the hidden mechanisms of perception that facilitate illusion. But, writes Robert Irwin (1985, 12): "[a]s one educated and practiced as a painter, my first hint (intuition) that the world of my perceptual and aesthetic concerns might not begin and end at the edge of my canvas was something that had no tangible reality. But my question would not go away and it was soon joined by others".

Without the unattended interstices of our perceptual world, things cannot be *things*. Both Odenwald and Irwin, in their own ways, are demonstrating means of attending to the gaps in our perception. Although not tangible, they play a significant role in perceptual experiences and are "things" to the extent that they have effects. Our relationship with the page depends upon that unseen void which Odenwald describes. The centrality of the canvas depends upon the fact that its context is unattended by the viewer. As proof of this phenomenon, we need only compare this "normalized" experience with that of one of Irwin and Turrell's *Art and Technology* subjects who is placed in a blacked-out anechoic chamber for a period of isolation no longer than 10 minutes. Upon being asked how the room felt, the subject (a 25-year old female student) answered: "Hard to put a shape to it. Flat in front of me. Hallucinations had shallow depth. On looking straight ahead, I felt light converging on the sides as if from behind, but when I turned it was even darker". The subjects repeatedly claimed to have feelings of "convergence" and "claustrophobia". The unseen void that maintains a healthy distance between the world of objects and us breaks down when there are no sensory referents to maintain it. One subject said she felt claustrophobic when she tried to look around. Without reverberation, sounds occur "in the head" or a sneeze "sticks" to the body. Without the transparency of light to see through, air cloaks us and weighs us down, pressing in (Tuchman 1971, 136).

The artists were especially interested in the relationship between sensory response *inside* the sphere and the experience upon stepping *outside* again. As Irwin told Weschler (1982, 129):

> There were all kinds of interesting things about being in there which we observed, but the most dramatic had to do with how the world appeared once you stepped out. After I'd sat in there for six hours, for instance, and then got up and walked back home down the same street I'd come in on, the trees were still trees and the street was still a street, and the houses were still houses, but the world did not look the same; it was very, very noticeably altered.

Irwin's "sharp-focus" walk down the street came after several hours in the anechoic chamber but even subjects who spent only minutes there reported that normal sound was sharply louder for some time afterward. Coherence, in a state of perpetually attended blackness, becomes something very different from what "makes sense" on the street. That our sensory organs adjust to circumstance should be obvious to anyone who has stood blinking in the glare when a light is suddenly turned on but that fact is too often conveniently forgotten in our need to stabilize what we see in order to orient ourselves.

Coherence does not therefore inhere in the anechoic chamber, but is a product of perceptual fine-tuning. It is constituted by the relationships between the viewer and her circumstances *and* between one experience and the next. This assertion is supported in encounters with a second important type of device made available to the artists and commonly used in sensory deprivation experiments: the ganzfeld sphere. The ganzfeld is the visual equivalent of an anechoic chamber insofar as it reduces sensory input as nearly to an absolute neutral as possible. The "whole field" of a sphere several feet in diameter sufficient to encompass the viewer's field of vision is finished in a uniform color and must be utterly smooth, as the ganzfeld relies upon a perfectly even distribution of light for effect. By looking at a stimulus of no color variation whatsoever, the experimental subject experiences the sensation of looking at nothing at all. The field of vision becomes utterly formless. There is no horizon or clearly defined object of any sort by which to orient oneself. The effect is one of a strange vast intimacy. In such circumstances color becomes a uniform presence. In the case of the Garrett experiments, the ganzfeld was white but it can be any one color in the spectrum: looking at a yellow ganzfeld for even a brief time will therefore have the "corrective" perceptual effect of making the subject see the world in magenta tones (its spectral opposite) for a time.

Turrell was particularly taken with the possibilities of ganzfeld, and has subsequently used it often in his work, in various colors. Ganzfeld technology and related light diffusion experiments transform the bare white cube of the exhibiting space, perhaps even rendering it wondrous: Turrell's light experiments reveal that its neutrality is an illusion. Turrell's *Virga* (1974) is one of a series of situational works, including work by Robert Irwin and others, commissioned by Count Giuseppe Panza for his private collection (now publicly held in trust by the Guggenheim Foundation). In the installation at Villa Panza in Varese, Italy the artist used ganzfeld effects to transform a plain white room into a situation for looking at light rather than with it. Two rhomboid veils of natural light appear to descend from thin, diagonal fissures hidden in the ceiling of a long, narrow, white room (12.25' wide × 14.67' high × 30.75' long; 3.73 × 4.47 × 9.37 m); the effect is one of separating the rectangle of the room into shrouded thirds. It does not serve to enhance the salience of an art object, but holds in the visitor's attention an awareness of her own sensibilities to light and space. As the light in Northern Italy changes in intensity throughout the day, and throughout the year, and as the viewer moves through the space, the effect is altered, so that the work is continually revised and renewed, a continual and nuanced reminder of the contingent nature of perception.

The strange sense of displacement brought about by looking into the ganzfeld makes it a popular tool in para-psychological research in addition to the kind of psycho-physical research performed at Garrett Corporation. There, like the anechoic chamber, the ganzfeld was used for experiments in sensory deprivation. Certainly the extent and nature of the group's experiments suggest that the work could at times be considered more para-psychological than psycho-physical, as it expanded to include experiments in alpha conditioning and Buddhist meditation practices (Tuchman 136, 137). Indeed, the extent to which these artists conflate sensory deprivation, meditation, and aesthetic experience proffers an important insight into their approach and practice. Ed Wortz asserted that Robert Irwin for one engaged in sensory deprivation as part of his artistic practice (Tuchman 1971, 139). When we consider Irwin's previously noted tolerance for tedium when making his line paintings, Wortz has a valid point. We come to understand that for these artists sensory deprivation constitutes a framework for creativity. The anechoic chamber and the ganzfeld are therefore not artistic media—that is perception—but the means of returning to a posture of inquiry. In this regard, they are of equal value to the artist as to the neurophenomenologist seeking a practical yet controlled means for studying first-person consciousness. The rise of cognitive psychology in Southern California meant that Irwin and Turrell could avail themselves of an experimental infrastructure wherein the subject perceived himself perceiving. This conflation of the phenomenology of the artist and the experimental discipline of the scientist closed a gap long held open by significant forces in both fields.

White Cube and Black Box: Exposing the Explanatory Gap in Modernism and Behaviorism

Having placed the artistic enterprise of Irwin and Turrell in historical and intellectual context, this section will demonstrate how their art challenges central tenets of both disciplines from which it draws. For in mid-twentieth-century American art, the material and immaterial realms of conscious reality were compartmentalized by Clement Greenberg and his formalist followers—by way of the white cube. Simultaneously, in the then-regnant school of B.F. Skinner's radical behaviorist psychology, these realms were commensurately bracketed by way of the black box. In casting embodied experience aside as either inscrutable or irrelevant, each methodology maintains its disciplinary autonomy: the behaviorist is only concerned with recorded actions; the formalist with visualizing ideals. To maintain this autonomy, however, is also to maintain an explanatory gap that has become a focus for cognitive studies in the past half century. Susan Blackmore offers a forthright description of the problem (2006, 261): "The gap in explanation between mind and brain, inner and outer, objective and subjective, or the physical world and consciousness, or the claim that facts about the physical world can never satisfactorily explain facts about consciousness". While behaviorism describes experience from the outside and formal analysis does the same from the inside (the eye as mind) the relationship between the two remains unconsidered. What this section will show is how Irwin and Turrell collapsed both disciplinary and inter-disciplinary boundaries by using the black box and white cube as artistic materials, rather than as theoretical frames.

In psychological circles, the "black box" was an effective metaphor for consciousness, inscrutable and isolated from observable behavior. A commonplace in psychological parlance, "black box" in this context refers specifically to the complex biological mechanism that includes the brain and its attendant inner workings (central nervous system, or "CNS") including conscious thought; it is invisible to the outside observer, operating for the most part beneath our conscious notice. It is "black because we cannot see inside it" (Hamlyn 1990, 3). The term "black box" is generally attributed to B.F. Skinner, but is more widely proliferated by those who oppose his views; nevertheless, it is he who insisted that objectively verifiable data is the sole concern of psychological research. In his words, radical behaviorism "does not deny the possibility of self-observation or self-knowledge or its possible usefulness, but it questions the nature of what is felt or observed and hence known" (Skinner 1974, 16). Introspection is environmental "collateral"; thus, what is empirically observed is effectively severed from what is intellectually thought within the observed subject.

Commensurately, the white cube fosters an environment in which the sensing individual plays only a supporting role to a disembodied and discerning "eye". So critic Brian O'Doherty has trenchantly observed:

> Art exists in a kind of eternity of display, and though there is lots of "period" (late modern), there is no time. This eternity gives the gallery a limbo-like status; one has to have died already to be there. Indeed, the presence of that odd piece of furniture, your own body, seems superfluous, an intrusion. The space offers the thought that while eyes and minds are welcome, space-occupying bodies are not—or are tolerated only as kinaesthetic mannequins for further study (1976, 15).

The "white cube", or high modern gallery, provided a pristine, even antiseptic, means of separating life from art. While interpreting very different material or data, with the black box and white cube both behaviorist psychologists and formalist critics nevertheless omitted the same element—embodied experience. Doing so enabled the professional observer, whether critic or scientist, to maintain a position that (his) description suffices as explanation.

In this environment, where science and art occupied mutually non-transgressable realms, Robert Irwin and James Turrell adopted a middle position. Coincident with cognitive psychology, they asserted that felt experience is the essential *matter* of art rather than the stuff of traditional artistic media. Works like Turrell's *Virga* undermine the idealizing potential of the exhibiting space by calling attention to its spatial, temporal, and human contingencies. *This* particular white cube cannot be *the* white cube. In this regard, the work of such "phenomenal" artists differs sharply from that of conceptualists, which is fundamentally propositional rather than experiential. As described by the conceptualist Joseph Kosuth: "For the artist, as an analyst, is not directly concerned with the physical properties of things ... [The] propositions of art are not factual, but linguistic in character ...; they express definitions of art, or the formal consequences of definitions of art".[6] Instead of doing away with art or "dematerializing" it, Irwin and Turrell shifted the notion of what constitutes "material". In order to take such a stance, however, they needed to accept a position that human consciousness is accessible rather than epiphenomenal (acting only upon itself in metaphysical isolation). The conditional art of Irwin and Turrell necessarily pries open the black box and in so doing, undermines, or even obliterates, the presumed neutrality of the white cube, revealing its profound contingency.

In piercing and vitalizing the pristine space of the white cube, Irwin and Turrell were simultaneously provoking engagement with formalist aesthetics and the black-boxing of experience it privileged. In an artists' statement given in conjunction with the *Art and Technology* experiments (Tuchman, 128) Irwin and Turrell wrote: "A problem may arise with this project in the minds of the art community who may regard it as 'non-art'– as theatrical or more scientific than artistic or as being just outside the arena of art. Although it is a strong alteration as far as methods, means, and intent, we believe in it as art, and yet recognize the possibility of a redefinition needed to incorporate it into the 'arena.'"

This can only be interpreted as a direct challenge to the formalist critic Michael Fried. In his epochal and still provocative essay "Art and Objecthood" of 1967, Fried had used the term "theatricality" to describe "non-art", and in particular the "literalism" of minimal sculpture. While Michael Fried's essay serves as my key example for pointing out some limitations of formalist criticism (specifically because of his well-known reading of minimal art as "literalism") he belongs to a larger and influential tradition that owes a great deal to the work of his onetime mentor, Clement Greenberg. Greenberg (1962) advocated a strict adherence to medium specificity: i.e. a painting must be evidently so, rather than posing as image, which is illusionistic. The miscegenation of arts such as sculpture and painting would therefore obscure the role of media. According to these criteria, Fried can rightly claim that minimal art is transgressive, wherein mere objects inappropriately pose as sculpture.

Tony Smith's *Die* (1962) is quintessentially minimal. It is unreadable, impermeable. *Die* is 72 inches (182.88 cm) cubed, made of steel, and painted black: factually, a black box. Yet oddly, in 1967 Fried asserted that it was the latent anthropomorphism of such a thing that made it so "literal". *Die* is the height of a large man; furthermore, Smith had it placed directly on the ground/floor rather than raised on a plinth or dais; it shares a space and scale with its viewers. These apparently human qualities in fact undermine its role *as art* as far as Fried is concerned (1967, 155–156): "... the entities or beings encountered in everyday experience in terms that most closely approach the literalist ideals of the nonrelational, the unitary, and the holistic are *other persons*". *Die*'s apparent muteness gets at the heart of the

---

[6] Joseph Kosuth, quoted in Alberro and Stimson (1999, xxxi n. 7).

problem of literalist art for Fried. The sculpture remains tethered to its circumstances, offering nothing to supersede them as far as the critic is concerned, a mere object rather than a work of art.

To Fried, the intransigence of the object deflected viewer attention onto its context, including the viewer's own immediate experience (which, as behaviorists would point out, is unreliable as evidence). In the same essay, Fried recounts an anecdote from Smith where the artist suggests that an epiphanic experience on the New Jersey turnpike may well be "the end of art" for him. For the critic, the described experience is no more than an "empty, or 'abandoned' *situation*"; lacking an art *object*, it cannot be art at all. The situation, he asserts, "reveal[s] the theatrical character of literalist art, only without the object, that is, without the art itself—as though the object is needed only within a *room*" (Fried 1967, 159). In this essay Fried clearly delineates what he considers to be the risks of "dematerializing" art and offers a means by which we might distinguish art from non-art (or art from "theater"). Thus, the essay becomes an important point of rupture between Greenbergian formalist criticism and artistic practices in the 1960s and 1970s that refused to acknowledge modernist parameters (Lee 2006).

Fried does not specifically mention the work of any of the California artists upon which this study is based; "Art and Objecthood" references minimal art being made in 1960s New York. But, though Fried (1964) at one time wrote positively about Irwin's early work, we can infer from his critique of minimalist work by artists such as Smith, Robert Morris, and Donald Judd that Irwin's later work as well as Turrell's, would be interpreted as "theatrical". However, the limitations of Fried's methodology show up in any world that accepts Robert Irwin's *1°2°3°4°* (1997) as art. *1°2°3°4°* can be aptly described in Fried's terms as a situation that is not only "empty" but continuously emptying. In this work, Irwin facilitates interplay between the museum gallery and external conditions of its coastal Southern California setting (the Museum of Contemporary Art in La Jolla). The gallery space is an odd off-shoot from the main museum structure; at the end of a corridor, the visitor finds herself in a room surrounded by picture windows, two of which intersect the corners on either side of the facing wall (room dimensions: 9.6 × 26.7 × 18.41 ft; 29.21 × 81.28 × 56.13 m). Through the windows one views a panorama of the Pacific Ocean, the rocky coast, and the museum gardens immediately below. Three precise, apparently square, cuts in the heavy glass windows release the stale museum air while admitting unfiltered light and the sound of the surf below, intermingled with the sights and sounds of human activity both beyond the walls and within them. Two of the apertures on either side (24″ h × 30″ w; 60.96 × 76.2 cm) intersect the windows at the corners of the room while the center cut mitered (24″ h × 26″ w; 60.96 × 66.04 cm) is flush with the glass. In the mid-1960s, while experimenting with the perceptual properties of canvas size, Irwin discovered that a perfect square will appear slightly elongated to a typical viewer. To achieve perceived square*ness*, he stretched canvases that were slightly rectangular. With *1°2°3°4°* he evidently used the same principle. Though the cuts are neither square nor equal in dimensions, they appear as such to the museum visitor. The room, transformed into an aesthetic situation by whomever views it as such, requires no object. *1°2°3°4°*, derived from negation, owes its fluctuating presence to the conscious attention of its viewer/listeners, and ceases to exist *as art* in their absence. In this sense *1°2°3°4°* is a late manifestation of what Andrew Perchuk (2006) has deemed Irwin's "refusal of the gestalt", the artist's persistent concern with the dynamic immediacy of the work.

Irwin's insistence that the properties of a work of art are experientially contingent departs from the behaviorist psychological model that had dominated in American laboratories and universities for the first half of the twentieth century, and which had been implicit in Fried's critique. Positing *behavior* as the appropriate subject for scientific testing, behaviorist psychology dismissed the idiosyncrasies of subjective experience as the purview of psychoanalysis and philosophy. B.F. Skinner (1974, 207) explains: "A person is first of all an organism . . . The organism becomes a person as it acquires a repertoire of behavior under the contingencies of reinforcement to which it is exposed during its lifetime. The behavior it exhibits at any moment is under the control of a current setting. It is able to acquire such a repertoire under such control because of processes of conditioning which are also part of its genetic endowment". Non-behavioral factors in human development such as thought, feelings and ideas belong (to Skinner's way of thinking) to a mentalist viewpoint: because it is manifest, behavior is the only aspect of human learning not "locked in" the black box. It is not that Skinner does not acknowledge the phenomenology of the situation, but its idiosyncratic messiness and apparent immateriality force him to leave it aside. The behaviorist's means of dealing with this problem is to conduct his experiments in a laboratory where phenomenal nuances can at least be constrained if not controlled outright.

The prevalence of behaviorist paradigms in mid-century America is evident in the ways in which its methods even seep into mid-twentieth-century American art critical discourse. In Fried's analysis, artistic properties inherent to the work of art elicit recognition in the beholder: *if* there is any alteration it is on the part of the viewer. This is apt if your model is a behaviorist one, which charges that input and output—more commonly referred to in psychological circles as "stimulus" and "response"—constitute the measurable (and therefore appropriately material and scientific) content of human experience. Fried's oddly passive approach opens up a gap between input—the observable qualities of an object—and output—response. How does this transaction occur? The possibility of conscious *input* on the part of the viewer is foreclosed; meaning is seen as residing within the work of art rather than emerging through the engaged attention of a living brain/body. In the formalist artworld, this interaction between art stimulus and responsive viewer takes place in a carefully prescribed environment meant to maximize stimulation: the white cube. Like its scientific counterpart, the behaviorist lab, the museum gallery was designed to be as ideally "neutral" as possible (its own salience minimized by white paint and muted lighting).[7] The white cube was meticulously tended to in order to prevent

[7]The laboratory analogy is by no means isolated, nor was it particularly new by the middle of the twentieth century. In 1905, for example, the trustees of the Boston Museum of Art specially

viewers from being distracted; by declining to include an object/stimulus Irwin and Turrell instead direct viewer attention to the cube/space itself.

From the artists' perspective, there is little if any difference between a gallery, a laboratory, and an artist's studio: each is a site for ongoing experimentation. For Turrell in particular the use of scientific experimental devices was a natural extension of much of his previous work, since he had studied experimental psychology in college. He was also well versed in the terminology and methodology and (perhaps as importantly) in phenomenological philosophy. His *Mendota Stoppages* (1969–1974) was an on-site installation in his Santa Monica studio where the artist emptied out the space and then "stopped up" the windows except for carefully controlled apertures which allowed the ambient external light (sunshine in the daytime, streetlights and passing cars at night) to animate the space. The work of art was utterly temporal; the light played upon the walls as the sun set, as the streetlights came on, and then more urgently as passing headlights breached the stoppages and crisscrossed the interior walls. Turrell's interventions operated in an analogous way to the stops in an organ, by simultaneously suppressing and admitting light. Night-time was the lively movement that followed a sedate daytime pattern. In direct contrast to the expressive object, then, this work allowed salience to "leak in".[8]

To think of conception as something physical and contingent is to undermine the formalist ideals of art with experiential immediacy and to muddy the science of behaviorism with philosophy. To a formalist, the work of art emanates or communicates its secrets to an attentive but otherwise passive viewer; likewise, in a behaviorist laboratory, the stimulus acts to evoke response. Meaning, understood as immeasurable and atemporal, is therefore set apart from the immediate situation in which it is encountered. Although far from taking any psychological stance in art criticism, Fried shares with behaviorists an understanding of experience as off-limits to analysis. I think this is what makes something like Tony Smith's *Die* so "human" to him (apart from its human scale). The sculpture "hides its thoughts": "[T]he apparent hollowness of most literalist work—the quality of having an *inside*—is almost blatantly anthropomorphic. It is, as numerous commentators have remarked approvingly, as though the work in question has an inner, even secret life ..." (1967, 156). Whereas for Fried (and for behaviorists) an *art* object—or stimulus—is expressive, while the beholder—or test subject—absorbs its qualities: what she brings to the situation (or more specifically, how she constructs it) is less important

commissioned an experimental gallery for the purpose of testing conditions especially lighting conditions in a scientific manner (Gilman 1905, 1906). Excessive glare and shadow especially were to be avoided. The Boston trustees' prejudice in favor of the eye (rather than brain) as the critical viewing organ continued in the preparation of gallery spaces elsewhere in twentieth-century America. Seventy years later Brian O'Doherty (1976, 29) asserted how this prejudice had come to dominate curatorial practice: "It is now impossible to paint up an exhibition without surveying the wall like a health inspector ..." The irony of discussing these experiments in this context is that light and space art often consists of *nothing but* glare and/or shadow.

8 James Turrell, in conversation with Jan Butterfield in Butterfield (1993, 69–71); Turrell (1980, 27–29).

than what she understands the work of art to be "saying". By contrast, situational art such as Irwin's and Turrell's shows the character of any object to be dependent upon what goes on inside the observer's own "black box."

As a formalist, though, Fried is also a keen observer of the phenomenology of his own experience. He therefore provides well-articulated evidence for anyone seeking to explain how aesthetic engagement takes place. In establishing minimal art as "non-art" Fried effectively describes its contingent nature, pointing out the ways in which situational art reveals to us our own conscious roles in constructing aesthetic experience (Fried 1967, 155):

> It is, I think, worth remarking that 'the entire situation' means exactly that: *all* of it—including it seems, the beholder's *body*. There is nothing within his field of vision—nothing that he takes note of in any way—that declares its irrelevance to the situation, and therefore to the experience in question. On the contrary, for something to be perceived at all is for it to be perceived as part of the situation. Everything counts—not as part of the object but as part of the situation in which its objecthood is established and on which that object at least partly depends.

Regarding Fried's mistrust of the temporality of minimal art in "Art and Objecthood", Pamela Lee (2006, 38) writes: " . . . no text articulates the particular mechanics of minimalism's reception as brilliantly as its does, in spite of its antagonism toward the work in question". For Fried's assessment to make sense to a "literalist" however, the definition of "situation" must include conscious thought. Fried stops short of doing so by drawing a line at perception, attributing it only to the body; understanding is not likewise situated. He thus effectively bisects thought into two spheres: literal and abstract.

Robert Irwin made a discovery similar to Fried's—that everything counts—but in his case it turned him toward an integrated situational approach: site *conditioned* rather than site specific. By taking this stance he opened up for dialogue the spheres of experience closed off by modernist ideals and behaviorist restrictions (1985, 26): "*Being and circumstance*, then, constitute the operative frame of reference for an extended (phenomenal) art activity, which becomes a process of reasoning between our mediated culture (being) and our immediate presence (circumstance)".

Light and space work continues to make use of the essential function of the white cube as a space apart; to foster an environment that suspends perceptual certainty this must be so. In everyday life we are far too reliant on our swiftly adjusting faculties to be aware of what they are accomplishing. In *Inside the White Cube* (1976, 78), Brian O'Doherty claims that in the 1970s the white cube was being challenged in an understated way by what I have been terming situational art (including a cross-section—and cross pollination—of minimalism, performance, video, and site-specific work): "[Seventies art] is not in search of certainties, for it tolerates ambiguity well". The critique of the white cube implicit in Irwin and Turrell's investigations is made on purely—and deeply—aesthetic grounds. Thus, they differ from interpretations of the "white cube" that see it as excluding social discourse from the space, instead choosing to embrace and reveal the experiential possibilities of the white cube, and in so doing undermining its assumptive in-transience. A work like Irwin's *1° 2° 3° 4°* achieves this by interrupting the viewer's thoughts with sensations—the smell and sound of the ocean, the coolness of the breeze—calling attention to the museum's situatedness. Similarly, Turrell with the *Mendota Stoppages* and with *Virga*, perforated the gallery walls and ceilings, engaging the vicissitudes of urban life and weather to create a perpetually conditional art.

In contrast, Michael Fried requires that art have "presentness", a quality superseding circumstance. Nevertheless, he relies on his own situated experience to make this claim: his only means of determining that such an atemporal quality belongs to a work of art is to indeed engage with the art in question in real time. In this case, *what* he feels is proffered as the explanation for why he feels it: a behaviorist error. By citing "presentness" as an *a priori* characteristic of artistic (as opposed to "theatrical") phenomena Fried "[ignores] first-person phenomenological as well as third-person empirical constraints in the formation of [his] basic conceptual tools" (Metzinger 2003, 3). He has transformed a conditional physiological response into an objective criterion of aesthetic judgement, and so "abstracts logical principles from incarnate inquiry and attempts to safely ensconce them in the Museum of Eternal Forms" (Johnson 2007, 106). He expects that another viewer will recognize "presentness" when she comes across it; but by doing so, Fried must himself extort complicity from the reader.

In an interview Irwin described how the *Art and Technology* experiments loosened the hold of such abstract constructs (Weschler 1982, 129):

> I think that what happens is that in our ordinary lives we move through the world with a strong expectation-fit ratio which we use as much to block out information which is not critical to our activity. . . .So that what the anechoic chamber was helping us to see was the extreme complexity and richness of our sense mechanism and how little of it we use most of the time. We edit from it severely, in time to see only what we expect to see.

A description of the stimulus does not explain how or why we are stimulated. This is the key limitation in Fried's methodology as it is of behaviorism. Looking back on his decision to begin to explore a conditional art Irwin (1985, 23) wrote: "It takes a peculiar kind of compounded belief to plan, proselytize, or thrust your abstractions onto the world". If we take a formalist's approach the question of what constitutes aesthetic experience is necessarily set aside for the sake of a rigorous determination of what constitutes the correct properties of an art stimulus. Max Kozloff described the same problem with regard to Fried's one-time mentor Clement Greenberg (Newman 2000, 168):

> The will to convince is not the same as earning your convictions. Now, Greenberg made sure to separate his descriptions from his judgments; they really didn't evolve from his explaining, though they gave the illusion that they did. All other contentions against him are secondary, compared with this one. He had decided the worth of an artist "off stage", according to his scheme, rather than by virtue of the particular artistic "phenomena".

A similar problem arises with regard to the explanatory gap in cognitive science. There, the neuronal processes of the brain are observed in ever-increasing detail and yet, as Francisco Varela pointed out, that work takes place in circumstances, both literal and theoretical, that alienate them from first-person, individuated, circumstantially-contingent human life. Varela (1996) asserted that the gap could only be bridged by inserting "disciplined, first-person accounts" of experience into the scientific study of consciousness. In neurophenomenological terms: "... explaining what is happening inside the black box is not explaining what is happening *for* the black box, so to speak. It is one thing to try to account for what is going on in the brain—at whatever level of explanation—... and another one to try to account for what we feel or think is going on..." (Petitot et al. 1999, 12). Varela's proposed solution was to "naturalize phenomenology", establishing a "scientific study of the processes of the phenomenalization of reality". Varela clarifies: "I will not provide a naturalized account in the sense of 'explaining away' or 'giving substance' to the phenomenological description. My aim is just as much to naturalize phenomenology as it is to phenomenologize cognitive science" (Petitot et al. 1999, 577, n.1). In this regard, he shares a posture of inquiry with Irwin and Turrell, who wanted to study the process of perception *as it occurs*, rather than from the standpoint of theories for how it ought to, or is understood to, occur.

Behavior does not necessarily require consciousness to guide it; we conduct our lives to a large extent via unconscious means (Damasio 1999). But experience is another matter; it is shaped by our thoughts and memories, and aesthetic experience in particular is delineated by way of attention, an alert and directed form of interest. A critic necessarily hones his attentional faculties on works of art, providing him with more material from which to articulate the experience. Consciousness follows attention slavishly. As William James (1890, 381) puts it: "Only those items which I *notice* shape my mind—without selective interest, experience is an utter chaos. Interest alone gives accent and emphasis, light and shade, background and foreground—intelligible perspective, in a word". To be sure, the sensory properties of the art object serve to snare and possibly hold that attention, but the experiments of light and space artists undermine the role of salience by presenting situations where all the perceptual 'snags' have been smoothed out, leaving us with no *thing* to attend to, and yet there is no denying the intensity of somehow "un-stimulated" viewer response. Asked to attend to a vacuum, the viewer can make art of anything, or even of nothing. What Robert Irwin and James Turrell show so well is that this murky transitional realm allows us to get beyond the categorizing "what" questions of behaviorism and formalism (i.e. of what do they consist?) to the "how" questions of art and consciousness (i.e. how and whence do they emerge?).

Conclusion

Until the 1960s, rationalizing schemas had enabled a body-mind divide in twentieth-century American art and psychology. For the New York-centered art world, such a schema meant the "neutral" white cube. For the behaviorist, it was the off-limits "black box" of consciousness. The work of modernist art critics such as Michael Fried on the one hand and behaviorist psychologists like B.F. Skinner on the other allowed for ideas (the purview of philosophers and critics) to be abstracted from action (the realm of behaviorist psychologists), effectively exacerbating the explanatory gap. In each of these cases, the analytical method precludes the viewer-subject from recognizing her role as the agent in which actions and ideas integrally emerge. In Southern California, however, developments in quantum physics and space exploration were rapidly undoing Newtonian paradigms of a stable, measurable, and atomized world. Astronautics necessitated a psychology that could accommodate novel and disorienting states. Cognitive psychology in turn assumed an embodied mind for which consciousness was understood to be a material process. In this setting Maurice Tuchman launched the Los Angeles County Museum of Art's *Art and Technology* Program (1968–1971) where Irwin and Turrell investigated the parameters of perceptual thresholds with sensory deprivation devices such as the anechoic chamber and ganzfeld sphere. Both in effect were like the white cube (absent an art "object"), providing undifferentiated spaces for contemplation. But Irwin and Turrell's experiments pushed the logic of the gallery space to such an extreme that they proved the impossibility of its presumed neutrality. To understand how these circumstances become artistically meaningful singularities, the behaviorist input-output model must be replaced by one more robustly phenomenological. Current work in neuropsychology that recognizes the contingencies of conscious experience provides that insight. An insistence upon the conditional nature of experience displaces the presentness of autonomous art with absences, leaving (quite literally) nothing to which one can attach attributes and opening the door to a situational approach.

References

Adcock, C., (ed.) (1990), *James Turrell: The Art of Light and Space*. Berkeley, CA: University of California Press.

Alberro, A. and Stimson, B. (eds.) (1999), *Conceptual Art: A Critical Anthology*. Cambridge, MA: MIT Press.

Baars, B. (1997), *In the Theater of Consciousness*. New York: Oxford University Press.

Berthoz, A. (2000), *The Brain's Sense of Movement*. Cambridge, MA: Harvard University Press.

Blackmore, S. (2006), *Conversations on Consciousness*. Oxford: Oxford University Press.

Butterfield, J. (1993), *The Art of Light and Space*. New York: Abbeville Press.

Cage, J. (1961), *Silence: Lectures and Writings by John Cage*. Hanover, NH: Wesleyan University Press.

Churchland, P. (1986), *Neurophilosophy*. Cambridge, MA: MIT Press.

Compton, M. (1970), *Larry Bell, Robert Irwin, Doug Wheeler*. London: Tate Gallery of Art.

Damasio, A. (1999), *The Feeling of What Happens*. New York: Harcourt, Brace & Company.

Davis, M. (1992), *City of Quartz*. New York: Vintage Books.

Dewey, J. (1934), *Art as Experience*. Chicago, IL: University of Chicago Press.

Emmerik, P. V., in collaboration with Herbert Henck and András Wilheim (2003–2007), *A John Cage Compendium*. http://www.xs4all.nl/~cagecomp/

Flores-González, L. M. (2008), "Phenomenological Views on Intersubjectivity: Towards a Reinterpretation of Consciousness", *Integrative Psychological and Behavioral Science* 42: 187–193.

Fried, M. (1964), "New York Letter", *Art International* 8: 81–82.

Fried, M. (1967), "Art and Objecthood", reprinted in Michael Fried (1998), *Art and Objecthood*. Chicago, IL: University of Chicago Press.

Gilbert-Rolfe, J. (1993), "Expression: Lines, Dots, Discs; Light", in R. Ferguson (ed.), *Robert Irwin*. Los Angeles: Museum of Contemporary Art, Los Angeles, NY: Rizzoli Publications, 93–111.

Gilman, B. (1905), "The Museum Commission to Europe", *Communications to the Trustees*, III. Boston, MA: Boston Museum of Fine Arts.

Gilman, B. (1906), "The Experimental Gallery", *Communications to the Trustees*, IV. Boston, MA: Boston Museum of Fine Arts.

Greenberg, C. (1962), "After Abstract Expression", *Art International* 1: 24–32.

Hamlyn, D. W. (1961), *Sensation and Perception*. New York: Humanities Press.

Hamlyn, D. W. (1990), *In and Out of the Black Box: on the Philosophy of Cognition*. London: Basil Blackwell.

Husserl, E. (1991), *Ideas Pertaining to a Pure Phenomenology and a Phenomenological Philosophy, First Book: General Introduction to a Pure Phenomenology*, trans. Fred Kersten. Dordrecht and Boston, MA: Kluwer Academic Publishers.

Irwin, R. (1970–2004), Robert Irwin Papers, Getty Research Institute, Research Library, Accession no. 940081.

Irwin, R. (1977), *Robert Irwin*. New York: Whitney Museum of Art.

Irwin, R. (1985), *Being and Circumstance*. Larkspur Landing, CA: The Lapis Press.

James, W. (1890), *Principles of Psychology*. Cambridge, MA: Harvard University Press.

Johnson, M. (2007), *The Meaning of the Body*. Chicago, IL: University of Chicago Press.

Judd, D. (1969), "Complaints: Part I", *Studio International* 177: 166+.

Koppes, C. R. (1982), *JPL and the American Space Program*. New Haven, CT and London: Yale University Press.

Leider, P. (1966), *Robert Irwin*. Los Angeles, CA: Los Angeles County Museum of Art.

Lee, P. (2006), *Chronophobia: on Time in the Art of the 1960s*. Cambridge, MA: MIT Press.

Magnifico, M. and Lucia B. D. (2001), *Villas Menafoglio Litta Panza and the Panza di Biuomo Collection*. Geneva, Milan: Fondo per L'Ambiente Italiano.

Mandler, G. (1975), "Consciousness: Respectable, Useful, and Probably Necessary", in R. L. Solso (ed.), *Informational Processing and Cognition: The Loyola Symposium*, New Jersey: Erlbaum, 229–254.

Metzinger, T. (2003), *Being No One*. Cambridge, MA: MIT Press.

Miller, G. A. (2003), "The Cognitive Revolution: A Historical Perspective", *Trends in Cognitive Sciences* 7: 141–144.

Neisser, U. (1967), *Cognitive Psychology*. New York: Appleton-Century-Crofts.

Newell, H. (1980), *Beyond the Atmosphere: Early Years of Space Science*. Washington, DC: Scientific and Technical Information Branch, National Aeronautics and Space Administration.

Newman, A. (2000), *Challenging Art: Artforum 1962–1974*. New York: Soho Press.

Odenwald, S. F. (2002), *Patterns in the Void*. New York: Westview Press.

O'Doherty, B. (1976), *Inside the White Cube* (expanded edition, 1986). San Francisco, CA: Lapis Press.

Palmer, S. E. (1999), *Vision Science: Photons to Phenomenology*. Cambridge, MA: MIT Press.

Perchuk, A. (2006), "From Otis to Ferus: Robert Irwin, Ed Ruscha, and Peter Voulkos in Los Angeles", Ph.D. dissertation, Yale University.

Petitot, J., Varela, F. J., Pachoud, B. and Roy, J.-M. (eds.) (1999), *Naturalizing Phenomenology: Issues in Contemporary Phenomenology and Cognitive Science*. Stanford, CA: Stanford University Press.

Pinker, S. (2002), *The Blank Slate*. New York: Viking Press.

Robertson, W. G. and Wortz, E. C. (1969), *The Effects of Lunar Gravity on Metabolic Rates*. Washington, DC: National Aeronautics and Space Agency.

Schrödinger, E. (1935), "The Present Situation in Quantum Mechanics", trans. J. D. Timmer, *Proceedings of the American Philosophical Society* 124: 323–338.

Skinner, B. F. (1964), "Man", *Proceedings of the American Philosophical Society* 108: 482–485.

Skinner, B. F. (1974), *About Behaviorism.* New York: Alfred A. Knopf.
Smith, D. W. (2007), *Husserl.* London: Routledge.
Thompson, E., Lutz, A., and Cosmelli, D. (2005), "Neurophenomenology: An Introduction for Neurophilosophers", in Andrew Brook and Kathleen Akins (eds.), *Cognition and the Brain: The Philosophy and Neuroscience Movement*, Cambridge, UK: Cambridge University Press, 40–97.
Tuchman, M. and Livingston, J. (1971), *A Report on the Art & Technology Program of the Los Angeles County Museum of Art 1967–1971.* New York: Viking Press.
Turrell, J. (1980), *James Turrell, Light and Space.* New York: Whitney Museum of American Art.
Varela, F. J. (1996), "Neurophenomenology: A Methodological Remedy for the Hard Problem", *Journal of Consciousness Studies* 3: 330–350.
Vitz, P. C. and A. Glimcher (1984), *Modern Art and Modern Science: The Parallel Analysis of Vision.* New York: Praeger.
Watson, J. B. (1925, rev. ed. 1930), *Behaviorism.* New York: W.W. Norton.
Weschler, L. (1982), *Seeing is Forgetting the Name of the Thing One Sees.* Berkeley, CA: University of California Press.
Whiting, C. (2006), *Pop L.A.: Art and the City in the 1960s.* Berkeley, CA: University of California Press.

Art and Neuroscience

John Hyman

1. I want to discuss a new area of scientific research called neuro-aesthetics, which is the study of art by neuroscientists. The most prominent champions of neuro-aesthetics are V.S. Ramachandran and Semir Zeki, both of whom have both made ambitious claims about their work. Ramachandran says boldly that he has discovered "the key to understanding what art really is", and that his theory of art can be tested by brain imaging experiments, although he does not describe these experiments, or explain what results the theory predicts (Ramachandran and Hirstein 1999, 17). Zeki, who originally coined the term "neuro-aesthetics", claims to have laid the foundations for understanding "the biological basis of aesthetic experience", and to have formulated a "neurobiological definition of art" (Zeki 1999, 2, 22).

If these claims are true, we are at the dawn of a new age in the study of art. Up to now, most of the people studying art have been historians, some of whom can read Latin, but hardly any of whom have mastered even the rudiments of brain science. And aesthetics has been in the hands of philosophers, who still disagree among themselves about ideas that were stated in the fourth century BC. Neuro-aesthetics is different. As Ramachandran (2000, 19) says: "These ideas have the advantage that, unlike the vague notions of philosophers and art historians, they can be tested experimentally". So, is neuro-aesthetics the next big thing? I want to assess its prospects, starting with Ramachandran.

2. As I have said, Ramachandran claims to have discovered "the key to understanding what art really is". He also calls this key "[a] universal rule or 'deep structure', underlying all artistic experience" and "a common denominator underlying all types of art" (1999, 16). He writes as follows:

> The purpose of art, surely, is not merely to depict or represent reality—for that can be accomplished very easily with a camera—but to enhance, transcend, or indeed even to *distort* reality. . . . What the artist tries to do (either consciously or unconsciously) is to not only capture the essence of something but also to amplify it in order to more powerfully activate the same neural mechanisms that would be activated by the original object (1999, 16f).

J. Hyman (✉)
University of Oxford, Oxford, UK
e-mail: john.hyman@queens.ox.ac.uk

R. Frigg, M.C. Hunter (eds.), *Beyond Mimesis and Convention*, Boston Studies in the Philosophy of Science 262, DOI 10.1007/978-90-481-3851-7_11,

By "the original object" Ramachandran means the object represented by an artist: for example, a man or a woman, the interior of a room, a landscape, and so on. His hypothesis is that the works of art we enjoy activate the neural mechanisms that are normally activated when we see the kinds of objects which they represent, but they activate these mechanisms more powerfully.

But why should a distortion of reality have this effect? Ramachandran's answer, which he describes as "the key to understanding what art really is", is that this is an example of a psychological effect called "peak shift". He writes as follows:

> If a rat is taught to discriminate a square from a rectangle (of say, 3:2 aspect ratio) and rewarded for the rectangle, it will soon learn to respond more frequently to the rectangle. Paradoxically, however, the rat's response to a rectangle that is even longer and skinnier (say, of aspect ratio 4:1) is even greater than it was to the original prototype on which it was trained . . . this principle holds the key for understanding the evocativeness of much of visual art (1999, 18).[1]

Ramachandran's favorite example of peak shift in art is the way in which the female figure was represented by classical Indian sculptors. Figure 1 shows an example from the twelfth century, a sculpture of the goddess Parvati. This kind of sculpture, Ramachandran says, is essentially "a caricature of the female form". And he adds this:

> There may be neurons in the brain that represent sensuous round feminine form as opposed to angular masculine form and the artist has chosen to amplify the "very essence" of being feminine by moving the image even further along the male/female spectrum. The result of these amplifications is a "super stimulus" in the domain of male/female differences (1999, 18).

So Ramachandran proposes a generalization about art and then postulates a mechanism to explain the generalization. The generalization is that "the purpose of art . . . [is] to enhance, transcend, or indeed even to *distort* reality. . . . not only capture the essence of something but also to amplify it". More pithily: "all art is caricature" (1999, 18). And the mechanism which explains the biological function of art is peak shift. In combination, these things explain a profound and pervasive part of human life in terms of a simple physiological mechanism, which can be demonstrated in the laboratory with a rat, square, a rectangle and some cheese.

This is quite enough to damn the theory, in some people's eyes. It is brazenly reductionist, and that, some people think, is a bad thing. This is of course has been a well-established view of modern science since the Romantic movement. For example, it is expressed in following lines by William Blake (1982, 478):

[1]Following the description of peak-shift as "a common denominator underlying all types of art" (1999, 16) and "the key to understanding what art really is" (1999, 17), the more cautious phrase "the evocativeness of *much* of visual art" may signal a quiet step in reverse. (As the philosopher J.L. Austin once said, there's the bit where you say it and there's the bit where you take it back.) However, even the qualified claim is an exaggeration. The trouble is that the claim that *some* art is caricature is neither very exciting nor very new.

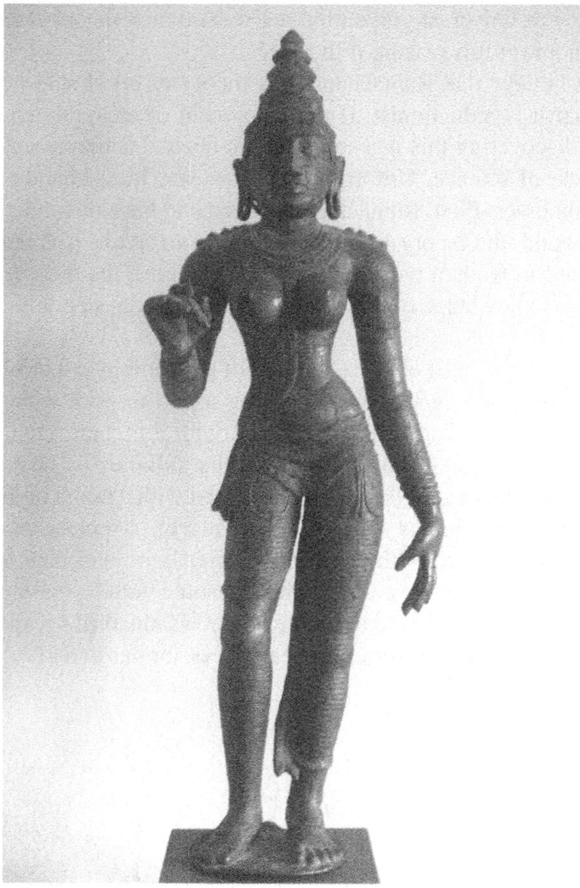

Fig.1 *The goddess Parvati*, Chola, twelfth century AD Private collection

The Atoms of Democritus
And Newtons particles of light
Are sands upon the Red sea shore
Where Israels tents do shine so bright

For my own part, I love Blake's poetry but I do not accept his anti-scientific world-view. I do not believe that modern science drains enchantment from the world, or (as Keats put it) that all charms fly at the touch of cold philosophy. In my view, explaining complex phenomena in terms of simple mechanisms, or explaining a variety of phenomena in terms of a single mechanism, is a good thing. Furthermore, Blake's verse reminds us that we cannot accept reductionism in science that is more than a century old, and reject it in more recent scientific work. This is not an intellectually defensible position. Anyone who uses the word "reductionist" as a term of abuse should ask themselves whether Darwin's theory of natural selection and

Newton's theory of universal gravitation are reductionist theories, and whether we should reject them for this reason, if they are.

So I do not believe that Ramachandran's theory of art should be dismissed on the grounds that it is reductionist. If the enjoyment of art really can be explained by peak shift, discovering this is a stunning intellectual achievement, a formidably impressive piece of science. Unfortunately, however, Ramachandran's theory has three fatal weaknesses. First, Ramachandran seems to have misunderstood the peak shift effect. Second, the theory is not really about art at all. It is really about why men are attracted to women with big breasts. And third, the theory is based on an extremely limited knowledge of art. I shall comment on these points in turn.

3. I begin with the peak shift effect. One kind of psychology experiment which was popular about fifty years ago involved training a bird to peck when it saw a light with a certain color or when it heard a sound with a certain pitch. The bird was rewarded each time it responded to this particular stimulus by pecking at a target, and then once it had learned to do this, it was tested with a range of different stimuli including the training stimulus. The solid line in Fig. 2 represents an experiment where pigeons were trained with a 550 nanometer light, and then tested with different lights, some with higher wavelengths and some with lower wavelengths. The training stimulus is called S+, and as the figure shows, the bird's response decreased more or less uniformly as the stimulus became less similar to S+.

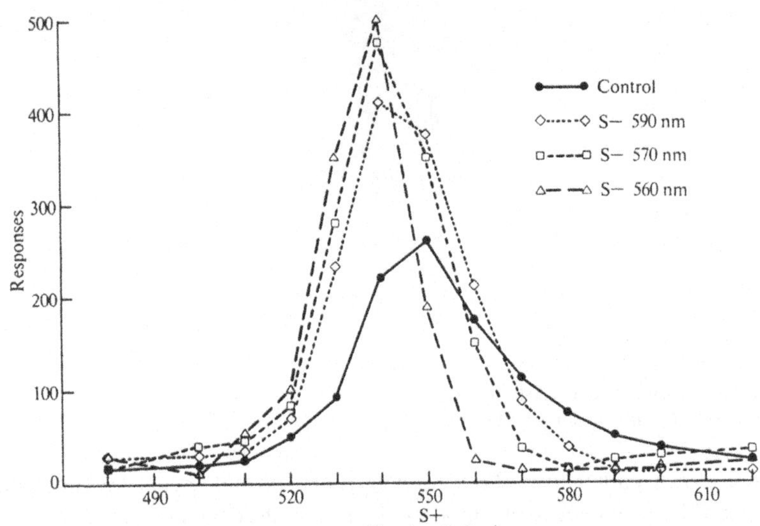

Fig. 2 The peak-shift effect

Now in the experiments represented in Fig. 2 by broken lines, pigeons were rewarded for responding to S+ but they were also shown another stimulus in the training period, which is called S–, which they were not rewarded for responding to. When the pigeons were tested after this kind of training, they responded more vigorously to a stimulus that is different from S+, in the direction away from the S−, than they did to S+ itself. So the peak of the distribution was shifted to about 540 nanometers. This new peak is sometimes called S++. Hence the term "peak shift".

Now we can see that in this experiment S+ and S− are very similar to each other. S+ is a 550 nm light, which is bright yellow. And S− varies from 560 nm, which is yellowy-orange, to 590 nm, which is orangey-yellow. In fact peak shift *only* occurs when S+ and S− are very similar, and the commonest theory of peak shift explains why. The theory, which is represented by Fig. 3, is that peak shift is the result of an interaction between an excitatory gradient around S+ and an inhibitory gradient around S−. What is thought to happen is that when the animal is trained to respond to S+, it also acquires a tendency to respond to stimuli that resemble S+, both to the left and to the right. But if it is also trained not to respond to S−, and S− is similar to S+, this part of its training also inhibits its tendency to respond to S+. So the net effect is that the animal responds more vigorously to a stimulus that resembles S− a bit less than S+ does.

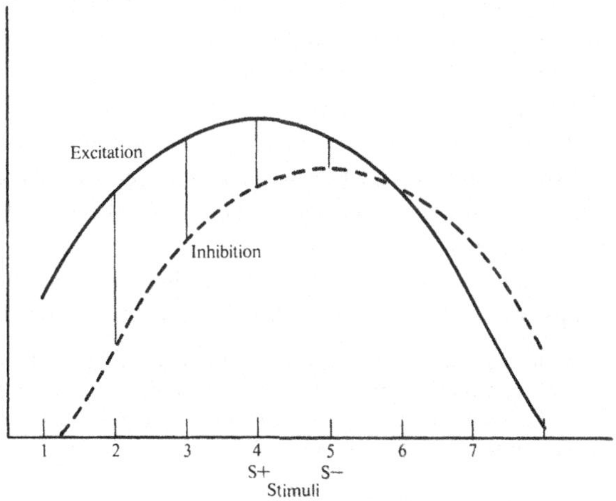

Fig. 3 An interaction between an excitatory gradient around S+ and an inhibitory gradient around S−

The lessons of Fig. 3 are, first, that S− has to be very close to S+ in the subject's quality space to produce the peak shift effect; and, second, that if the effect does occur, the new peak stimulus, S++, will be even closer to S+ than S+ is to S−. This means that if male subjects were predisposed to respond positively to a stereotypical female body at reproductive age—in other words, if that was the S+ body shape—the peak shift towards wider hips and larger breasts would only occur if an inhibitory gradient was created around a female body with slightly narrower hips and smaller breasts than average, and the predicted effect would be that the subject's response would peak at a female body with *very* slightly wider hips and *very* slightly larger breasts than average.

It follows that Ramachandran's explanation of the beauty of the Indian sculpture does not work. It is obvious that classical Indian sculptors gave goddesses such as Parvati prominent breasts and narrow waists—as Ramachandran put it, they "amplify the 'very essence' of being feminine". But there is no evidence that male spectators who find these sculptures beautiful have innate or learned stereotypes that interact to produce a peak shift in their response to female body-shapes. Besides, the body shape of the goddess deviates too far from the norm to be an example of peak shift. Peak shift is simply the wrong mechanism to explain how a "'super stimulus' in the domain of male/female differences" affects the male brain.[2]

4. That is the first reason for rejecting Ramachandran's theory of art. The second is that the theory is not really about art at all. It is really a theory about why men are attracted to women with big breasts.

Remember: the theory is meant to be giving us "the key to understanding what art really is". But the fact that the Indian sculpture is a work of art is completely irrelevant to this theory. It could just as well be a theory about Pamela Anderson. The theory would be that Pamela Anderson has amplified the "very essence" of being feminine—in other words, she has had her breasts enlarged—and the result is a "super stimulus" in the domain of male/female differences. And of course this is more or less true, although it cannot be described as a cutting-edge piece of science.

The point I want to underline is that Ramachandran's theory of art (we can call it the *Baywatch* Theory of Art) doesn't distinguish between a work of art and the kind of object that it represents. For example, if it doesn't distinguish between a sculpture that represents a woman with big breasts and a woman with big breasts. And it follows that the theory cannot be telling us what "the key to understanding what art really is".

This is something every undergraduate who studies aesthetics learns in the first couple of weeks. In Plato's *Republic*, Socrates says that everyone can be an artist:

> Don't you see that you yourself could make all these things in a way? ... Take a mirror and carry it about everywhere. You will quickly make the sun and all the things in the sky, and quickly the earth and yourself and the other animals and artefacts and plants and all the objects of which we just now spoke (*Republic*, 596d).

We can't be sure how seriously Plato meant us to take the comparison between painting and mirroring. But every student learns how to criticize it. Every student learns that understanding "what art really is" means understanding first and foremost

[2]This line of criticism is elegantly advanced in Martindale (1999). In response, Ramachandran acknowledges that he is "not using the phrase 'peak shift' in its original, strict technical sense" (1999, 73), and he has added (in correspondence with me) that he isn't "much concerned with the exact meaning of words and phrases like 'peak shift'" and that he deplores "excessive preoccupation with purely semantic issues". But these comments are not reassuring. For how nonchalant we can afford to be about the definition of a term depends on the term. "Peak shift" is a technical term, so it means nothing until it has been explained. And if it is not being used in its original, strict technical sense, no alternative sense has been introduced. Furthermore, scientists do need to think about semantic issues, i.e. about the concepts they use and the language in which these concepts are expressed. This is an indispensable part of the most serious and challenging work in science—try to imagine twentieth-century physics without Einstein's analysis of the concept of simultaneity—and there is no reason for thinking that neuroscience is exempt.

that it is art. Ramachandran seems to have grasped half of this lesson. He seems to have grasped that a work of art isn't a true mirror image of the world. Remember, he says that the purpose of art is to enhance and to distort reality. But if a work of art isn't a true mirror image of the world, it isn't a silicone-enhanced mirror image of the world either. This is the part of the lesson he seems to have missed.

5. That is the second reason for rejecting Ramachandran's theory of art. The third is that it is based on a very limited knowledge of art. There are really two points here. First, as we saw earlier, Ramachandran begins with following observation: "The purpose of art, surely, is not merely to depict or represent reality—for that can be accomplished very easily with a camera—but to enhance, transcend, or indeed even to *distort* reality" (1999, 16f). When E.H. Gombrich (2000, 17) was asked to comment on Ramachandran's theory of art, he made the following remark:

> To the historian of art, it is evident that the authors' notion of "art" is of very recent date, and not shared by everybody ... They do not explain how one could photograph Paradise or Hell, the Creation of the World, the Passion of Christ, or the escapades of the ancient gods—all subjects that can be found represented in our museums.

I cannot improve on this remark. I would only add that photography itself is one of the visual arts, and has become increasingly important during the last hundred years. So even the kind of representation of reality that *is* accomplished with a camera cannot be excluded from the domain of art.

The second point is that even if we limit ourselves to erotic images made by male artists, and presumably in conformity with male taste, it is obvious that Ramachandran's idea about the distortion of reality, the idea that all art is caricature, is quite unconvincing. Here are a few examples, which were made in very different societies and with different techniques.

The first is a small red-figure jug made in Athens in about 430 BC, which is an unusually touching image of a boy and a girl making love by the standards of Greek art (Fig. 4). The boy is leaning back in his chair, his arms at his sides and his hands gripping the seat, his mantle pushed down around his knees. A young girl, naked except for a wide band around her hair, is about to straddle his uncovered lap. Their foreheads touch and they gaze into one another's eyes with a tenderness which is rare in this period. (I don't mean that tenderness between lovers was rare. I mean that it was rarely represented in art.)

My second example is a woodblock print by Utamaro made in the 1780s (Fig. 5). Several things contribute to the subtle eroticism of this image: the intense concentration of the couple, which is expressed in their hands and the man's eye, just visible below his lover's hair; the confusing tangle of limbs which is partly hidden by the man's delicate silk kimono; and the powerful contrast between the fabrics, with their intense colors and lively designs, and the graceful forms of the woman's bottom and neck, the clear white of her skin divided by two similar curves.

My final example is one of Rembrandt's last etchings, *Jupiter and Antiope*, which he made in 1659 (Fig. 6). The composition is based on an etching by Annibale Carracci which is dated to 1592 (Fig. 7). But the figure of Amor, the curtain in the

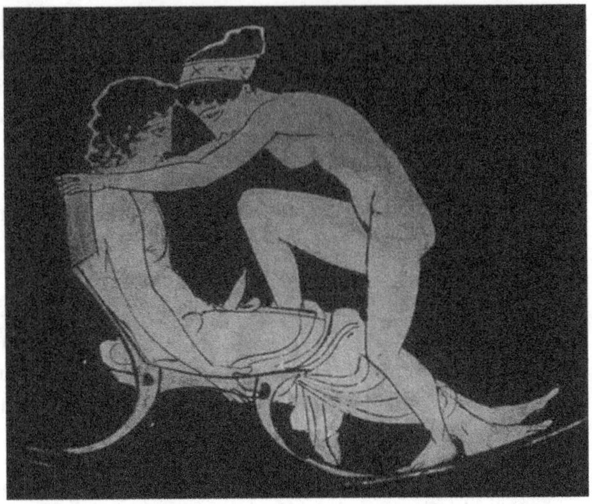

Fig. 4 Attributed to the shuvalov painter, *Attic Red-Figure Oinochoe*, ca. 430 BC (Altes Museum, Berlin)

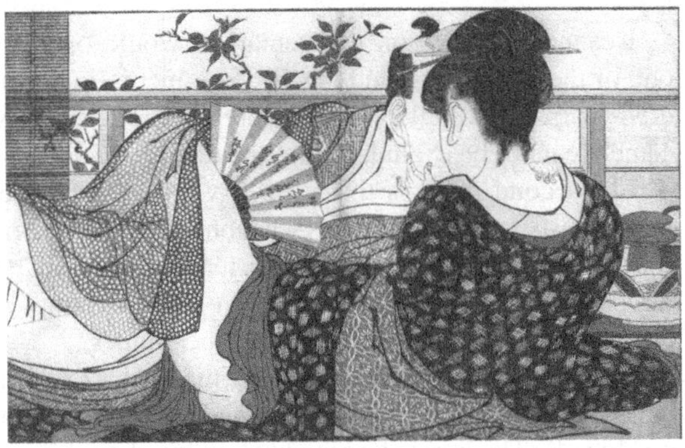

Fig. 5 Utamaro, *Lovers*, from *The Poem of the Pillow*, 1788. Woodblock print

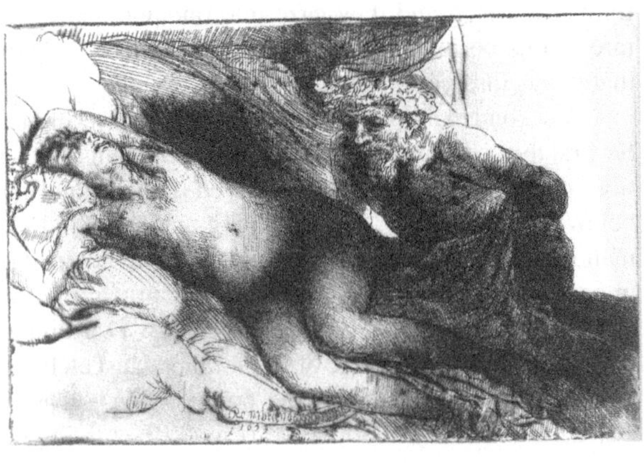

Fig. 6 Rembrandt van Rijn, *Jupiter and Antiope*, 1659. Etching, drypoint and burin

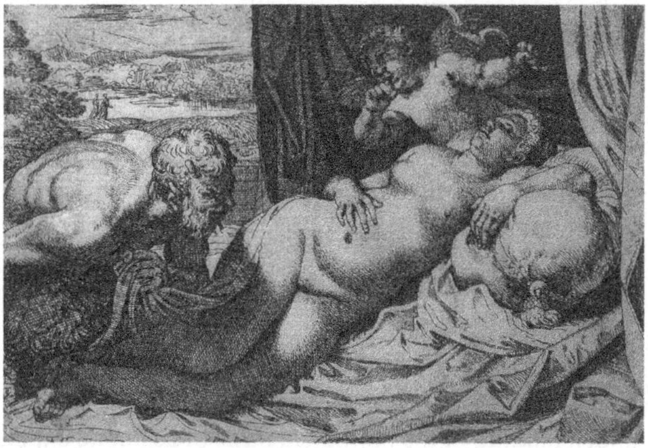

Fig. 7 Annibale Carracci, *Jupiter and Antiope*, 1592. Etching

foreground and the landscape have all been omitted, to concentrate on the two main figures, and Antiope has been given a more natural pose. Her arms are thrown back behind her head, and she is lost in a deep sleep.

None of these images confirms Ramachandran's generalization about art, and of course they stand here for many hundreds of others.

6. I said earlier that Ramachandran has missed the basic point that understanding "what art really is" means understanding that it is art. In other words, works of art are produced with specific tools, materials and techniques. A comparison between these two etchings will help to bring out the significance of this point.[3]

When we look at Annibale's etching, we can see that although he used the tools of an etcher he worked in the style of an engraver. So when he wanted to depict a shadow, he employed the regular cross-hatching of the engraver. The result is a competent print. But there is nothing personal about the technique: it is merely a useful means to an end.

By contrast, Rembrandt's use of drypoint and burin over an initial layer of etching gives his print an extraordinary depth and subtlety of light and shadow. The burin is used to provide an intermediate tone of shading on Antiope's stomach and thighs. And then various details on her arms and head are touched in with drypoint, as is the thick blanket of tone behind her, which sets off the bright upper part of her body. The mixture of these techniques yields a richness and variety of tone, and thereby a subtle atmosphere and register of feeling, far beyond anything Annibale's straightforward method could provide.

The lesson of this example is twofold. First, the comparison brings home how deeply involved Rembrandt was in printmaking. A skilful etcher could follow Annibale's design, but only Rembrandt himself could execute his plate, because the technique was so important to the final result. Second, it illustrates the fundamental

[3]This comparison is entirely derived from White (1999).

point that works of art are produced with specific tools, materials and techniques. Understanding "what art really is" has to involve understanding how the ability that works of art have to express meaning, and to communicate thoughts and feelings and perceptions, depends on these tools, materials and techniques.

Ramachandran's theory of art therefore fails three times over. It fails because he has missed this fundamental point about what art is; it fails because his generalization about what works of art represent is not borne out by the facts; and it fails because even if the generalization were true, the peak shift mechanism would not explain why.

7. I shall turn now to Semir Zeki, and in particular to the two key ideas in his book *Inner Vision*, the book in which he attempts to lay the foundations for "an understanding of the biological basis of aesthetic experience", and defends his "neurobiological definition of art". One of these ideas is about the visual arts in particular and the other is about the arts in general.

The first idea is expounded in a large part of Zeki's book. But it is expressed in the most striking way in his remark that "artists are in some sense neurologists, studying the brain with techniques that are unique to them" (1999, 10). Zeki happily concedes that this is a surprising thing to say. But although the formulation is surprising, the idea has been well established since the last quarter of the nineteenth century. I shall explain what is original about Zeki's version of it shortly. But first I shall quote a passage from its original source, which is a lecture given by Helmholtz in 1871:

> We must look upon artists as persons whose observation of sensuous impressions is particularly vivid and accurate, and whose memory for these images is particularly true. That which long tradition has handed down to the men most gifted in this respect, and that which they have found by innumerable experiments in the most varied directions, as regards means and methods of representation, forms a series of important and significant facts, which the physiologist, who has here to learn from the artist, cannot afford to neglect. The study of works of art will throw great light on the question as to which elements and relations of our visual impressions are most predominant in determining our conception of what is seen, and what others are of less importance. As far as lies within his power, the artist will seek to foster the former at the cost of the latter (1995, 280).

In this passage, Helmholtz combines the idea that artists test and explore the visual system with a theory of vision whose broad outlines he inherited from Locke and Kant. The theory is that visual perceptions occur when the unconscious mind interprets "sensuous impressions". Sensuous impressions are raw patterns of color, without any intrinsic meaning. Artists, he claims, are particularly good at observing their sensuous impressions, and at figuring out which patterns trigger which interpretations.

Most visual scientists have abandoned Helmholtz's theory of vision. They no longer talk about sensuous impressions, or about the unconscious mind interpreting sensuous impressions. Instead, it is generally held that different parts of the brain are simultaneously performing various highly specialized tasks, reacting to form, or to motion, or to color; and that somehow or other the results of these processes are combined to form a unified visual perception, although nobody is sure yet how this synthesis occurs.

But abandoning Helmholtz's theory of vision does not entail abandoning the idea that artists test and explore the visual system. On the contrary, it allows for a more detailed and discriminating version of the same idea, since different kinds of art can now be shown to correspond to different parts of the visual system. For example, kinetic art specializes in V5, the part of the visual cortex that reacts to motion. Fauve art specializes in V4, which reacts to colors. A painting by Mondrian will excite V1, which reacts to horizontal and vertical lines. And so on. The message is that in some cases different kinds of art excite different groups of cells in the brain. This is the principal idea that Zeki defends in his book.

I want to make two comments about this idea. First, it is undeniable that we could not appreciate a painting by Mondrian if the cells in our brains which are excited by vertical and horizontal lines were not functioning properly. But this does not explain why the painting is pleasing or interesting to look at, or what it means. In fact, it reveals nothing whatever specifically about art. Because it is equally true that I could not see the text on a page or the railing in a fence if the cells in my brain which are excited by vertical and horizontal lines were not functioning properly.

The second comment I want to make is this. It may be an amusing paradox to describe painters as neurologists, studying the brain in their own special way. But the real substance of this claim is, *first*, that paintings are designed to have specific kinds of psychological effect on viewers; and *second*, that specific kinds of psychological effect are produced by specific kinds of activity in the nervous system. I do not want to dispute either of these ideas. They have been commonplace for more than a hundred years, and they are both surely true. But if we can think of paintings in this way, the same is true of many other things. For example, hamburgers and ice cream are designed to produce a specific kinds of psychological effect on consumers: the experience of tasting hamburgers in one case and the experience of tasting ice cream in the other. And these specific psychological effects are produced by specific kinds of activity in the nervous system.

So there are two reasons for doubting whether the claim that artists are in some sense neurologists is a useful one to make. First, it does not say anything distinctive about artists. It tells us nothing about Picasso and Cezanne that doesn't apply equally to Häagen Dazs and MacDonalds. And second, it skates over many interesting differences between artists, for example, the difference between painters who are interested in geometrical optics, such as Piero della Francesca, and painters who are interested in the psychology of perception, such as Seurat and Bridget Riley. Or the difference between painters who are interested in the character of visual experience, as Monet and Bonnard seem to have been, at least in theory; and painters who regard themselves as being more like naturalists, and are therefore uninterested in visual experience, but very interested in the visible word—for example, Constable and Turner.

8. The second idea I want to comment on is the boldest and most speculative in Zeki's book. "Aesthetic theories", Zeki maintains, "will only become intelligible and profound once based on the workings of the brain" (1999, 217). Encouraged by this thought, he proposes what he calls a "neurobiological definition of art" (1999, 22). He writes as follows:

> Great art can thus be defined, in neurological terms, as that which comes closest to showing as many facets of the reality, rather than the appearance, as possible ... The inestimable quality [of great art] is the opportunity that the brain is offered to give several interpretations, all of them valid (1999, 22f).

Zeki sometimes calls this inestimable quality of great art "ambiguity". This may not be the right word for it. But I shall not quibble about terminology. The important question is whether the opportunity it affords us to give several valid interpretations *is* what we value in great art. There is also the interesting and contentious question of whether it is brains that do the interpreting, or whole animals. But I shall not address this question here.

So, is Zeki's claim about the value of great art plausible or not? I am not sure that the category of great art is a useful one, in the history of art or in philosophy or science. But I think we all know roughly what properties make art repay serious and sustained attention. We think about these properties when we use the concepts to which criticism constantly returns—among them the concepts of imagination, truth, beauty, form and emotion. But we face two difficulties when we theorize about art. First, none of these concepts is pellucid. They all need careful study. And second, their significance lies in the very particular uses that we put them to, in criticism, so merely identifying them by name does not get us very far.

Take the concept of imagination.[4] Every serious work of art, every work of art that deserves close critical attention, is imaginative, at least in some respects. But the idea of imaginativeness works by contrast. To describe a work as imaginative is to say what it is *not*. It to say both that it is not banal, conventional or academic, and that it is not gimmicky or fanciful or kitsch. Of course it is sometimes hard to decide whether a work of art, or part of one, falls on one or the other side of imaginativeness. For example, consider the famous opening lines of T.S. Eliot's poem, "The Love Song of J. Alfred Prufrock":

> Let us go then, you and I,
> When the evening is spread out against the sky
> Like a patient etherised upon a table ... (1980, 3)

Is the simile "like a patient etherised upon a table" imaginative, or is it meretricious? In many cases, opinions vary and it is hard to know. But part of the business of criticism is make these hard decisions, and to back them with convincing reasons.

So we cannot hope to assess the idea that imaginativeness is a central concept in the theory of art without considering examples. And the same is true of the idea that the value of great art—or art that repays serious attention—lies in "the opportunity that the brain is offered to give several interpretations, all of them valid". Zeki acknowledges this, and in fact he offers many examples of the kind of art he thinks of as having several valid interpretations. He mentions, for instance, Vermeer's painting *Woman in Blue Reading a Letter* and comments that there is no way of telling what the letter she is reading is about (Fig. 8). True. As it happens, there is an earlier painting by Vermeer entitled *A Girl Reading a Letter by an Open Window* (Fig. 9),

[4] My comments on imaginativeness are entirely derived from Passmore (1998).

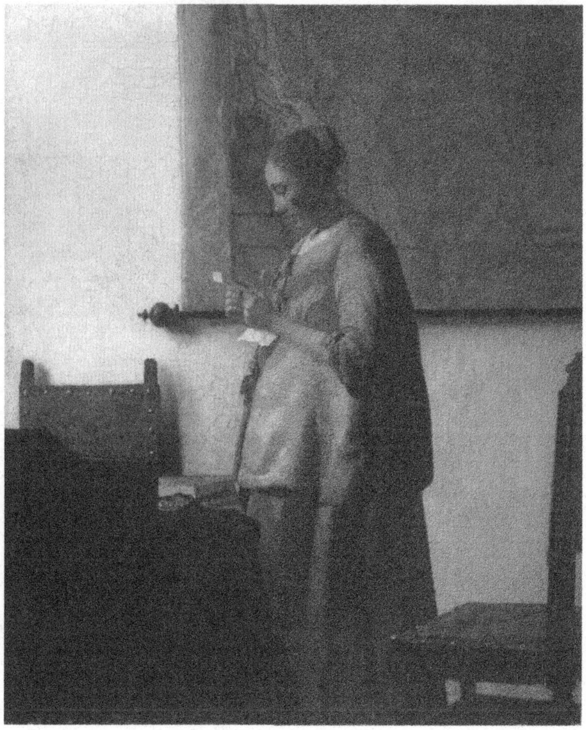

Fig. 8 Jan Vermeer, *Woman in Blue Reading a Letter*, ca. 1662–1665. Oil on Canvas. (Rijksmuseum, Amsterdam)

and in this case an x-ray revealed a painting of Cupid hung on the wall behind the woman, the very same one we see in *A Lady Standing at a Virginal* (Fig. 10). This was a clue that the girl is reading a letter from a suitor or a lover. But Vermeer decided in the end to hide the clue, and I think we can see why.

But the difficult question is what kinds of indefiniteness or multiplicity can contribute to the value of a work of art. Multiplicity is not always a good thing, and more multiplicity is not always better than less multiplicity. For example, consider one of Chardin's still-life paintings of a hare, a partridge or a duck (Fig. 11). I doubt whether it would have been a greater painting if Chardin had managed to paint a duck-rabbit hanging on the wall instead, although this would have introduced an ambiguity which is not there now (Fig. 12).

Perhaps what we want is *imaginative* multiplicity—multiplicity that it is not banal, conventional or academic, and that is not gimmicky or kitsch. But if that is right, the idea that great art is art that has many valid interpretations boils down the idea that great art is art that is imaginative, in this specific way.

My last comment on this idea is that this kind of imaginativeness seems to me to be one property among many which sometimes contributes to the value or interest of a work of art. It certainly is not a definition of great art in neurological terms.

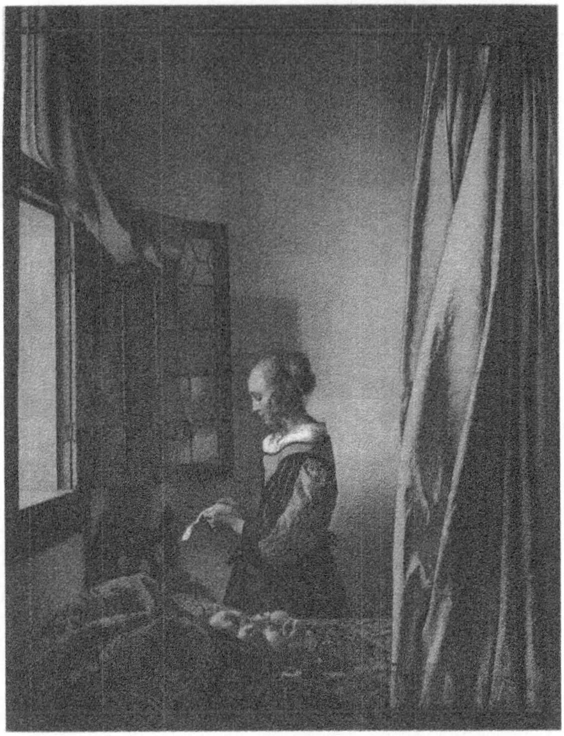

Fig. 9 Jan Vermeer, *A Girl Reading a Letter by an Open Window*, ca. 1657–1659. Oil on Canvas. (Staatliche Kunstsammlungen, Gemäldegalerie, Dresden)

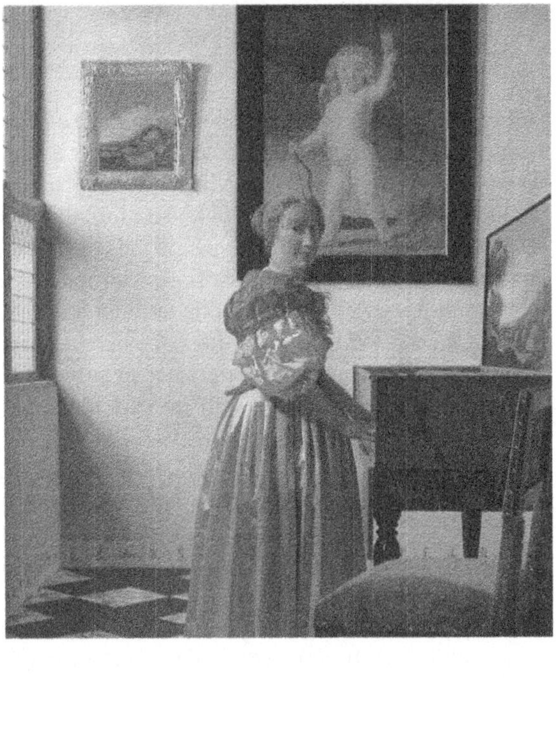

Fig. 10 Jan Vermeer, *A Lady Standing at a Virginal*, ca. 1670–1673. Oil on Canvas (National Gallery, London)

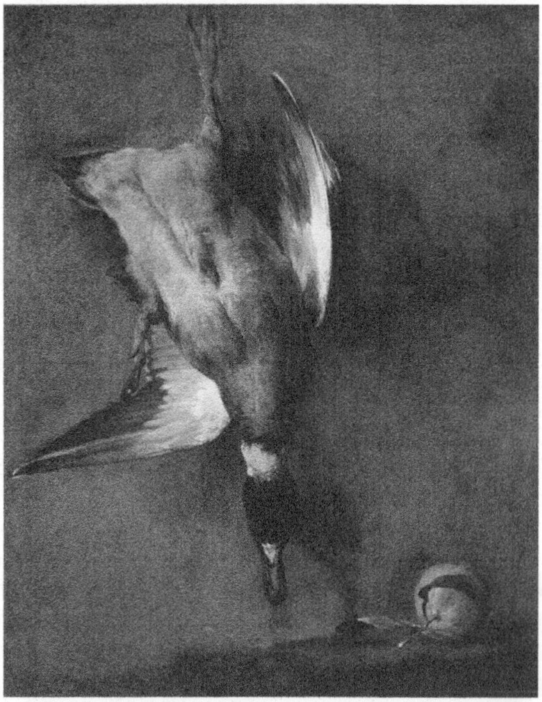

Fig. 11 Jean Siméon Chardin, *Un canard col-vert attaché à la muraille et une bigarade*, ca.1730. Oil on canvas. (Musée de la Chase et de la Nature, Paris)

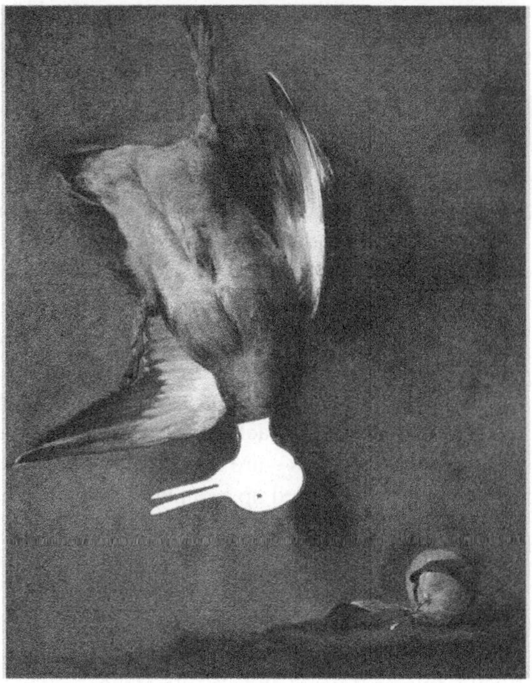

Fig. 12 After Chardin, *Un canard-lapin attaché à la muraille et une bigarade*

And there are many other reasons for admiring art. Here, for variety, is a literary example.

It is a well-known fact that the most erotic line in English poetry—which is in Donne's elegy "To his Mistris Going to Bed"—consists entirely of prepositions:

> License my roving hands and let them go
> Behind, before, above, between, below (1965, 15)

It is true that this line leaves room for different interpretations, or at least for different ideas about what exactly the author has in mind. For example, there are several things that Donne could want his roving hands to get between, although one doesn't imagine that he has her teeth in mind. But I doubt whether this explains the line's extraordinary effect. Surely it is the economy of means and the combination of imagination and technical control that enables Donne to pack such an erotic charge into a string of prepositions. In other words, it is sex and syntax, not ambiguity.

This example reminds us, again, that works of art are produced with specific materials, and it is often the relationship—in some cases the surprising relationship—between these materials and the thoughts and feelings they communicate that matters most. But please do not think this is my universal theory of art. It is just one kind of example. And it is meant to show that we should be pluralists about artistic value. It is a mistake to think that ambiguity or caricature or anything else of this kind defines all art, or all "great" art. Philosophers and historians have for a long time considered it a truism that there is no single source of value or single overarching motive in art. Why should there be such a thing in art, any more than in human life generally?

9. Zeki maintains that "aesthetic theories will only become intelligible and profound once based on the workings of the brain". So, does neuroscience at last hold out the promise of an intelligible and profound aesthetic theory, or one that will provide "the key to understanding what art really is"? And will we soon be able to discard the unintelligible and shallow aesthetic theories proposed by Plato, Aristotle, Hume and Kant?

Extrapolating from the present, the answer must be no. Neuroscience can explain some features of some paintings. For example, some of the color effects of impressionist paintings are explained by lateral inhibition. But the idea that there is a neurological theory of art in prospect is utterly implausible, in my view. The eye-catching paradoxes Ramachandran and Zeki propose—that all art is caricature, that artists are neurologists—are in fact very weak ideas. And in Ramachandran's case, this weak idea is dressed up as a piece of science by misleadingly associating it with the peak shift effect. This, in particular, gets a black mark in my book.

I shall add two final comments. First, the main defect in the work I have discussed is that both authors propose extravagant generalizations about art—all art is caricature; all great art is ambiguous—and then discuss a small number of examples, which are chosen to *illustrate* the generalization they favour and not to *test* it. Would Zeki or Ramachandran tolerate this procedure in their own subject? I expect they'd laugh at it. How easily we shrug off our academic training when we take the brave step outside the furrows we were taught to plough!

Second, I firmly believe that neuroscience can contribute something to our understanding of the visual arts. But progress is only possible if we build on the intellectual tradition we have inherited. This is especially true of neuroscience, which is a nineteenth-century subject rooted in the philosophy of Locke and Kant. In neuroscience, and in psychology in general, philosophy is unavoidable; and if we ignore the philosophy of the past, we shall simply reinvent the wheel. In other words, our ideas will be based on mediocre and amateurish philosophy of our own.

References

Blake, W. (1982), "Mock On, Mock On, Voltaire Rousseau", in D. V. Erdman (ed.), *The Complete Poetry and Prose of William Blake*, Garden City, NY: Anchor Books, 477–478.

Donne, J. (1965), "To his Mistris Going to Bed", in H. Gardner (ed.), *The Elegies and The Songs and Sonnets*, Oxford: Clarendon Press, 14–16.

Eliot, T.S. (1980), "The Love Song of J. Alfred Prufrock", in *The Complete Poems and Plays 1909–1950*. London: Harcourt Brace & Company.

Gombrich, E. H. (2000), "Concerning 'The Science of Art'", *Journal of Consciousness Studies* 7, 8/9: 17.

Helmholtz, H. (1995), "On the Relation of Optics to Painting", in D. Cahan (ed.), *Science and Culture: Popular and Philosophical Essays*, Chicago, IL: University of Chicago Press, 279–308.

Martindale, C. (1999), "Peak Shift, Prototypicality and Aesthetic Experience", *Journal of Consciousness Studies* 6, 6/7: 52–54.

Passmore, J. (1998), *Serious Art*. London: Duckworth.

Ramachandran, V. S. (2000), "Reply to Gombrich", *Journal of Consciousness Studies* 7, 8/9: 19.

Ramachandran, V. S. and Hirstein, W. (1999), "The Science of Art: A Neurological Theory of Aesthetic Experience", *Journal of Consciousness Studies* 6, 6/7: 15–51.

White, C. (1999), *Rembrandt as an Etcher*. London: Yale University Press.

Zeki, S. (1999), *Inner Vision: An Exploration of Art and the Brain*. Oxford: Oxford University Press.

Index

R. Frigg, M.C. Hunter (eds.), *Beyond Mimesis and Convention*, Boston Studies in the Philosophy of Science 262, DOI 10.1007/978-90-481-3851-7,

The manufacturer's authorised representative in the EU is Springer Nature Customer Service Centre GmbH, Europaplatz 3, 69115 Heidelberg, Germany. If you have any concerns regarding our products, please contact ProductSafety@springernature.com

Printed and bound by CPI Group (UK) Ltd, Croydon, CR0 4YY

28/09/2026

02224088-0014